Mathematical Modelling of Zombies

TO MATT,

ENJOY USING YOUR BRAAAIIINNNSSS!
AS YOU DEVOUR THIS BOOK...

Robert?

Robert Smith?

Mathematical Modelling of Zombies

University of Ottawa Press|2014

u Ottawa

The University of Ottawa Press acknowledges with gratitude the support extended to its publishing list by Heritage Canada through the Canada Book Fund, by the Canada Council for the Arts, by the Federation for the Humanities and Social Sciences through the Awards to Scholarly Publications Program and by the University of Ottawa.

Copy editing: Bryn Harris
Proofreading: Trish O'Reilly-Brennan
Cover design: Llama Communications and Édiscript enr.

Library and Archives Canada Cataloguing in Publication

Mathematical modelling of zombies / edited by Robert Smith?

Includes bibliographical references.
Issued in print and electronic formats.
ISBN 978-0-7766-2210-1 (pbk.).--ISBN 978-0-7766-2168-5 (pdf).--
ISBN 978-0-7766-2167-8 (epub)

1. Zombies--Mathematical models. I. Smith?, Robert J. (Robert Joseph), 1972–, editor

GR581.M38 2014 398.2101'51 C2014-906565-5
C2014-906566-3

Printed in Canada

To Richard Tongue and Bon Clarke,
my high school and undergraduate mathematics teachers.
You showed me the beauty of mathematics
but, more importantly, also passed on your passion for teaching.
This is your opus.

Contents

FOREWORD: I RAN WITH A ZOMBIE

Andrew Cartmel

ONE of the hallmarks of a good horror film is that you leave the cinema retaining a residual sense of unease. Suddenly, there seems to be a chill to the sunlight and the world becomes an oddly-less-comfortable place. Then you pinch yourself and remember that you aren't actually living in a society beleaguered by perpetual onslaughts from hordes of the undead. Cue an embarrassed rush of relief.

And the only time such a sense of unease, perhaps blossoming into full-blown terror, is likely to recur is in your dreams. Or nightmares. I've personally lost count of the number of times that zombies, vampires (and, in the interests of full disclosure, occasional Nazi battle groups) have troubled my own sleep.

But there was always one source of comfort in zombie scenarios. As established by films—from the earliest efforts such as Val Lewton, Curt Siodmak and Jacques Tourneur's 1943 *I Walked with a Zombie* right through to George Romero's tipping-point classic *Night of the Living Dead*—the zombie has not been depicted as the most nimble or fleet of creatures. In fact, the lumbering, shambling, clumsy gait of the inadequately reanimated became their hallmark. Consequently, even in the most effective of films or the most personal of nightmares, there has always been a ray of hope. Escape is possible.

Ever since Romero's 1968 debut feature, zombies have gone hand in (cadaverous, rotting) hand with the destruction of the broader social order. This specific trope of a collapsed human society under siege by fiends actually goes back further still, to Richard Matheson's brilliant 1955 novel *I Am Legend*, which depicted a world of the vampiric undead where the last human scrabbles fearfully through the empty city streets, keeping a sharp eye out for sunset.

Matheson's book preceded—and prefigured—the urban and suburban zombie sieges of Romero with their outnumbered and outgunned (or out-toothed) protagonists, deserted metropolitan sprawl and underpinning scientific rationale for the nature of its menace. Indeed, Romero openly acknowledges his debt to Matheson's book and its early film incarnation, 1964's *The Last Man on Earth*.

With zombies now very much off the (Caribbean) plantation and on the march across the homeland in great swaying legions, destroying all social infrastructure and cohesion as they go, there was only one source of comfort. They didn't move quickly. As we sat enrapt en masse in our cinemas or awoke terrified and alone in our beds, we could remain reassured by one unshakeable tenet: the varmints can't skedaddle.

However, things changed.

In 2002, Alex Garland and Danny Boyle gave new life, inappropriately enough, to the undead. The highly successful, and influential, *28 Days Later* was a zombie movie in all but name and indeed became a game-changer. Intriguingly, one of its clearest influences was another film adaptation of Richard Matheson's classic novel: 1971's *The Omega Man*.

But *28 Days Later*'s snazzy, derelict cityscapes were populated by a distinctively new breed of the living dead. Shambling and lurching were out of vogue. They had been decisively replaced by dashing and running. These critters could move. Such new behaviour was justified by references to a rabies-like plague and its ensuing ramped-up aggression and animal ferocity, all calling to mind David Cronenberg's 1977 shocker *Rabid*. Of course, the notion of such a disease ironically moves us away from vampires, and towards lycanthropy and therefore the roots of the werewolf legend. But, crucially, it makes the zombies a still more dire and inescapable threat.

These days, to quote Bram Stoker's *Dracula*, "The dead travel fast."

In a curiously similar fashion, and at much the same time, zombies have accelerated and expanded into popular culture, rapidly occupying new niches. They have long since invaded the toy shop—my favourite example being the Cubes' Corporate Zombies, poseable figures that come complete with coffee mugs, clipboards and cell phones that they can clutch as they shamble around their office cubicles between management meetings.

Comic books had succumbed to the plague decades ago, with Stuart Kerr, Ralph Griffith and Vince Locke's highly successful *Deadworld* debuting in 1987. Most recently, zombie comics have made a splash with Robert Kirkman, Tony Moore and Charlie Adlard's *The Walking Dead*, which has lately lurched on and taken over television, thanks to the celebrated 2010 dramatization developed by Frank Darabont for AMC.

Now, clearly some decisive paradigm shift has taken place, and zombies infest the bestseller lists with such books as Seth Grahame-Smith's *Pride and Prejudice and Zombies*, and Max Brooks' *Zombie Survival Guide* and *World War Z*, the latter migrating to the big screen.

And then of course there is the book in hand, which represents a welcome new onslaught, with zombies crashing through the groves of academia. Invading the realm of mathematical modelling, to be precise.

In these pages, you will find an epidemiological analysis comparing the effect of zombies and their old rivals vampires as disease-causing agents, statistical inference to help you judge the severity of any zombie outbreak, and a study of whether you will be safer in the countryside fending off rural ghouls (*Night of the Living Dead*-style) or in the city battling urban undead (as in *Dawn of the Dead*). The city zombie also comes in for scrutiny using the discipline of agent-based modelling.

The dread epidemic is examined in terms of swarming biological organisms (for what else are zombies if not this?). Diffusion techniques are called into play to model the zombie population and provide an estimate for that all-important question: how long is it before you're likely to run into one of the flesh-eating bastards?

Elsewhere, a consideration of the implications of the relative population densities of humans and zombies in shifting ratios gives insights into adaptive strategies for dealing with zombie attacks. You will also learn how best to defend yourself in a situation where, perhaps not surprisingly, availability of firearms is suddenly at a premium. Meanwhile, government policy in response to the zombie outbreak is simulated using an evolvable linear representation.

Systems for warning of zombie infections are proposed based on fuzzy decision-making. Echoing the viral spread of zombie infection, we also consider the viral spread of a media story, based on an appropriately undead example. Statistical regression methods address that pressing question of the demographics of zombies in the United States. And the zombie goes social, with the use of social networks to model outbreaks.

At last science—or at least mathematics—is taking the zombie threat seriously. The articles in these pages constitute an admirably varied, provocative and engaging bunch. Although, interestingly, none of them address the question of whether zombies are here to stay.

You do the math.

INTRODUCTION: WHAT CAN ZOMBIES TEACH US ABOUT MATHEMATICS?

Robert Smith?

IN 2009, I published an article I thought would amuse me and no one else: a mathematical model of zombies. The idea was to examine a hypothetical zombie apocalypse through the lens of disease modelling: that is, using the same types of differential equations I use every day to examine the spread of infections such as HIV, malaria, human papillomavirus and a variety of tropical diseases. What we found was that zombies would overwhelm a mid-sized city like Ottawa in just four days.

Much to my surprise, the media appeared to share my sense of humour. The zombie paper was the number one story on the BBC for 24 hours. Stories about it appeared in the *Wall Street Journal*, the *Daily Telegraph*, Dutch and Finnish papers. It was also on television (TVO's *The Agenda* with Steve Paikin) and on radio, such as CBC and NPR. For about four years, it was the top-ranked pdf on Google.

The result was an enormous increase in awareness of disease modelling as a concept. Many people who had never heard of disease modelling became aware of it through this paper. High-school teachers reported that their students became interested in mathematics for the first time. Popular science books on mathematics (*The Calculus Diaries: How Math Can Help You Lose Weight, Win in Vegas and Survive a Zombie Apocalypse*) were written that showcased both my zombie modelling and also my more serious work on AIDS.

In Winter 2012, I taught a class called "Mathematical Modelling of Zombies," which illustrated a variety of disease-modelling techniques through the angle of a zombie apocalypse. It received significant attention, with CBC coming to class to film the lectures and also interview students for a news segment. The fact that zombie modelling received so much attention, despite involving mathematics—one of the most impenetrable disciplines and one that engenders fear-based reactions in many people—is still astonishing to me.

But zombies aren't just a piece of frivolity. They present important lessons that can help us understand just how useful mathematics can be—and that could potentially save our lives someday.

No, really! When the next pandemic comes, what happens? First we need to understand the biology, then create accurate mathematical models that simulate the mechanics of the spread of disease. Once we have these models in place, we use mathematical analysis to develop a conclusion. The great thing about mathematics is that the conclusion is absolutely robust, based on the premise (i.e., the model). This gives us a way to see through the darkness that avoids the messiness and uncertainty of the real world. Every step of the process involved in developing a zombie model is conceptually the same step we want to take when the next pandemic comes (as it surely will).

Furthermore, by using pop culture and social media, mathematics can be made accessible to non-mathematicians. Where physics has a popular following—concepts such as the Higgs Boson, black holes or quarks excite the popular imagination, even among people who may never understand the details—mathematics struggles to make its concepts similarly appealing beyond a niche audience. Zombies break this trend, allowing people who will never solve an equation to be engaged and excited by mathematical modelling.

Zombies also have inherent properties that lend themselves very nicely to various mathematical techniques. They swarm, they act together without being conscious of it and they move about pseudo-randomly. There are also an enormous number of them! On the flipside, when faced with an outbreak, the things we need to do also lend themselves to mathematics: we need to estimate parameters, make decisions based on imperfect information and develop government policy that needs to evolve as the situation changes.

Swarming can be modelled by integro-differential equations, which helps us understand how people (or zombies) cluster together and what the effects of that are on an infectious disease. Pseudo-random movement (of a staggering zombie or an airborne virus) can be described by diffusion, which requires partial differential equations. With so many zombies, each with specific desires (to eat our braaaiiinnnsss), individual-based modelling can step in to numerically simulate the effects of many individuals in an urban environment. Then there are networks, which describe the interactions that people and zombies may have (for example, infection can only spread to those nearby), showing how a disease might spread in a neighbourhood or via international travel.

As for our own response, to estimate parameters we use statistical inference (filling in the gaps when we have some information, but not everything we'd like). We can also use fuzzy decision-making to make logical choices even when information is only "somewhat" true. For example, if we spy three zombies, does that guarantee an epidemic? We also have techniques to develop an evolvable linear system to simulate government responses to an emergency. Some of these techniques are analytical, some computational and others involve a combination of the two. All of them apply both to the fun thought experiment of a zombie outbreak and to the more serious possibility of a major infection sweeping across the globe.

Thus from zombies we learn how to cope with the unknown. A pandemic might be terrifying, but if we know how many vaccines to stockpile in advance (or very quickly after it begins), then it becomes manageable. Knowing how a disease spreads through a population—which requires knowing the makeup of that population and how it might change—is immensely valuable information for knowing when to intervene. Mathematics can help us determine which control efforts might be most useful (and when to apply them) or it might tell us that a particular intervention will be functionally useless, thus saving lives, money and valuable time in an emergency.

So there you have it. Everything you ever needed to know to survive the next outbreak is right at hand. Just wait till the depraved undead arise from their graves and start munching on society's brains, unaware of the power of modelling to predict and thwart their every move. You'll be glad you took all those math classes then.

THE VIRAL SPREAD OF A ZOMBIE MEDIA STORY

Robert Smith?

Abstract

This chapter uses the case study of a popular media story—the 2009 coverage of a mathematical model of zombies—to examine the viral-like properties of a story's propagation through the media. The coverage of the zombie story is examined and then a model for the spread of a media story is developed. Stability conditions are derived and the model is refined to include multiple secondary hooks, a series of additional pieces of information that may reignite an existing story. Sample scenarios are investigated, under a variety of suboptimal provisions. Conditions under which a story goes viral include initial newsworthiness, the natural lifespan of the story, durability after the fact and at least one secondary hook that occurs early in the story's lifespan.

1.1 Introduction

1.1.1 A Media Invasion of Zombies

LIKE any huge event, it started small. In August 2009, an online blog for a newspaper [1] and an article in *National Geographic* [2] triggered a tidal wave of reports: a group of Canadian researchers had created a mathematical model of zombies [3]. The story was reported in *Wired* [4], which acted as a hub for spreading it significantly further afield. It was picked up in Canada's *Globe and Mail* [5] and then spread to *The Toronto Star* [6], *The Wall Street Journal* [7] and *BBC News* [8], where it was the number one story in the world for 24 hours. Twitter was a-flutter, blogs went into overdrive and searches in Google spiked.

The story gathered even more steam when it was discovered that the lead researcher had a question mark in his name [9]. From here it spread worldwide: *The Daily Mail* [10], *The Melbourne Herald Sun* [11], Finnish news [12]. The authors made appearances on *National Public Radio* [13] and participated in episodes of TV programs devoted entirely to the subject [14].

Upon reaching Australia, the story gathered another boost: the senior author was Australian, so the Australian media became particularly interested in covering it, thus extending the lifespan of the story even further [11]. Agents came calling. Book deals were offered [15]. The Hollywood Science and Entertainment exchange arranged a panel at the Director's Guild of America, putting the senior author in a discussion with George Romero and Max Brooks [16].

In every sense of the word, this story went viral.

August 2009 was a slow news month, with no natural disasters or political scandals. It also fell in the northern summer, which is usually the period where lighter stories can gain traction. Hallowe'en occurred a few months later, resulting in a brief reinterest in the story and it was discussed at the year's end in the summary of stories for the year (and decade). Occasional reports surfaced intermittently thereafter and quotes continue to be solicited to this day.

The story's timeline is shown in Figure 1.1. This is an underestimation of the true number of stories but illustrates repeated spikes in interest that have continued since. Figure 1.2 shows the Google trends for the word "Smith?" in 2009, illustrating the relative spike in searches that occurred in mid-August. The figure also shows Google Trends "zombies" in 2009, indicating their consistent popularity. Note that searching for the word "Smith" produces different results than searches for the word "Smith?" (such as many reports of the death of celebrity Anna Nicole Smith earlier that year).

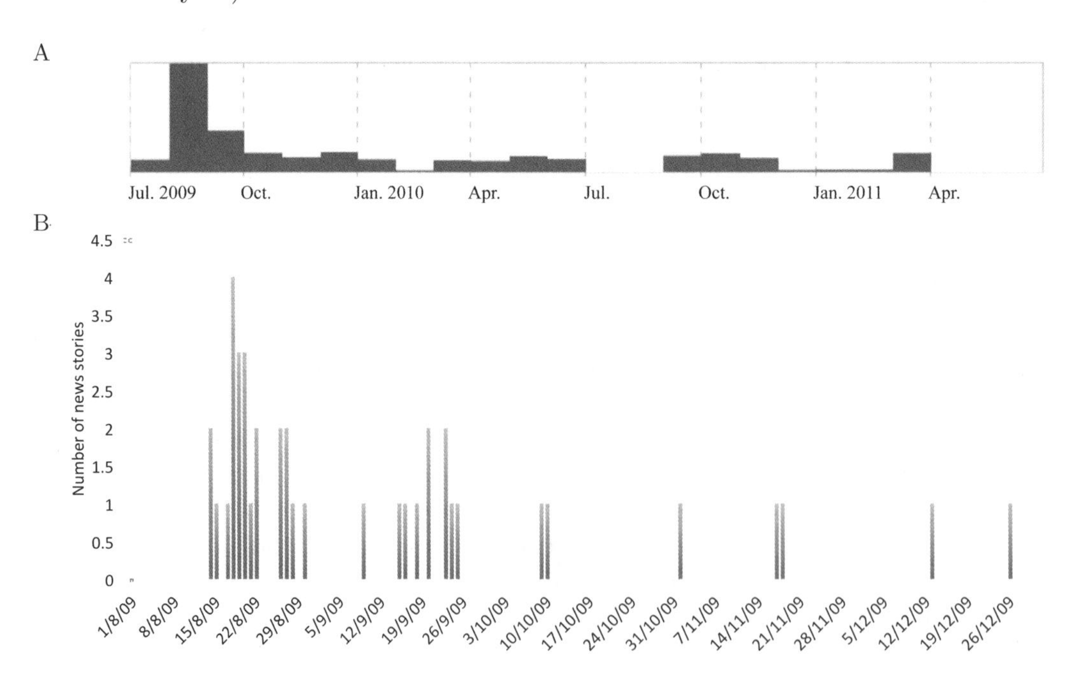

Figure 1.1: A: Timeline of stories in Google News archives featuring keywords "zombies" and "mathematics" since July 2009. B: Number of stories per day in Google News archives featuring keywords "zombies" and "mathematics" throughout the latter part of 2009.

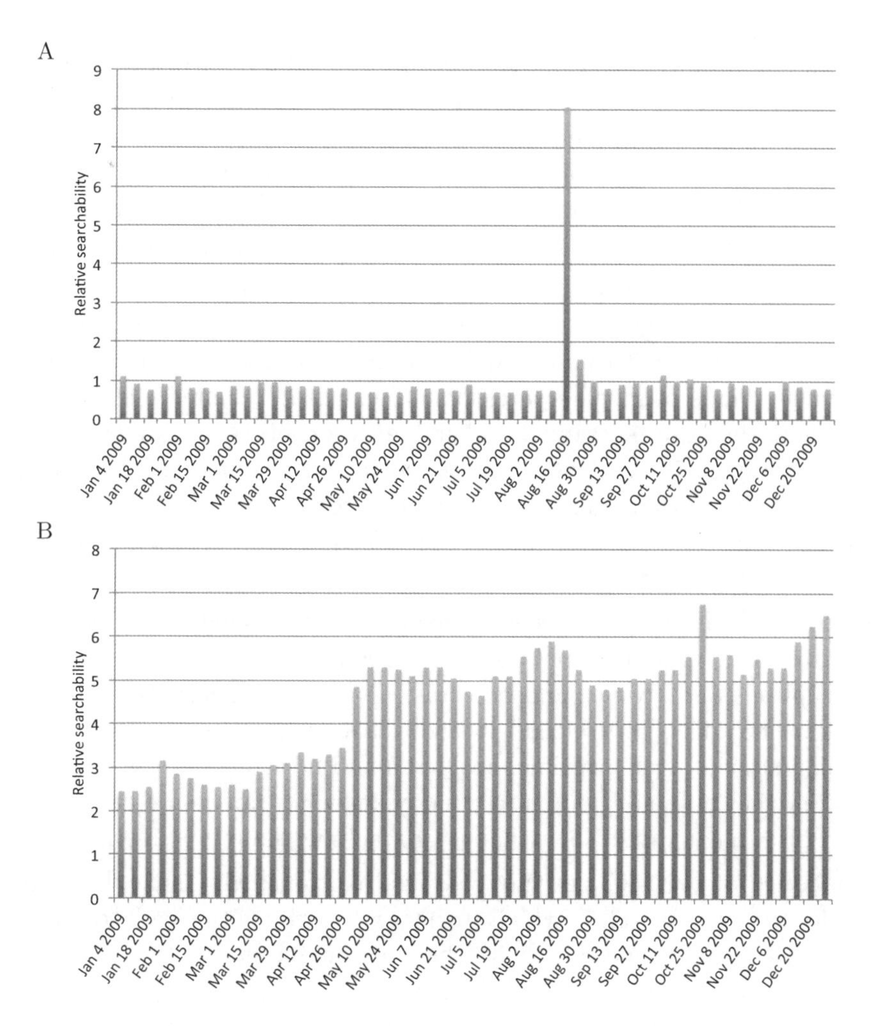

Figure 1.2: A: Google trends for the word "Smith?" in 2009. B: Google trends for the word "zombies" in 2009, as a comparison. Note that searches for "Smith?" produced different results than searches for "Smith."

Since the tail of Figure 1.1 is nonzero, it appears that the original zombie paper—or, specifically, its meme—may now have become endemic. This is fed by a constant recruitment of susceptible individuals who have not heard of the story because they were too young to be media savvy (or not born) at the time of the original outbreak. For example, there was a 2013 outbreak in the (relatively isolated) Canadian city of Kelowna, BC. The steady state prevalence will thus be nonzero, if low. As Andrew Cartmel said in the introduction to this book, zombies have invaded mathematics and appear to be here to stay.

1.1.2 The Effects of Media

The media influences individual behaviour, formation and implementation of public policy and perception of risk [17]. Media reporting plays a key role in the perception, management and even creation of crisis [18]. Since media reports are retrievable and because the messages are widely distributed, they gain authority as an intersubjective anchorage for personal recollection [19]. At times of crisis, non-state-controlled media thrive, while state-controlled media are usually rewarded for creating an illusion of normalcy [18]. Media exposure and attention partially mediate the effects of variables such as demographics and personal experience on risk judgments [20].

The original interpretation of media effects in communication theory was a "hypodermic needle" or "magic bullet" theory of the mass media. Early communication theorists [21, 22] imagined that a particular media message would be directly injected into the minds of media spectators. This theory of media effects, in which the mass media has a direct and rapid influence on everyday understanding, has been substantially revised. Contemporary media studies analyzes how media consumers might only partially accept a particular media message [23], how the media is shaped by dominant cultural norms [24, 25] and how media consumers resist dominant media messages [26, 27].

The choice of which stories receive coverage, and to what degree that coverage is emphasized, is a complex one, involving local, national and international concerns, perceptions of relevance and cultural effects embedded within the hierarchy of reporting. The adage "if it bleeds, it leads" is a journalistic shorthand for raising sensationalist stories (such as crime, accidents and disasters) to the top [28, 29]. A political story that does not dominate the day's news or fails to be the leading story from the capital has a diminished chance of reaching the public farther afield [30].

Journalistic practice includes the role of gatekeepers, who make decisions about which stories to cover, as well as which stories are worthy of the lead slot [31]. Access to media is access to influence [32], with the mass media serving the economic, social and political interests of the elite [33]. This involves persistent patterns of cognition, interpretation and presentation of selection, emphasis and exclusion through which symbol-handlers organize discourse [34]. For example, Canadian newspapers were three times as likely to have climate change or global warming stories as American ones [35]. Media coverage frequently conforms to cultural stereotypes involving gender and race [31, 29].

A handful of mathematical models have described the impact of media coverage on the transmission dynamics of infectious diseases. Cui et al. [36] showed that when the media impact is sufficiently strong, their model exhibits multiple positive equilibria. This poses a challenge to the prediction and control of the outbreaks of infectious diseases. Liu et al. [17] examined the potential for multiple outbreaks and sustained oscillations of emerging infectious diseases due to the psychological impact from reported numbers of infectious and hospitalized individuals. Liu and Cui [37] analyzed a compartment model that described the spread and control of an infectious disease under the influence of media coverage. Li and Cui [38] incorporated constant and pulse vaccination in SIS epidemic models with media coverage. Tchuenche et al. [39] examined the media impact on an influenza pandemic and showed that media overreactions could trigger a vaccinating panic and result in a significantly worse outcome than would occur without the media.

To examine the question of what makes a story go viral, this chapter will use the story of a mathematical model of zombies as a case study to determine the constituent elements of a media sensation. Although a story going viral is not a disease, it has the hallmarks of one and thus mathematical models for disease spread can be adapted to account for a story which is 'infecting' a variety of media outlets. The components of a media story (newsworthiness, durability, natural lifespan and hooks) are identified, in order to examine which aspects are most conducive to a story going viral and what the long-term outcome will be under a variety of suboptimal scenarios.

And, naturally, should you be a reporter reading this fine volume and discover that this meta-article on the spread of articles is itself so worthy of praise that it needs to be reported at once (perhaps under the headline, "Academic writes most egotistical article ever!!!"), then I welcome the chance to pen a future article discussing the spread of articles that discuss how other articles spread that you in turn can report on. If we do this right, we'll be on the gravy train for life...

1.2 The Model

A story that's currently running can be considered 'infectious,' in the sense that other media outlets may pick it up and run their own version. Stories that have recently run may also 'infect' susceptible media outlets, but this effect will lessen the more time passes. That is, unlike most diseases (with the possible exception of zombies!), susceptible media outlets can be infected by those who have already recovered from the infection. This is because journalists very often decide what to write about based on what their competitors have recently written about: not just what is hot right now, but what was recently hot.

Let S represent susceptible media, I represent media outlets that are currently running the story and R represent media outlets that have run the story. We define susceptible media to be those outlets (newspapers, television programs, radio programs, etc.) that have not yet run the story, but that have the potential to at some future time. Note that our active variables are media outlets, not news stories.

β measures how newsworthy the story is in the first place (based on the various criteria by which media outlets decide how 'interesting' a story is), α measures the durability of the story (driven in part by how good the interview subject was once the interview has run) and ν measures how quickly the story becomes old (so $1/\nu$ measures how long the story is news, or the story's natural lifespan). To capture the effect of distance from the story, α is time-dependent and eventually decreases to zero. We also assume that β could be time-dependent, although without the requirement that it necessarily approach zero.

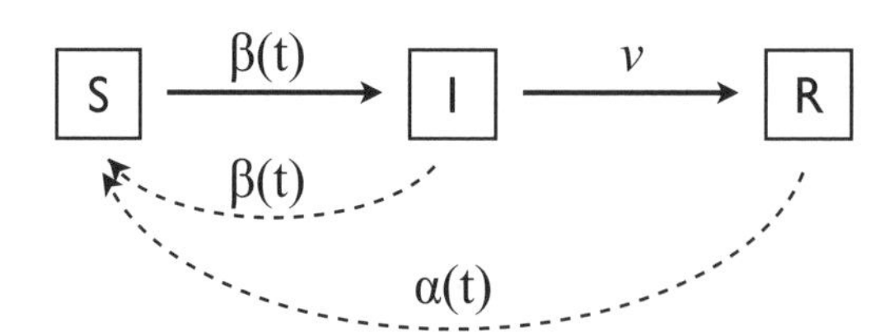

Figure 1.3: The model. Media outlets can be susceptible (S), can be currently running the story (I) or have run the story (R). The measure of a story's newsworthiness is β, the rate at which the story becomes old is ν (so that $1/\nu$ is the story's natural lifespan) and the story's durability is α.

The model is given by

$$\begin{aligned} S' &= -\beta(t)SI - \alpha(t)SR \\ I' &= \beta(t)SI + \alpha(t)SR - \nu I \\ R' &= \nu I. \end{aligned}$$

Mass-action transmission is assumed, since stories are easily accessible, thanks to the internet. The model is illustrated in Figure 1.3.

We say that a story goes viral if the infection rate initially rises, a classic epidemic wave appears and a majority of susceptible media outlets are infected.

1.2.1 The Durability of a Media Story

One technique for examining the dynamics of a model is the use of nullclines. These are places where one derivative is zero, but the others may not be (and usually aren't). In a two-dimensional phase plane representation, these would correspond to any time the tangent is either horizontal (the y derivative is zero at that moment) or vertical (the x derivative is zero at that moment). When the nullclines meet, we have an equilibrium.

The following conditions on the durability $\alpha(t)$ were assumed.

1. $\alpha(0) = 0$.
2. $\lim_{t\to\infty} \alpha(t) = 0$.
3. α is not uniformly zero.

If $\alpha \equiv 0$, then the S-nullclines are $S = 0$ and $I = 0$. The I-nullclines are $I = 0$ and $S = \frac{\nu}{\beta(t)}$. This suggests that, in the absence of a good interview subject, the story's peak would occur at $S = \frac{\nu}{\beta(t)}$ and then decrease until $I = 0$. See Figure 1.4.

A reasonable form for $\alpha(t)$ might be

$$\alpha(t) = \begin{cases} 0 & 0 < t < t_0 \\ \bar{\alpha} & t_0 < t < t_f \\ 0 & t > t_f, \end{cases}$$

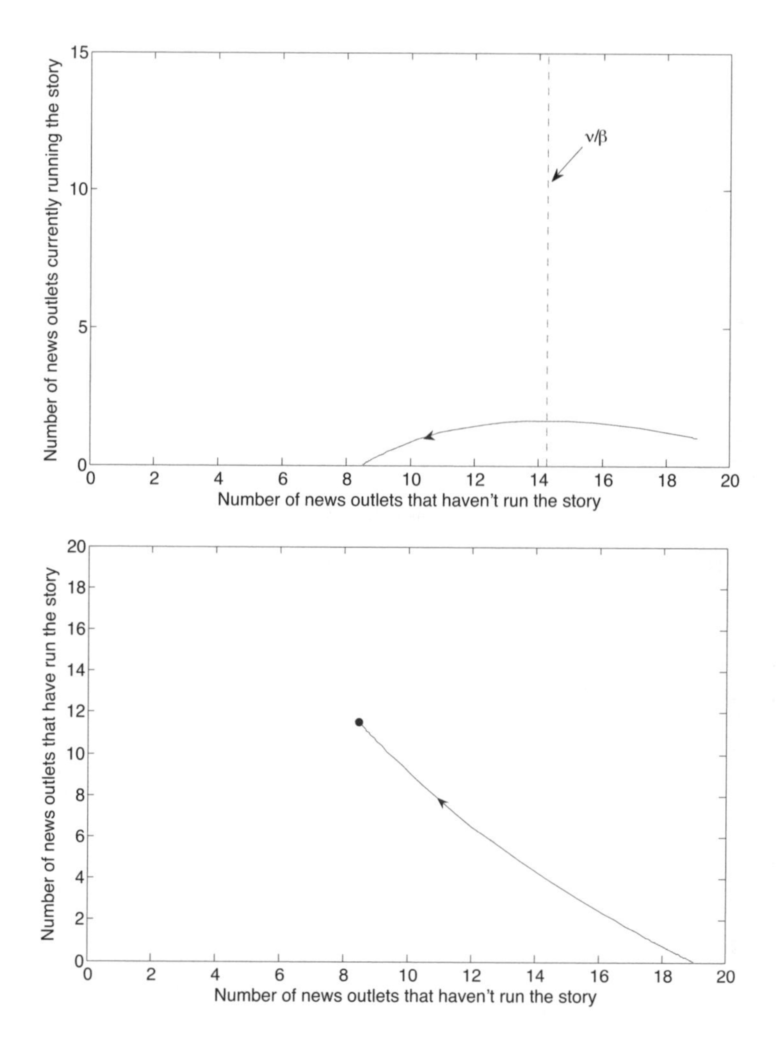

Figure 1.4: Nullclines and trajectories in the case $\alpha = 0$. Other parameters were $\beta = 0.01$, $\nu = 1/7$, $N = 20$, $S(0) = 19$ and $I(0) = 1$. In this case, the story has an initial rise, but does not go viral, since there is no significant epidemic wave.

where the interval $[t_0, t_f]$ is the time during which the interviewee provides added value to the story. That is, when the story breaks, a good interviewee initially adds no effect. However, at time t_0, the time of the first interview, the interviewee's skills are discovered. The interviewee remains a 'hot property' until time t_f, after which their interview skills are irrelevant. This is of course not the only form that such a function can take, but it's the example we'll consider.

In this case, the story receives an extra boost, as media outlets that have already run the story become infectious, since the interview subject has demonstrated a flair for interviews. As a result, more outlets run the story and the peak number of stories is higher. See Figure 1.5.

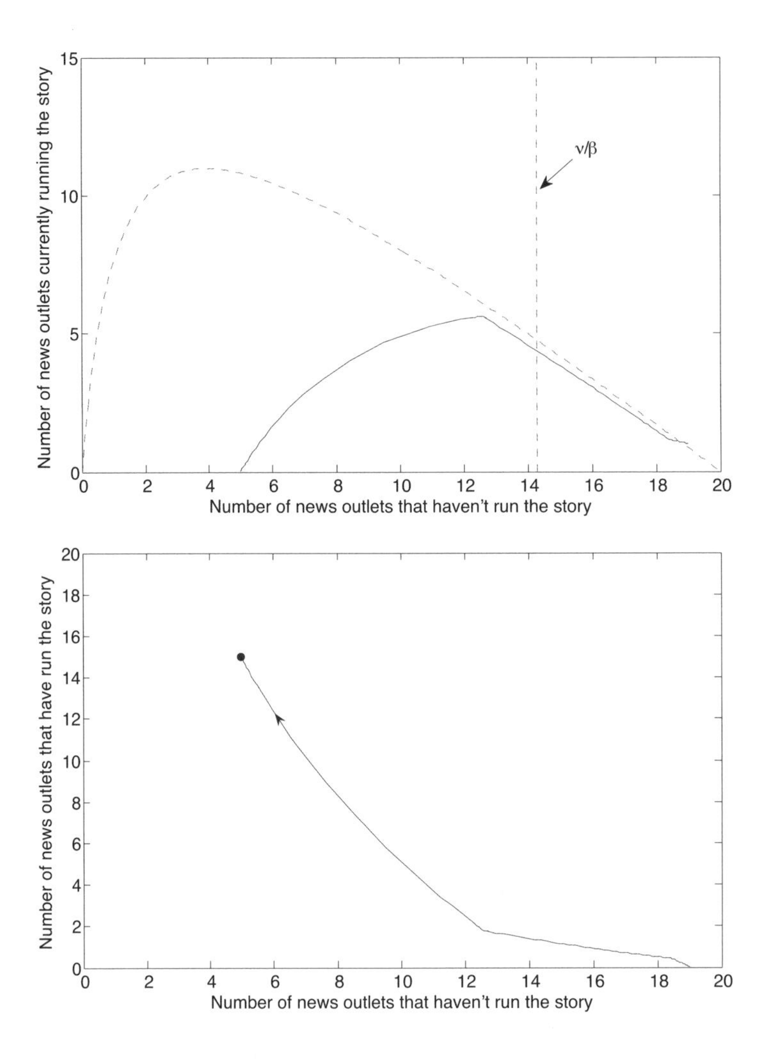

Figure 1.5: Nullclines and trajectories in the case $\alpha \neq 0$. Here, $\alpha = 0.1$ for $3 < t < 6$; otherwise, $\alpha = 0$. All other parameters are as in Figure 1.4. The dashed curve is the nullcline for $\alpha \neq 0$, which only applies in the region $S(t_f) < S(t) < S(t_0)$. The line $S = \frac{\nu}{\beta}$ is not a nullcline in this case, but is included for comparison. Having an interview subject who is 'hot' can significantly increase the number of media outlets that run the story. In this case, there is an initial rise and the majority of susceptible outlets are infected, but the epidemic wave is not significant so the story does not go viral.

1.2.2 The Newsworthiness of a Media Story

The following conditions on the newsworthiness $\beta(t)$ were assumed.

1. $\beta(0) > 0$.

2. $\lim_{t \to \infty} \beta(t) = \bar{\beta} \geq 0$.

Unlike durability, newsworthiness starts at $t = 0$ (whereas $\alpha(t) = 0$ for $0 \leq t < t_0$), so $\beta(0) > 0$. As we shall see, the form of β is less important as time varies, since β is primarily a parameter that applies at the start of an epidemic. Thus $\beta(t) = \bar{\beta}$ (i.e., β is constant) may be a reasonable form. This simply suggests that a story is newsworthy if anyone is currently reporting it.

1.2.3 The Natural Lifespan of a Media Story

Our third parameter is ν, a measure of how quickly a story becomes old, which we assume to be time-independent. The inverse, $1/\nu$, measures the natural lifespan of a story. This takes into account other news stories that may compete for a finite number of susceptible media outlets. For example, a sports story that would have been quite popular may have a significantly shorter lifespan if a tsunami has hit.

1.3 Analysis

1.3.1 Final Size Populations

One important concept in disease modelling is determining the final size of an epidemic: the number of noninfected outlets who are left when the disease has passed and the infection has done its damage.

First we demonstrate that solutions are bounded for large time. Define S_∞, I_∞ and R_∞ to be the final size populations of the susceptible, infected and removed outlets, respectively.

Suppose $t > t_f$. Then the model becomes

$$\begin{aligned} S' &= -\beta(t)SI \\ I' &= \beta(t)SI - \nu I \\ R' &= \nu I. \end{aligned}$$

Note that $I = 0$ is an equilibrium. Let $\Sigma = S + I + R$. Then $\Sigma' = 0$, so Σ is constant. In particular, since $R' = \nu I$, it is important to establish that R cannot blow up to infinity.

Next, notice that the R equation decouples from the system, so we can analyze the two-dimensional system in S and I. Since $S' \leq 0$, there are no limit cycles. If $S_\infty > 0$, then $I_\infty = 0$ and hence $R_\infty = \Sigma - S_\infty$ is finite.

Now consider the case when $S_\infty = 0$. In this case, every susceptible media outlet has run the story. Suppose $I(t) \neq 0$ for any t. Then the second equation can be divided by the first to derive

$$\frac{dI}{dS} = -1 + \frac{\nu}{\beta(t)S}.$$

Since we are interested in long-term dynamics, let $\beta(t) = \bar{\beta}$. Then, integrating, we have, for t large,

$$I(t) = I(0) + S(0) - S(t) + \frac{\nu}{\bar{\beta}} \ln\left(\frac{S}{S(0)}\right)$$

$$I_\infty = I(0) + S(0) - S_\infty + \frac{\nu}{\bar{\beta}} \ln\left(\frac{S_\infty}{S(0)}\right).$$

Since $S_\infty = 0$, this implies that $I_\infty = -\infty$. However, since $I(0) > 0$, there must exist a finite time t_a such that $I(t_a) = 0$, which is a contradiction. It follows that the assumption that $I(t) \neq 0$ is incorrect and hence $I_\infty = 0$ (since $I = 0$ is an equilibrium).

Thus, whether all media outlets run the story or not, the eventual outcome is that I reaches zero in finite time. Hence R cannot blow up.

1.3.2 Stability

Stability is an enormously useful concept in disease modelling (and beyond). If an equilibrium is stable, it means that small perturbations away from that equilibrium will return to it, or at least not stray far away. And equilibrium is unstable if small perturbations move away from it. So if you're sitting in a classroom with no zombies, you're at a disease-free equilibrium. If a single zombie enters the room and starts an epidemic, then that equilibrium is unstable. If, however, you and your classmates manage to hack the zombie to death before it infects anyone, then congratulations: not only have you saved your fellow students from becoming the living dead, you've also experienced a stable equilibrium.

Equilibria are of the form $(S, I, R) = (\hat{S}, 0, \hat{R})$. The Jacobian matrix is

$$J(S, I, R) = \begin{bmatrix} -\beta(t)I - \alpha(t)R & -\beta(t)S & -\alpha(t)S \\ \beta(t)I + \alpha(t)R & \beta(t)S - \nu & \alpha(t)S \\ 0 & \nu & 0 \end{bmatrix}.$$

We are interested in the stability of the disease-free equilibrium, where perturbations are applied at $t = 0$. Thus the Jacobian matrix is evaluated at $\alpha(0) = 0$ and $\beta(0)$. We thus have

$$\begin{aligned} \det(J(\hat{S}, 0, \hat{R}) - \lambda I) &= \det \begin{bmatrix} -\lambda & -\beta(0)S & 0 \\ 0 & \beta(0)S - \nu - \lambda & 0 \\ 0 & \nu & -\lambda \end{bmatrix} \\ &= \lambda^2(\beta(0)\hat{S} - \nu - \lambda) = 0. \end{aligned}$$

It follows that equilibria with $\hat{S} > \frac{\nu}{\beta(0)}$ will be unstable. That is, stories that are particularly newsworthy (high $\beta(0)$) or which have the potential to run for a long time (low ν) have the potential to go viral, since this condition predicts an initial rise in infections, which is a necessary (but not sufficient) characteristic of the viral spread of a story.

Stories with $\hat{S} > \frac{\nu}{\beta(0)}$ will have an initial rise in the number of infections, although it remains to be seen whether they will display the classic infection wave or whether a majority of susceptible media outlets can be infected. However, we can conclude that stories with $\hat{S} < \frac{\nu}{\beta(0)}$ cannot go viral.

Note in particular that this threshold only depends on the initial value of β. We could thus assume $\beta(t) = \bar{\beta}$ (as we will in numerical simulations).

1.3.3 A Competing Story

Next, we examine the case where two stories may compete for the resource of susceptible media outlets. Consider the example posed earlier, of a sports story versus a tsunami. For simplicity, suppose $\alpha = 0$ since we are only interested in the initial viral properties of the story.

The model thus becomes

$$\begin{aligned}
S' &= -\beta_1(t)SI_1 - \beta_2(t)SI_2 \\
I_1' &= \beta_1(t)SI_1 - \nu_1 I_1 \\
I_2 &= \beta_2(t)SI_2 - \nu_2 I_2.
\end{aligned}$$

The Jacobian is

$$J(S, I, R) = \begin{bmatrix} -\beta_1(t)I - \beta_2 I & -\beta_1(t)S & -\beta_2(t)S \\ \beta_1(t)I & \beta_1(t)S - \nu_1 & 0 \\ \beta_2(t)I & 0 & \beta_2(t)S - \nu_2 \end{bmatrix}.$$

At equilibrium, we have

$$J(\hat{S}, 0, 0) = \begin{bmatrix} 0 & -\beta_1(0)\hat{S} & -\beta_2(0)\hat{S} \\ 0 & \beta_1(0)\hat{S} - \nu_1 & 0 \\ 0 & 0 & \beta_2(0)\hat{S} - \nu_2 \end{bmatrix}.$$

Thus eigenvalues are $\lambda = 0$, $\beta_1(0)\hat{S} - \nu_1$ and $\beta_2(0)\hat{S} - \nu_2$. The disease-free equilibrium is unstable if $\max\{\beta_1(0)\hat{S} - \nu_1, \beta_2(0)\hat{S} - \nu_2\} > 0$.

Suppose that, independently, both stories are equally newsworthy. That is, the media is interested in each story. However, where they will differ is in their natural lifespan. Thus we assume $\beta_1(0) = \beta_2(0) = \bar{\beta}$ but $\nu_1 > \nu_2$ (so that Story 2, the tsunami, has a longer natural lifespan). It follows that $\bar{\beta}\hat{S} - \nu_1 < \bar{\beta}\hat{S} - \nu_2$.

In this case, if $\hat{S} > \nu_2/\bar{\beta}$ but $\hat{S} < \nu_1/\bar{\beta}$, then Story 2 can go viral, but Story 1 cannot. Thus Story 2 (the tsunami) would eat up the oxygen that might otherwise allow Story 1 (the sports game) to go viral.

1.4 The Power of a Right Hook

When a subsequent 'hook' appears (more information that makes the story more appealing), the effect is a near-instantaneous transformation in the number of susceptible media outlets: those that may not have thought the story newsworthy before, and were thus impervious to infection, may suddenly decide the story is newsworthy with the presence of the further hook. Alternatively, media outlets that ran the original story now have a new story to run. This may happen a number of times throughout the life of a story.

Such near-instantaneous changes to the system can be described using impulsive differential equations. These are equations that allow a rapid change to be approximated by an instantaneous one at certain impulse times. Applications include the rapid infusion of drugs in the body after taking a pill [40], the effect of spraying pesticides [41] or a pulsed vaccination strategy [42]. The interested reader is referred to [43, 44, 45, 46] for more details on the theory of impulsive differential equations.

The model then becomes

$$\begin{aligned}
S' &= -\beta(t)SI - \alpha(t)SR & t &\neq t_k \\
I' &= \beta(t)SI + \alpha(t)SR - \nu I & t &\neq t_k \\
R' &= \nu I & t &\neq t_k \\
\Delta S &= S_k & t &= t_k,
\end{aligned}$$

where t_k ($k = 1, 2, \ldots, n$) are the times at which the hooks occur and S_k is the strength of the kth hook. We assume there are only finitely many hooks. In particular, note that we do not assume the t_k's are fixed, nor that each S_k is equal for different k's.

Although a hook may increase a story's attractiveness, the net effect is that the initial conditions are reset. Since all trajectories are attracted to an equilibrium with $I = 0$, the result is that the story will still once again begin to die out. However, a series of hooks may prolong the story's lifespan and result in a significantly larger number of media outlets covering it than would otherwise be the case. Compare Figure 1.6 to Figure 1.7. Note that we chose fixed times for illustration, but that this is not required. We also chose the first hook to be significantly stronger than subsequent hooks.

1.5 Sample Scenarios

The model can now be used to examine a number of potential scenarios. News stories do not occur in isolation; they exist in the context of other stories happening at the same time, they have a degree of newsworthiness that media outlets determine and additional information may surface. While a combination of factors is clearly favourable to the viral spread of a story, cases where one or more factor is limited are examined.

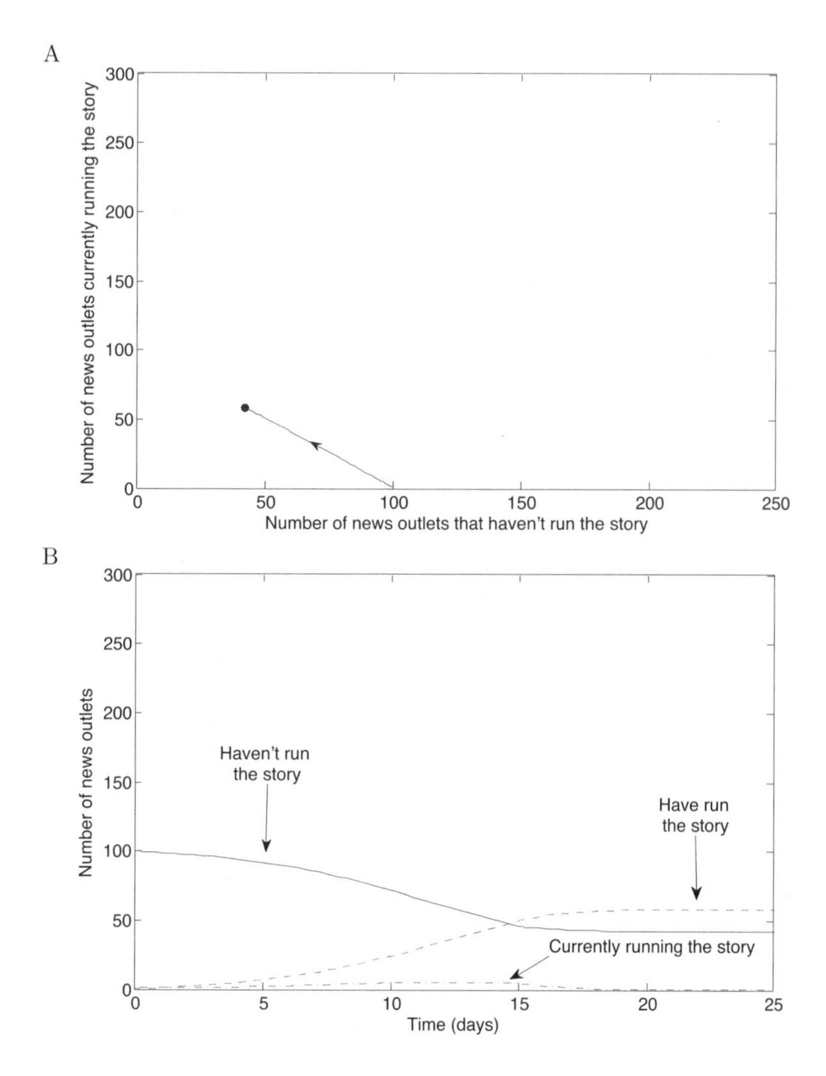

Figure 1.6: Behaviour of media outlets in the absence of a secondary hook. The story proceeds through its natural cycle in an orderly fashion. A: The phase plane (a representation where time is implicit) illustrating the tradeoff between media outlets that have run (or are currently running) the story versus those that have not. B: The time course of the story through the media. Parameters were as in Figure 1.4, except that $S(0) = 100$. This story does not go viral.

1. *Non-newsworthy story, good interview subject*

   In this case, the story never gets off the ground, regardless of the skills of the interviewee. This illustrates the power of the media to shape the cultural narrative, by determining what is or is not newsworthy. See Figure 1.8.

2. *Slow news week*

   In this case, the story remains in the infectious class for significantly longer than it otherwise would, allowing it to be sustained. The story can reap close

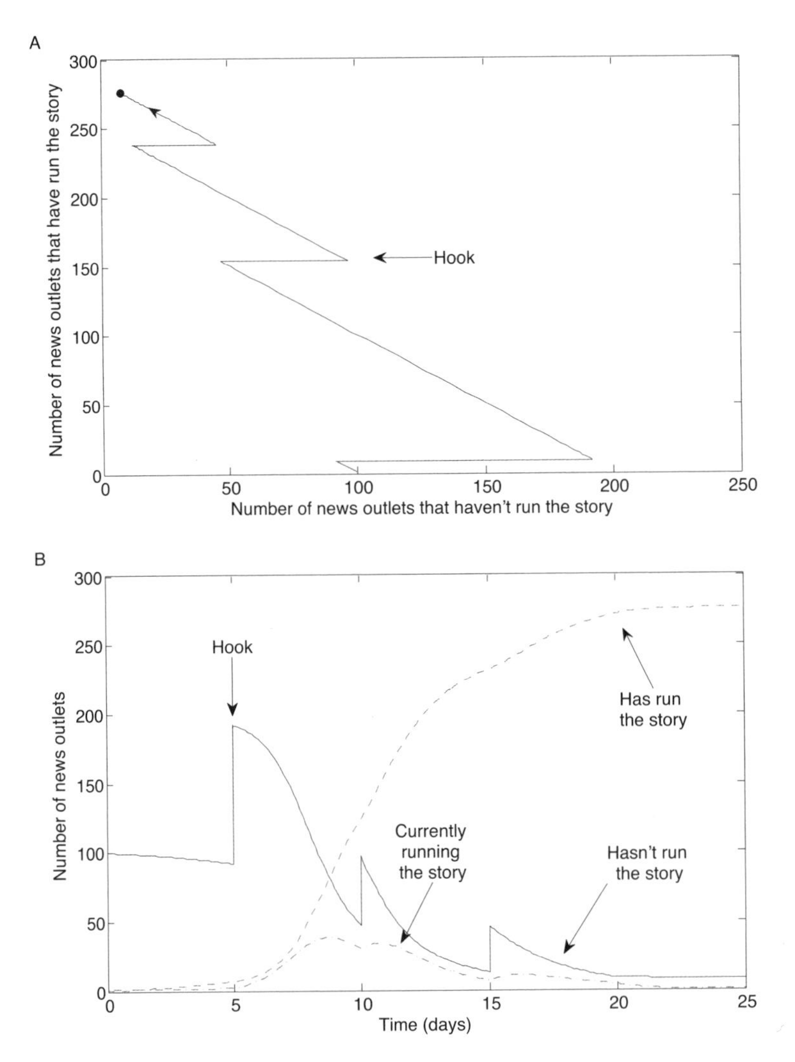

Figure 1.7: Behaviour of media outlets in the presence of multiple hooks. The story can be kept alive for significantly longer and be picked up by many more media outlets when successive hooks revive interest. A: The phase plane illustrating the trade-off between media outlets that have run (or are currently running) the story versus those that have not. B: The time course of the story through the media. Parameters were as in Figure 1.6, except that hooks of decreasing strength were added at regular intervals. In this case, there is a classic epidemic wave (double-peaked due to the impulses) and the majority of susceptible media outlets are infected, meaning that the story has gone viral.

to its maximum potential, with almost all susceptible media outlets running the story. See Figure 1.9.

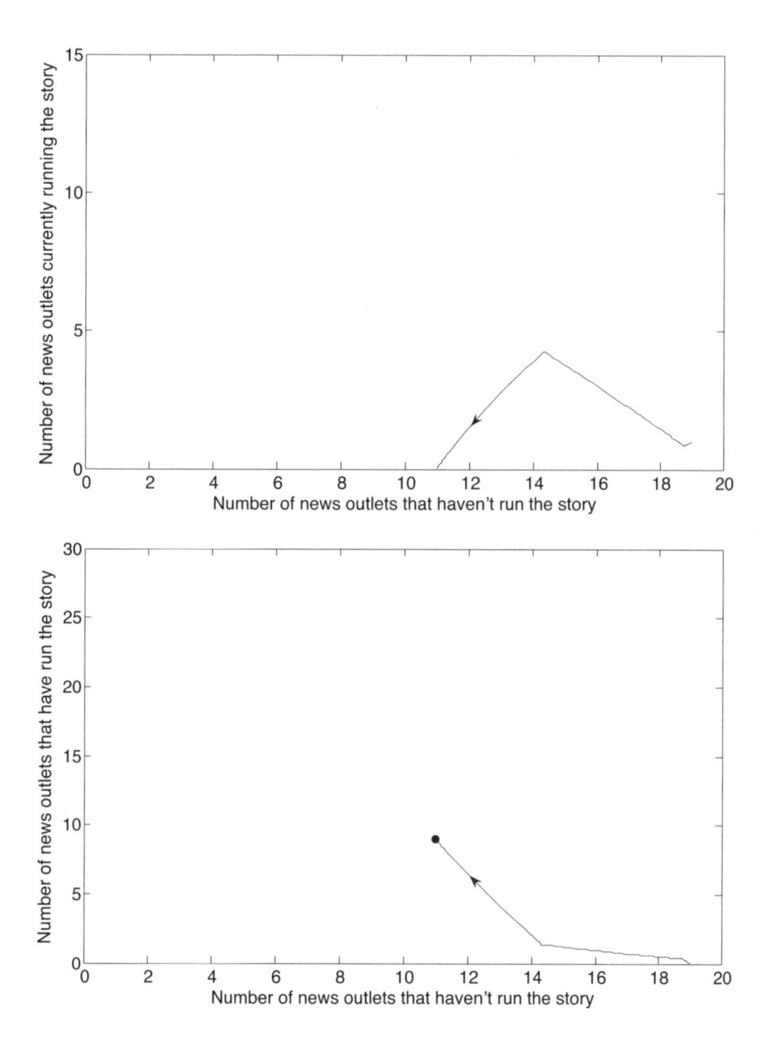

Figure 1.8: A good interviewee cannot compensate for a story not deemed newsworthy, resulting in the story barely registering in the media. Parameters were as in Figure 1.5 except that $\beta = 0.005$.

3. *Good topic, bad interview subject*

   If the the story has sufficient initial interest, then it can reap close to its maximum potential, with almost all susceptible media outlets running the story, even if it has no durability. See Figure 1.10.

4. *Secondary hooks, bad interview subject*

   In this case, secondary hooks mean that a story on its way out can receive new life. If a hook occurs early enough, this may result in a significant revival of the story, even in the absence of any long-term durability. See Figure 1.11.

It follows that a story can reap its original potential if the topic is sufficiently newsworthy or the story occurs during a slow news week. A good interview subject

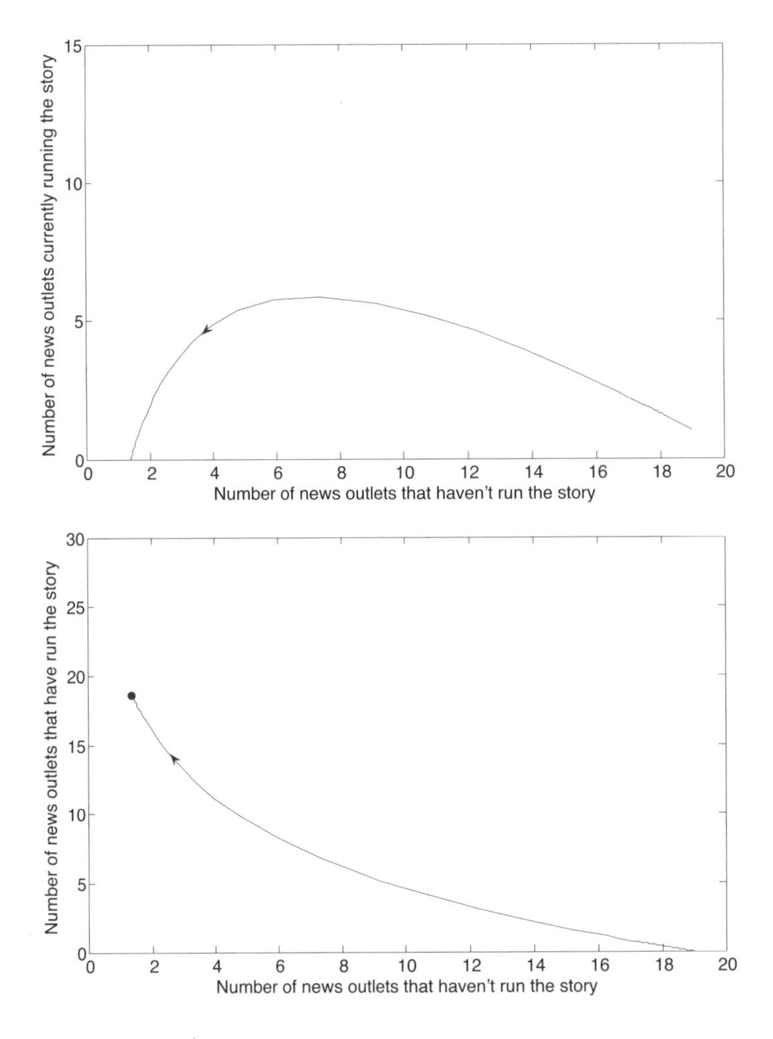

Figure 1.9: A slow news week can sustain a story, even in the absence of good interview subject. Parameters were as in Figure 1.4 except that $\frac{1}{\nu}$ was doubled.

can increase the power of a story. However, a series of secondary hooks, occurring at discrete times, can significantly expand a story's appeal beyond its original potential. This is especially true if the hooks occur early in the story's life.

1.6 Discussion

Just like zombies themselves, articles about zombies are the gifts that keep on giving: every time you think they're finally dead, they seem to come back to life.

Ultimately, the story of a mathematical model of zombies going viral was a confluence of circumstances: a diverting topic that happened to occur in a slow news week, which came with media-savvy interview subjects and had a major secondary

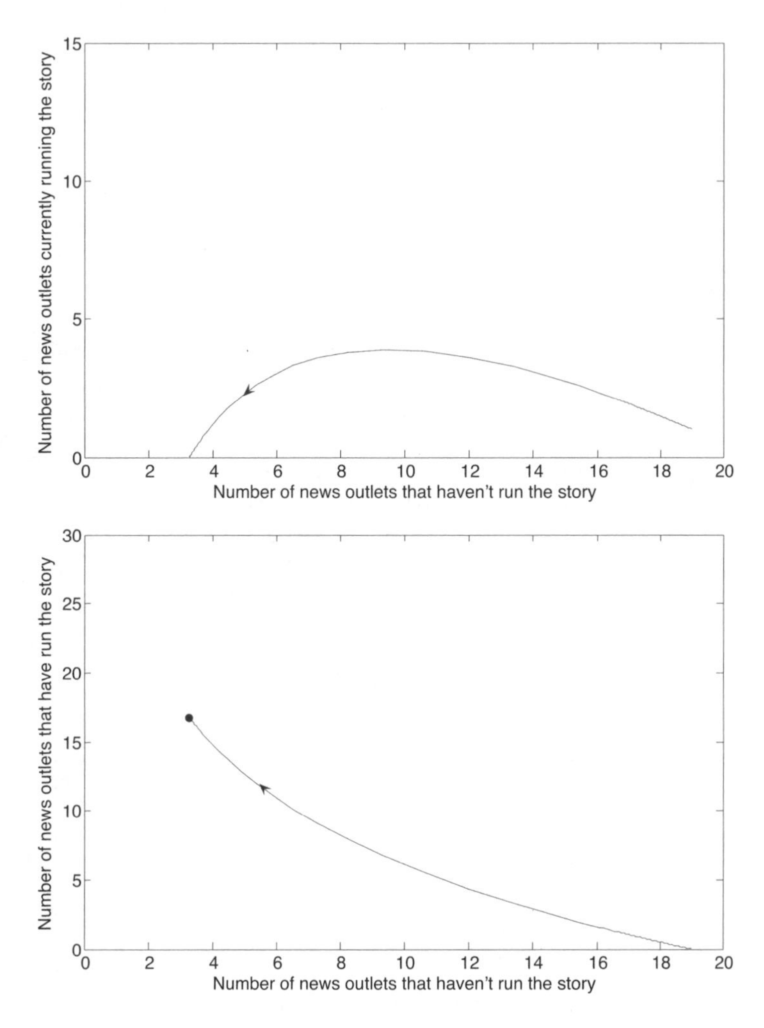

Figure 1.10: A fascinating topic can overcome a poor or nonexistent interview subject. Parameters were as in Figure 1.4 except that β was increased by 50%.

hook that occurred early in the story's lifespan. Furthermore, since it was a fairly self-contained phenomenon, it forms a useful case study for the effects of media.

For a story to go viral, it needs to be newsworthy and needs to have a long natural lifespan. Once a story is under way, a good interview subject can extend the lifespan of a story. However, a series of further hooks that increase a story's appeal have the potential to breathe new life into the story by creating new outlets for it to appear in. The effect is particularly significant if such a hook appears early in the news cycle. These hooks may occur randomly and may have different strengths when they do occur.

As an example of a story that did not go viral, but was studied intensely, consider Jensen's 1977 account of the lack of focus on lead in the blood of children living

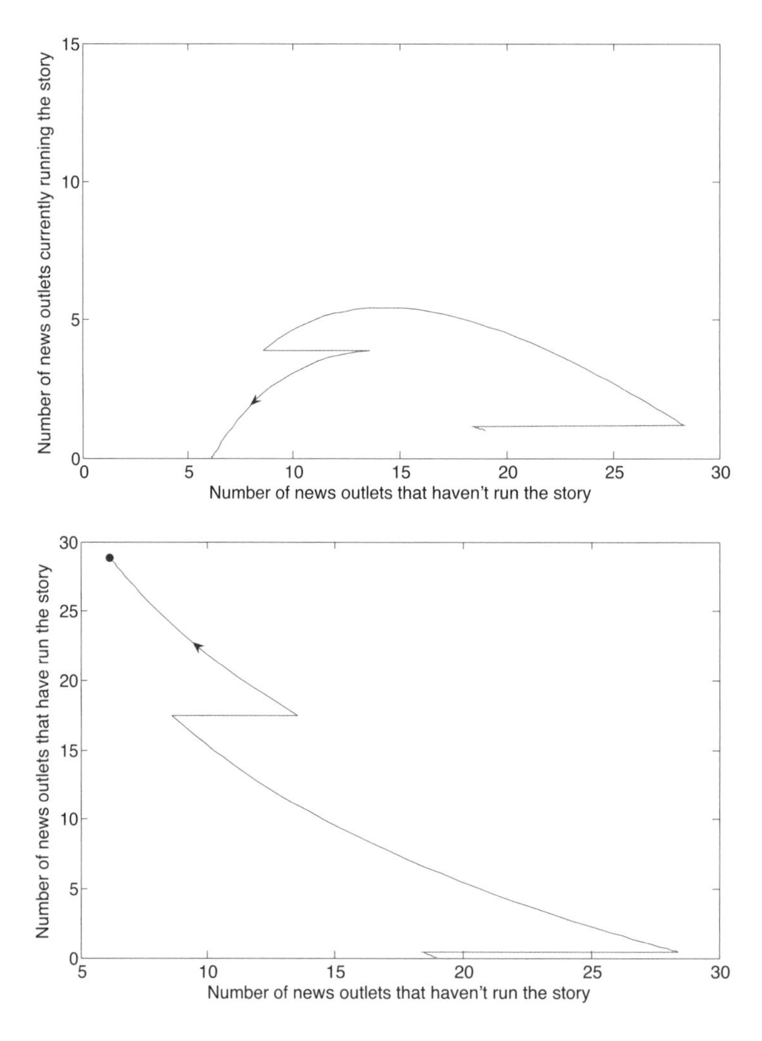

Figure 1.11: A series of hooks can sustain the story beyond its original lifespan, even in the absence of a good interview subject. However, hooks that occur early in the story's lifespan can breathe significant new life into the story, whereas those that occur late will not. Parameters were as in Figure 1.4 except that two hooks were added. Note that, despite an initial rise and the majority of susceptible media outlets being infected, there is no significant epidemic wave, despite the presence of secondary hooks. Thus this story does not go viral.

near a lead and zinc smelter in Kellogg, Idaho [47]. This story was featured briefly on national television, but thereafter had no national followup. A reporter for an independent local newspaper, Cassandra Tate, followed the story for two years, writing 175 articles on the subject. However, the story never received much attention beyond her direct employer, despite the presence of several hooks: the lead poisoning was not limited to Idaho and lead was later found in the newspaper-publishing industry. Jensen concluded with a general summary he called Cassandra's Law: the odds against comprehensive coverage of an environmental story are high, and are increased by complexity, proximity and distance. These effects are aggravated by the more recent dominance of corporate media.

In the context of our model, this story was somewhat newsworthy to begin with, but had no durability, further hooks failed to take hold and it was swamped by coverage of Evel Knievel's September 1974 motorcycle leap across Snake River Canyon, which was covered by hundreds of reporters. Thus, although β was relatively high, α was low (or zero), $\frac{1}{\nu}$ was low and there were no substantial secondary hooks.

A major difference that arises in this model, as compared to other epidemic models, is a counterpoint to cross-immunity. Many diseases provide some immunity to further infection, but in this model previous infection actually primes some media outlets for secondary outbreaks. Thus 'removed' media outlets are still infectious. However, since this form of infection decreases with time, the results are highly dependent on the speed of the epidemic. That is, if a story isn't going to go viral immediately, then it is much harder for it to take off later.

The model has several limitations, which should be acknowledged. Media 'infection' may focus more on an outlet's specific competitors than general reports, making the mass-action transmission factors less accurate. It would be instructive to generalize the model to include more details about the factors that determine 'newsworthiness,' in all its complication. Furthermore, the negative effects of media were not considered; if an interview subject was particularly off-putting, so that $\bar{\alpha} < 0$, then this could stop even a potentially fascinating story in its tracks.

Other effects that could be considered include the effects of unofficial media, such as blogs or Twitter (which were a factor in gaining initial attention in the zombie story and then sustaining a 'buzz' for it throughout its intense phase). However, while these may sustain a story that is under way, they do not, as yet, have the sheer reach that official media has and lack some of the factors that have been identified here, such as an interview subject. Furthermore, the lifespan of stories in these incarnations is significantly shorter, suggesting that stories would pass through their natural life cycle at an accelerated rate in the absence of traditional media.

It should also be noted that the factors affecting a story's newsworthiness, durability and natural lifespan are culturally specific. Thus what makes a story newsworthy in one media market may not in another. Media may also be limited in some places by government restrictions, the need to sell ads, etc.

This media model was informed by the news story of the zombie model going viral. Data from that story was used to inform and parameterize the media model and also to compare this news story with other news stories that did not go viral. Thus the zombie news story gives insights into how the media works, while also quantifying several factors.

In summary, a story can go viral, but it needs a perfect storm of events to do so. It must be deemed newsworthy in the first place and it needs room to breathe, a good interview subject and at least one hook. Given an arbitrary topic, the only factors external to the media itself that are potentially controllable are the skills of the interviewee (through media training) and perhaps the timed release of further information that acts as a secondary hook. Otherwise, the viral nature of a story is at the mercy of the randomness inherent in the media.

Appendices

A Acknowledgements

The author is grateful to Graeme Burk, Shoshana Magnet, Penelope Ironstone-Catterall, Sandra Robinson, Seth Sazant and Anna Kwik for technical discussions. The author also acknowledges two anonymous reviewers, whose comments greatly improved the manuscript. Handling editor for this chapter was Jane Heffernan. The author is supported by an NSERC Discovery Grant, an Early Researcher Award and funding from MITACS. For citation purposes, note that the question mark in "Smith?" is part of his name.

B Glossary

Durability. The essence of a media story's staying power, largely driven by how fascinating and insightful the interview subject is. For anyone wishing to interview me in greater, but entertaining, depth about this topic, I am easily contactable at rsmith43@uottawa.ca or by telephone after a quick google search of my now-quickly-identifiable name.

Epidemic wave. A steep rise in infections that reaches a peak and then starts to dissipate, eradicating the epidemic but leaving a trail of devastation and horror in its wake. A bit similar to the emotions you'd feel if you walked in on your parents in the act.

Equilibria. A state where all dynamic forces are balanced so that the net effect is no change in the system. A bit like being high, really.

Final size. The number of survivors of an epidemic. Finally, a mathematical definition that sounds suitably zombie-like!

Hook. Further information that increases a story's appeal, either to media outlets that were not previously interested or that causes those that have already run the story to run it again. They also use these things to catch fish, you know.

Impulsive differential equations. A fairly new type of mathematical model that contains continuous solutions punctuated by short, sharp shocks. These reset the system so that it starts again at new initial conditions that are related to the final conditions from the previous cycle. You know, before I was the first person to apply this theory to infectious diseases, I used it to examine sewage treatment for my Ph.D. That's right, my thesis was shit.

Infectious. A media outlet is infectious if it is currently running the viral story and has the potential to infect other media outlets by virtue of its cutting-edge insights, sheer depth of coverage and fascinating reporting. So Fox News is quite safe, then.

Jacobian. Just about the most massively useful thing that mathematics has ever invented. It combines multivariable calculus with linear algebra to create a matrix of partial derivatives whose eigenvalues almost completely determine the stability properties of an equilibrium. Okay, I realize this doesn't tell you what the Jacobian actually is, but some things defy a pithy one-paragraph summary, you know.

Mass-action transmission. A form of infection modelling that assumes every infected media outlet has equal chance of infecting every susceptible outlet. Mass action means that your cohort has to be either small or very well connected. So if you're modelling a serious disease in Africa, you'd have to restrict yourself to a small village where—Oh, I'm sorry, was I boring you? Oh look, here are some imaginary zombies!

Media. The collective apparatus for reporting the news. Mainstream media consist of newspapers, television, radio, etc., while other media include blogs, Twitter, tumblr and so forth. Between the writing and publication of this article, six entirely new media propagation vehicles have undoubtedly been invented.

Natural lifespan. The length of time that a story remains newsworthy, in the absence of further disturbances. Thus the inverse of this parameter is the rate at which a story becomes old. See also: Grandpa, and the precise point at which he stopped being sexy.

Newsworthiness. That mystical and capricious property that assigns some media stories value over others, seemingly at random, to the complete bafflement of anyone who has ever thought about it. Rupert Murdoch, you have a lot to answer for.

Nullcline. The place where just one variable is constant, although the others can usually change. Why my first-year calculus class can never grasp this simple concept, I'll never know.

Perturbations. Small changes to a system that might destabilize it. Either that or kinky sexual practices, I forget which.

Removed. A media outlet is removed/recovered if it has already run the story but may be capable of still infecting susceptible outlets. Any resemblance to zombies is purely coincidental.

Slow news week. Those rare times when nothing is taking the editor-in-chief's fancy. There you are, in the whiskey-filled back room, chain-smoking your way through packs of cigarettes, while the typewriters lie idle and your fedora with the word "press" stuck in it hangs sullenly on the hatstand. I think you know what I'm talking about.

Stable. A place where you keep a horse. No, wait, that's not right. It's when small perturbations away from an equilibrium return you to that equilibrium. Nothing to do with horses. Except for that one type of equilibrium that's called a saddle. But that type of equilibrium is never stable, so I really don't know why I brought it up.

Susceptible. A media outlet is susceptible if it is not infected but has the potential to be at some point in the future, regardless of whether that eventually happens. Also, one that is prone to jumping on the bandwagon of whatever the cool kids are doing. I'd tell you to see CNN, but I'm not that cruel.

Threshold. A definitive line between one thing and another. Not just the airy-fairy sense that today you're in the mood for waffles when yesterday you liked pancakes, but a real, honest-to-goodness change, like losing an arm and not being able to play the guitar any more. Don't mix these things up, because mathematicians take them very seriously indeed. Indeed, somebody once took something that wasn't a threshold and called it "the reinfection threshold." Much merriment was had in academia as a result.

Viral. A media story is viral if it is able to expand its reach significantly, with a virtual explosion of reports across multiple media outlets and types. Unless it uses a condom, of course.

Zombies. Hideous inhuman creatures who exist in a state of half-existence, shambling from location to location with outstretched hands and moaning audible pain, despite not feeling any. Or do I mean politicians?

References

[1] J. Gardy. Science tackles zombie attacks. *Globe Campus*, 10 August 2009.

[2] B. Scriber. What Can Zombies Teach Us About Swine Flu? *National Geographic*, 12 August 2009. http://blogs.ngm.com/blog_central/2009/08/what-can-zombies-teach-us-about-swine-flu.html

[3] P. Munz, I. Hudea, J. Imad, R.J. Smith?. When zombies attack!: Mathematical modelling of an outbreak of zombie infection. In *Infectious Disease Modelling Research Progress*, J.M. Tchuenche and C. Chiyaka, eds., Nova Science, Happuage, NY 2009, pp. 133–156.

[4] B. Mason. Mathematical Model for Surviving a Zombie Attack. *Wired Science*, 14 August 2009. http://www.wired.com/wiredscience/2009/08/zombies/

[5] S. Chase. Scholars put braaaaains together to thwart zombies. *The Globe and Mail*, 15 August 2009. http://www.theglobeandmail.com/news/national/scholars-put-braaaaains-together-to-thwart-zombies/article1253006/

[6] K. Wallace. Mathematicians use zombies to learn about swine flu. *The Toronto Star*, 18 August 2009. http://www.thestar.com/article/682691

[7] J. Goldstein. What's the Best Way to Fight Zombies? Someone Did the Math. *The Wall Street Journal*, 18 August 2009. http://blogs.wsj.com/health/2009/08/18/whats-the-best-way-to-fight-zombies-someone-did-the-math/

[8] P. Ghosh. Science ponders 'zombie attack.' *BBC News*, 18 August 2009. http://news.bbc.co.uk/2/hi/science/nature/8206280.stm

[9] K. English. The man in question is Smith? *The Toronto Star*, 22 August 2009. http://www.thestar.com/comment/article/684873

[10] M. Hanlon. Forget swine flu—could we cope with a plague of the Undead? Scientists ponder the threat of a zombie attack. *The Daily Mail*, 26 August 2009. http://www.dailymail.co.uk/sciencetech/article-1209052/Forget-swine-flu--cope-plague-Undead-Scientists-ponder-threat-zombie-attack.html

[11] K. Rose. Professor Robert J Smith? does maths on surviving oubreak of zombies. *Melbourne Herald Sun*, 20 August 2009. http://www.heraldsun.com.au/news/professor-robert-j-smith-does-maths-on-surviving-oubreak-of-zombies/story-e6frf7jo-1225764287719

[12] S. Leppänen. Tiedemiehet pohtivat zombien hyökkäystä. *Iltalehti*, 18 August 2009. http://www.iltalehti.fi/ulkomaat/2009081810098216_ul.shtml

[13] Who Will Win in Human, Zombie War? *All Things Considered*, National Public Radio, 20 August 2009. http://www.npr.org/templates/story/story.php?storyId=112075098

[14] N. Hayden. Zombies. *ABC TV*, December 2, 2009. https://www.youtube.com/watch?v=AjBxxKR5Uvg

[15] R. Smith?, ed. *Braaaiiinnnsss!: From Academics to Zombies.* University of Ottawa Press, Ottawa, 2011.

[16] Science of the Living Dead. *The Science and Entertainment Exchange*, 23 October 2009. http://blog.scienceandentertainmentexchange.org/2009/10/science-of-living-dead.html

[17] R. Liu, J. Wu, H. Zhu. Media/psychological impact on multiple outbreaks of emerging infectious disease. *Comp. Math. Meth. Med.*, 8(3):153–164, 2007.

[18] R. Ma. Media, Crisis, and SARS: An Introduction. *Asian J. Comm.*, 15(3):241–246, 2005.

[19] C.R. Simpson. Nature as News: Science Reporting in *The New York Times* 1898 to 1983. *Internat. J. Polit. Cult. Soc.*, 1(2):218–241, 1987.

[20] M.D. Slater, K.A. Rasinski. Media Exposure and Attention as Mediating Variables Influencing Social Risk Judgments. *J. Comm.*, 55(4):810–827, 2005.

[21] H. Lasswell. *Propaganda Technique in the World War.* MIT Press, Cambridge, MA, 1971.

[22] P. Lazarsfeld. *Radio and the Printed Page: An Introduction to the Study of Radio and Its Role in the Communication of Ideas.* Duell, Sloan and Pearce, New York, NY, 1940.

[23] J. Radway. *Reading the Romance.* University of North Carolina Press, Chapel Hill, NC, 1984.

[24] S. Hall. *Cultural Representations and Signifying Practices.* Sage Publications, London, UK, 1997.

[25] R. Williams. *Television: Technology and Cultural Form.* Fontana, London, UK, 1974.

[26] D. Hebdige. *Subculture: The meaning of style.* Methuen & Co., London, UK, 1979.

[27] P. Treichler. *How to Have Theory in an Epidemic.* Duke University Press, Durham, NC, 1999.

[28] L.C. Owens. International Accident, Disaster Stories Generate Greater Interest Among Students. *Newspaper Research J.* 28(2):107–113, 2007.

[29] T.C. Trautman. Concerns about Crime and Local Television News. *Comm. Research Reports* 21(3):310–315, 2004.

[30] F. Kuhn et al. What the readers see: How a sample of newspapers treats Washington news. *Columbia Journalism Rev.* 1:21–23, 1962.

[31] C. Burke, S.R. Mazzarella. "A Slightly New Shade of Lipstick": Gendered Mediation in Internet News Stories. *Women's Stud. Comm.* 31(3):395–418, 2008.

[32] K. Ross. The journalist, the housewife, the citizen and the press: Women and men as sources in local news narratives. *Journalism* 8(4):449–473, 2007.

[33] E. Herman, N. Chomsky. *Manufacturing consent: The political economy of the mass media.* Pantheon, New York, NY, 1988.

[34] T. Gitlin. *The whole world is watching: Mass media in the making and unmaking of the new left.* University of California Press, Los Angeles, CA, 1980.

[35] J.E. Good. The Framing of Climate Change in Canadian, American, and International Newspapers: A Media Propaganda Model Analysis. *Canad. J. Comm.* 33:233–255, 2008.

[36] J. Cui, Y. Sun, H. Zhu. The impact of media on the spreading and control of infectious disease. *J. Dynam. Differential Equations*, 20:31–53, 2003.

[37] Y. Liu, J. Cui. The impact of media coverage on the dynamics of infectious disease. *Int. J. Biomath.*, 1:65–74, 2008.

[38] Y. Li, J. Cui. The effect of constant and pulse vaccination on SIS epidemic models incorporating media coverage. *Commun. Nonlinear Sci. Numer. Simul.*, 14:2353–2365, 2009.

[39] J.M. Tchuenche et al. The impact of media coverage on the transmission dynamics of human influenza. *BMC Public Health* 11(Suppl 1):S5, 2011.

[40] O. Krakovska, L.M. Wahl. Optimal drug treatment regimens for HIV depend on adherence. *J. Theoret. Biol.* 246(3):499–509, 2007.

[41] B. Liu, Z. Teng, L. Chen. The effect of impulsive spraying pesticide on stage-structured population models with birth pulse. *J. Biol. Systems* 13(1):31–44, 2005.

[42] A. d'Onofrio. On pulse vaccination strategy in the SIR epidemic model with vertical transmission. *Appl. Math. Lett.* 18(7):729–732, 2005.

[43] V. Lakshmikantham, D.D. Bainov, P.S. Simeonov. *Theory of Impulsive Differential Equations.* World Scientific, Singapore, 1989.

[44] D.D. Bainov, P.S. Simeonov. *Systems with Impulsive Effect.* Ellis Horwood Ltd, Chichester, UK, 1989.

[45] D.D. Bainov, P.S. Simeonov. *Impulsive differential equations: Periodic solutions and applications.* Longman Scientific and Technical, Burnt Mill, UK, 1993.

[46] D.D. Bainov, P.S. Simeonov. *Impulsive Differential Equations: Asymptotic Properties of the Solutions.* World Scientific, Singapore, 1995.

[47] D. Jensen. The loneliness of the environmental reporter. *Columbia Journalism Rev.*, 15(5):40–43, 1977.

THE UNDEAD: A PLAGUE ON HUMANITY OR A POWERFUL NEW TOOL FOR EPIDEMIOLOGICAL RESEARCH?

Jane M. Heffernan and Derek J. Wilson

Abstract

Zombies and vampires have long been considered an existential threat to humankind that should be eliminated by any means possible. (If such a means exists, which it does not.) The Umbrella Corporation Centre for Research on the UnDead (CRUD) recognizes the inconvenience to humanity posed by the undead, and emphasizes that there is no conclusive proof of its involvement in the current tide of zombie-based destruction that is presently spreading across the globe. However, it must be recognized that the undead, while certain to bring about our destruction, provide a unique and powerful mathematical model for disease epidemiology. In this work, we hope to develop three simple mathematical models: one that describes the spread of zombies and zombie-like pandemic diseases such as bubonic plague and Spanish flu; one that describes vampires and vampire-like endemic diseases such as HIV; and one that describes werewolves and episodic diseases such as herpes. However, we doubt that we will have the time before the zombie horde is at our door. The zombie model will provide critical insights into zombie/disease dynamics, identifying the reasons we survived the plague but will not survive zombies. An additional benefit of our work is that the development of these models provides a pleasant distraction from the more mundane activities associated with staying alive while surrounded by hordes of flesh-eating undead.

2.1 Introduction

In recent years, the undead have advanced considerably, both in numbers and in their adeptness at flesh-eating and blood-sucking. Indeed, there is a growing list of cities completely overrun with zombies (e.g., London [1]; Milwaukee suburbs [2]; and New York [3]) or infested with vampires (e.g., Los Angeles [4]; Beaumont, Louisiana [5]; and Forks, Washington [6]). The viral spread of zombies through the media has already been documented. Things are looking grim . . .

. . . or are they? While some may feel a twinge of concern over these recent developments, we at the Umbrella Corporation [7] Centre for Research on the UnDead (CRUD) prefer to focus on the positive potential of our reanimated brethren. As an example, we discuss here how zombies and vampires provide us with an invaluable tool for studying certain less-mystical human afflictions.

In particular, we note similarities between the way zombies rapidly devour human populations, and epidemic diseases such as bubonic plague and Spanish flu. Vampires, in contrast, are more subtle in their infestation of humanity, providing an excellent model for endemic diseases such as HIV. Thus, by carefully monitoring the progress of the undead in their inexorable march toward the annihilation of the living, and by closely tracking our pathetic attempts to stop them, we can obtain key epidemiological insights. We expect that this research will have significant impact over the entire future of the human race which, depending on the ability of isolated groups of armed civilians to barricade themselves inside shopping malls, may extend for up to several months.

2.2 Creeping up on Us: The Origins of Zombies and the Bubonic Plague

Prior to developing our 'undead' model of disease, it is necessary to understand the nature of undead species and their similarities to the illnesses in question. As with the beginnings of many catastrophic plagues, zombies were known but not immediately recognized as an international threat by world leaders. While some might view our leaders' paralytic inaction as a colossal misstep leading ultimately to the annihilation of the human race, this must be viewed in the context of early zombie development. It must be remembered that zombies were initially:

1. *Limited in number*, probably due to the small number of Voodoo sorcerers available for their creation.

2. *Principally confined to the Caribbean* (specifically Haiti), though they were apparently occasionally exported (e.g., in 1945 a zombie was exported to New York to perform in the nightclub Zombie Hut) [8].

3. *Without the desire to consume human flesh or brains.* Flesh eating by zombies wasn't reliably reported until 1964 [9].

4. *Apparently uninfectious.* There are no records of zombie-to-human transmission of zombie-ism until 1968 [10].

Similarly, the bacterium *Yersinia pestis* existed prior to being responsible for three of the most devastating plagues suffered by humanity (including the Black Death that would ultimately kill up to 30% of all Europeans) [11], but was not recognized as a threat because it was:

1. *Limited in number.* There is no compelling evidence of large scale infestation of humanity by *Y. pestis* until the Justinian plague around 550 AD, though the bacterium has a genetic history stretching back on the order of 20,000 years [11].

2. *Principally confined to northern Africa*, although it was probably exported along trade routes with the Middle East [11].

3. *Without the ability to consume human flesh.* Prior to the Justinian plague, there is no conclusive evidence that *Y. pestis* was particularly virulent in humans, although this may reflect a low number of infections (see point 4 below) and/or poor records from the pre-Justinian plague era [11].

4. *Relatively uninfectious in humans.* The apparent lack of bubonic plagues prior to 550 AD may also result from inefficiency in infecting humans upon being transmitted by plague-carrying fleas.

Thus, in the origins of zombies and the bubonic plague, we find much common ground. Both were present prior to their decimation of the human race, but were not perceived as a future threat. Indeed, early zombies were evidently much sought after as performers in avant-garde night club entertainment [8]. It is a sad irony that this latent period in which we are blind to the coming dangers may be the only one in which effective steps can be taken to stem the tide of destruction.

A related commonality is the necessity for some kind of change, or mutation—to borrow a term from the biology of the living—that facilitates transition from the docile latent infestation to the virulent form. In zombies, the flesh-eating variety that appeared in 1964 was scary, certainly, but not in itself a viable candidate for harbinger of humanity's doom [9]. The critical change was the one that occurred just four years later [10], which allowed these shambling monsters to transmit their condition to living beings by eating their brains. The critical mutation in *Y. pestis* yielding increased virulence and/or infectiousness probably occurred immediately preceding the Justinian plague. Modern 'molecular clock' genetic experiments suggest that *Y. pestis* may have undergone a significant evolutionary change around 1,500 years ago, which agrees quite well with the $\sim$ 550 AD date of the Justinian plague [11].

Beyond the historical similarities, however, it must be noted that zombies and the bubonic plague bear little resemblance to each other today. This can be appreciated simply by looking at the relevant mortality rates. In modern times, there are just 1000–3000 cases of bubonic plague reported to the World Health Organization (WHO) each year (although the actual global number is probably somewhat higher) and, with early treatment, the mortality rate is approximately 1–15% [12]. Minor epidemics of plague continue to occur, mainly in sub-Saharan Africa, but the disease is generally not considered a global threat. Zombies, in contrast, have if anything become even more deadly over the past 40 years. Unlike *Y. pestis*, which is currently less transmissible than during the great plagues, zombies have evolved new ways of spreading their contagions and appear to be faster, stronger and more malevolently intelligent than their flesh-eating forebears. The current varieties have taken over London and by now most of the United Kingdom [1], New York [3], Milwaukee [2] and indeed probably most of the continental United States [13] in a short period, corresponding to a new infection rate of roughly 500,000 per day.

Another substantial difference is the complete lack of effective treatment for zombies. The mortality rate when infected by a zombie continues to be 100%. (Note: we recognize that 'mortality' may be the wrong word in this instance, given that the infected 'dead' are subsequently reanimated.) A few isolated incidences of resistance [14] and a potential curative serum [3] have been reported for some zombie strains, although emphatically not for the most recent and deadly phenotypes. Reanimated or infected corpses can be stopped using small-arms fire [2]; however, this is at best a temporary strategy as one typically runs out of bullets before running out of zombies. Based on data from the outskirts of London, cricket bats are a highly effective counter-zombie weapon [15]; however, this information may be of limited utility in North America where the great game of cricket never really caught on.

2.3 The Rising Tide: Modelling Epidemics

Having outlined the historical similarities between zombies and a typical epidemic disease, we will now show how the current zombie infestation can provide a model for future epidemics of new diseases. The first thing we must consider is how we want to express the answers that our model will generate. For example, it is possible to build a model that is designed to predict exactly how many people in, say, Alaska will die from an outbreak of bubonic plague, but to do this the exact population of Alaska would need to be known. In light of the unstable (i.e., rapidly shrinking) living human population at the moment, it is probably better to think instead of proportions of a population, so that it is not necessary to focus only on one group of survivors. From such a model, we will be able to say things like, "At the peak of the outbreak, which will occur thirty-seven days after the plague first appears, 50% of the population of will have been infected and 30% will have died." Relative predictions may not be as satisfying as hard numbers, and may be less useful for small populations, but thinking in relative terms frees us from the requirement of knowing the initial conditions (i.e., total population size, number of infected, number uninfected, number in recovery, etc.) at the onset of the outbreak.

Even without the need for hard numbers on population size, however, it would be helpful if we could know something about the composition of this population (such as gender ratio, age distribution, and general health) or their living conditions (such as how spread out they are and their level of sanitation). While we cannot know these parameters directly under the current circumstances, the zombie outbreak can simplify our guesswork. Evidence suggests that the constitution of zombie attack survivor groups will be eerily similar to the original populations from which they are drawn including, among other members, at least one representative of a racial minority, a woman and a child who may hold the key to human survival. We can also be quite sure that, over the next few months, surviving populations will have barricaded themselves inside shopping malls [2] or the local pub [15], which gives us some indication of population density. Groups of survivors tend to encounter at least one other, nearly identical group prior to their annihilation, giving us an idea of population mixing. Including these factors explicitly in our model would greatly increase its complexity and may only slightly improve our predictions. To keep things simple, we will instead 'roll' this information into the most important parameters which are the rates at which disease-related events, such as transmission, occur.

Here then, we have a basis for designing our model. We have a population of living humans, which we can separate into three classes:

1. *Susceptible* (S) are individuals who are ostensibly healthy, living humans who can get the disease. At the start of an outbreak, the vast majority of the population will be in this class.

2. *Infected* (I) are members of the population who have contracted the disease, and who can spread it. At the peak of an epidemic, a significant fraction of the population will be in this class.

3. *Removed* (R) are individuals who have had the disease and survived. For most epidemic diseases, including bubonic plague, cholera and zombies, the removed class is a small proportion of the population.

Movement between the classes is governed by the following rates:

1. *Infection* moves individuals between the 'Susceptible' and 'Infected' classes. Transmission occurs when an S and an I meet, with rate β.

2. *Recovery* moves individuals between the 'Infected' and 'Removed' classes. This occurs with rate σ.

3. *Loss of Immunity* moves individuals from the 'Removed' to 'Susceptible' class. This occurs with rate ω.

4. *Birth/Death.* Each class has its own birth and death rates, resulting in the addition or removal of individuals from any class. Death occurs with rates

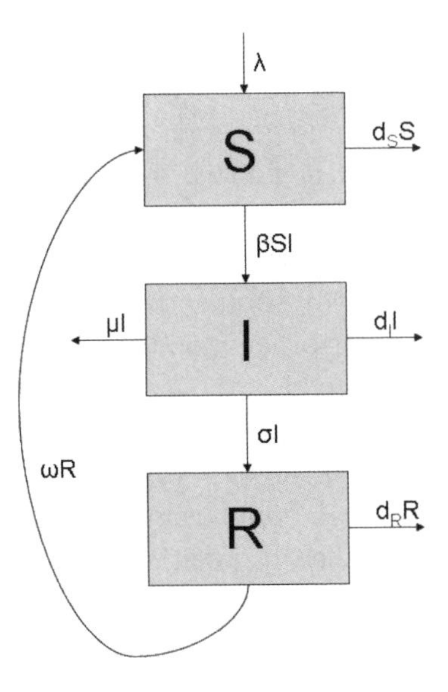

Figure 2.1: A schematic depiction of a model suitable for studying zombies and plague. Boxes represent classes of individuals: S = susceptible, I = infected, R = removed. Arrows indicate rates associated with movement into and out of classes; λ = the baseline birthrate, βSI = infection, σI = recovery, ωR = loss of immunity, $d_S S$ = natural death of susceptibles, $d_I I$ = natural death of infecteds, $d_R R$ = natural death of removed individuals, μI = unnatural death of infecteds.

d_S, d_I, d_R and μ (death due specifically to disease); births arrive with rate λ. For simplicity, we assume that everyone is born susceptible.

Thus, in diagrammatic form, our model will look like that shown in Figure 2.1. For the more mathematically inclined, we can represent these models as systems of differential equations as follows:

$$\frac{dS}{dt} = \lambda - \beta SI - d_S S + \omega R$$
$$\frac{dI}{dt} = \beta SI - d_I I - \mu I - \sigma I$$
$$\frac{dR}{dt} = \sigma I - d_R R - \omega R.$$

These equations depict the change in the S, I and R populations over time with respect to birth, death, infection, recovery and waning immunity.

But how can the current zombie plague help us refine our model? As described in Section 2.2, outbreaks of bubonic plague and zombies have much in common in their initial stages. The main difference, of course, is that humanity survived the plague while zombies will in all likelihood destroy us all. We can therefore look to zombies as a special case of plague where the rate parameters are such that the population of the living (the susceptible class) ultimately goes to zero. From this we can predict boundaries for these parameters which mark the 'tipping points' between ultimate survival of the human race and utter destruction. If in our (brief) future, a pathogen appears whose rate parameters lie on the 'destruction' side of those boundaries, we may be able to take action soon enough to ensure that our annihilation comes at the hands of a devil that we know (zombies), rather than some unfamiliar superbug. The next issue that we must tackle is how to determine our rate parameters.

For historical plagues, which we will use to define parameters that do not lead to the destruction of the human race, we can acquire rates of infection, recovery and loss of immunity from the scientific literature or historical records. As it happens, CRUD has retained historical papers on the bubonic plague during the Black Death that provide reasonable estimates of infection and recovery rates [16, 17, 18, 19, 20]. The most common epidemic studied is that of the plague in a town called Eyam which lost 76% of its population. Based on data from the Eyam plague [20], we find that:

- The rate of infection (β) was 0.1507 infections by each infected individual per day.
- Those infected recovered at a rate (σ) of 0.063 each day.
- The death rate in the infected class (μ) was approximately 0.027 each day so that the mortality rate is approximately 30% (0.027/(0.027 +0.063)).

The 'loss of immunity' rate is more challenging to determine experimentally, and so there is less literature evidence to work with. From the general literature, there is documentation that some individuals were infected by the plague more than once; however, others seemed immune to further exposure for life [21]. Here, we will assume that the rate of loss of immunity (ω) is about 0.0005 per day and thus immunity is lost in few years.

Given the rapid pace at which zombies are devouring the human race, and the hopeless preoccupation of the vast majority of remaining scientists with finding a cure, it is unsurprising that there is little in the scientific literature from which to acquire zombie-associated rate parameters for our model. However, we can rely on recent historical records from London that provide estimates for the zombie-generating 'rage' virus [14]. Under normal circumstances, the rage virus would not be highly transmissible, as it requires direct blood, saliva or mucous-membrane contact with infected blood or saliva. However, those infected with the rage virus are compelled to consume the flesh of those uninfected, providing an exceedingly efficient route for saliva-to-blood transmission. Based on the depopulation of uninfected individuals in London, which occurred over a period of 28 days, we estimate the infection rate (β) of the rage virus (and similar zombie viruses) to be approximately 0.5048 infections per infected individual each day. During this time, there was at least one case of 'recovery' insofar as the subject was bitten, but not subsequently consumed with the desire to eat human flesh. A reasonable rate of recovery (σ) would thus be $1/(7000000 \times 28) = 5 \times 10^{-9}$/day.

With zombies, it is difficult to define 'death' in the infected class, since the subjects are already biologically dead. To circumvent this difficulty, we will define zombie death as cessation of the ability to transmit the contagion. By this definition, the death rate is variable depending on the particular zombie phenotype. Zombies generated by the rage virus require the continuation of their normal biological functions and are thus susceptible to starvation, poison and, presumably, old age. Zombies prevalent in North America, such as those generated by the T-virus [7], are immune to most biologically disruptive agents and are also highly resistant to physical attacks. This means that zombies could live for hundreds or thousands of years. We thus assume that the death rate of zombies ($d_I + \mu$) is very small and can be approximated by zero since it is negligible compared to the rates of infection and death in humans.

We have not yet addressed the birth rate (λ) and the death rates of the susceptible and removed classes (d_S, d_R). In general, these parameters are very small compared to the rates of infection, recovery and death due to infection. Also, in the span of an epidemic, the number of births and deaths is very small compared to the actual size of the population. Therefore we can simply ignore these parameters and set them to zero.

With the addition of rates from the literature, our models are now in their final form. Figure 2.2 shows the results of the SIR model for both the plague (top) and zombie (bottom) cases. Here, we see that the plague increases but dies out a few months later. In contrast, zombies can take over the population in a very short period (approximately one month). Why this change in dynamics? The change depends on the values of the death rate (μ) and the recovery rate (σ). When these are very small (i.e., close to zero), infected individuals spend an infinite amount of time in the infected class. In other words, zombies take over because they don't 'recover' and they don't 'die.' This leads to a counterintuitive suggestion as to why humanity survived the bubonic plague: it was too lethal. The death rate in the infected class was so high that those infected died faster than they could transmit the disease. In the case of the Spanish flu, the opposite was true: it wasn't lethal enough. The recovery rate in the infected class moved people rapidly into the removed class (which had a low rate of return to the susceptible class through loss of immunity).

There are times when an infectious disease will not cause an epidemic. If, for example, infected individuals do not successfully transmit the infection, then the disease will die out. This is related to the parameter

$$R_0 = \frac{\beta}{\mu + \sigma} \tag{2.1}$$

which is called the basic reproductive ratio. In general, if $R_0 < 1$ the disease will die out. However, if $R_0 > 1$ the disease will cause secondary infections and will cause an epidemic (for more information on R_0 see [22, 23]).

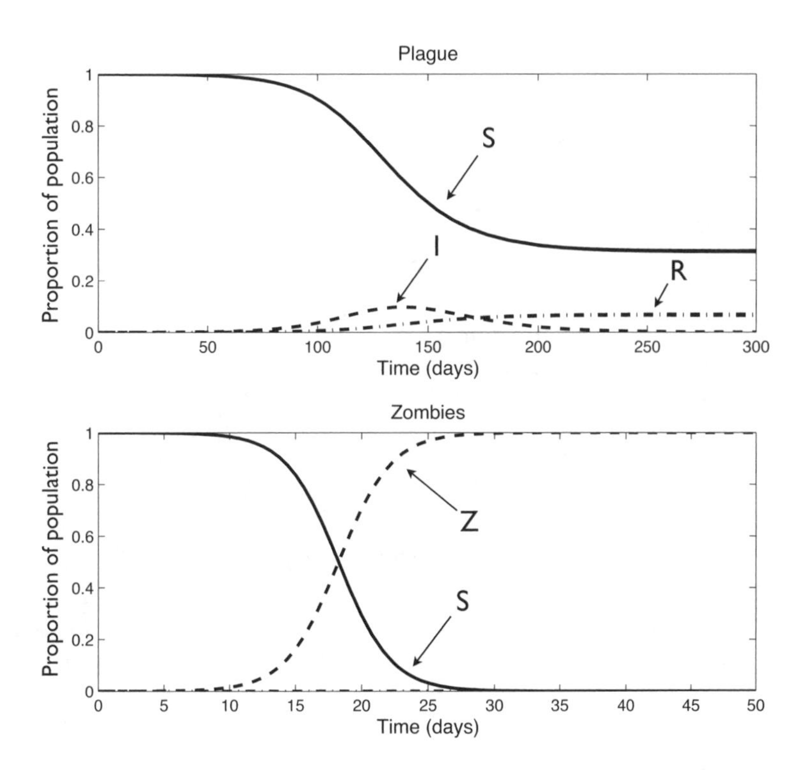

Figure 2.2: Epidemiological dynamics of the bubonic plague (upper) and zombies (lower). In the case of bubonic plague, there is a substantial outbreak that peaks and subsides over the course of a few months. In the case of zombies, the human race is entirely consumed over roughly 28 days.

Finally, now that we have a model and parameter values, we can achieve our objective of predicting the properties of a disease that distinguish it as a classical plague (one that we ultimately recover from) or an apocalyptic plague (one that will destroy us all). In Figure 2.3, we plot death rate vs. recovery rate in the infected class, identifying regions where the disease will: i) die out ($R_0 < 1$, above solid line), ii) cause a classical epidemic ($R_0 > 1$ and moderate, below solid line) or iii) take over the population (R_0 very big, small region around $\mu = 0$, $\sigma = 0$, circle). We also show the approximate positions of the bubonic plague (triangle) and the Spanish flu (star), which have moderate R_0 values that are greater than 1.

Using the final size relation, which can be derived from the SIR model [20], we can determine the number of susceptibles that will never be infected in the epidemic:

$$\log(S_0) - \log(S_\infty) = R_0 \left[1 - \frac{S_\infty}{K}\right], \tag{2.2}$$

where S_0 is the number of susceptibles at the beginning of the epidemic, S_∞ is the number of susceptibles after the epidemic is cleared, and K is the total population at the beginning of the epidemic which we will assume is S_0 plus one infected ($K = S_0 + 1$). This relation demonstrates that if R_0 is large, S_∞ becomes small; thus, in the case of zombies, which have a very large R_0, S_∞ will be almost zero.

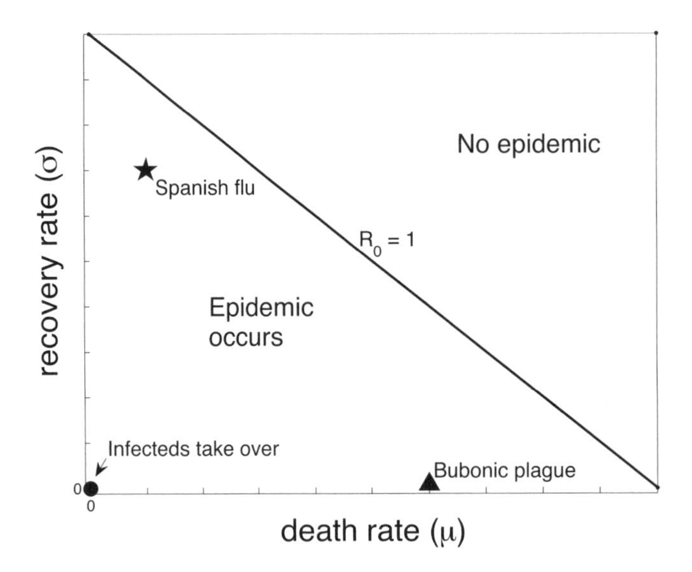

Figure 2.3: A phase diagram plot of disease-induced death vs. recovery with the boundary $R_0 = 1$ indicating the 'tipping point' between the occurrence and non-occurrence of an epidemic. The star corresponds to the Spanish flu (high rate of recovery and low death rate), the triangle denotes the bubonic plague (low rate of recovery and high death rate) and the circle represents zombies (very low rate of recovery and very low death rate).

We hope that these results should provide a small measure of comfort to those concerned about being killed by something other than a zombie. Fortunately, the low death-rate requirement for an apocalyptic plague virtually ensures that, even if some unknown illness besets what remains of humanity, those infected will survive long enough to be killed by the undead.

2.4 Living with Vampires: Modelling Endemics

Confronted with the current unstoppable tide of 'zombified' undead, it is sometimes difficult to remember that not too long ago the major threat to humans from the deceased came from another quarter entirely. While zombies have appeared globally since the mid-1940s, and have only very recently initiated their annihilation of the human race, vampires have been feeding on humanity for all of recorded history and probably beyond [24]. Our perceptions of vampires have changed over the years, from the bloodsucking demons of Mesopotamia to the charismatic, romantic vampires of the gothic era [25] to the present-day New World Order businessmen [4] or teen heartthrobs [6]. Scholars generally agree, however, that vampires have in fact changed very little throughout history, being always well-organized socially, and quite romantic and alluring, particularly when trying to suck your blood. Nonetheless, it is the characteristic of establishing a long-lasting equilibrium within human populations that interests us here. This tendency to persist amongst humans without sudden, large population fluctuations makes vampires an excellent model for endemic diseases such as HIV. In this section, we will aim to determine what properties of infection, recovery and immunity are required for the establishment of a long-lasting equilibrium within a human population, be it an intractable retrovirus (i.e., HIV) or a bloodsucking ghoul (i.e., vampires).

As far as general structure, our model for endemic vampires and HIV will look similar to that developed above for epidemic zombies (Figure 2.4). The susceptible S class corresponds to humans who have not been infected by a vampire (or HIV). Individuals in the infected class I are vampires (or HIV infected) and movement of individuals from S to I is governed by a rate of infection. The removed R class is missing in this case because there are no recorded cases of recovery from being a vampire, nor HIV (although potentially complete remission has been reported with multi-drug therapy). Note that a negligible (or missing) R class is not in itself a hallmark of endemic disease, as evidenced by epidemic zombies, where the R class population is essentially zero. In our vampire model, the R class is replaced by an abstinent A class, which consists of vampires or HIV-infected individuals who refrain from activities where transmission can occur (i.e., biting of humans and condom-less sex, respectively). This also includes individuals that are too sick to participate in such activities. The equations that correspond to Figure 2.4 are

$$\frac{dS}{dt} = \lambda - \beta SI - d_S S$$
$$\frac{dI}{dt} = \beta SI - d_I I$$
$$\frac{dA}{dt} = pd_I I - d_A A$$

and the basic reproductive ratio is

$$R_0 = \frac{\beta S_0}{d_I},$$

where S_0 are the number of susceptibles in the population when an initial infected individual is introduced. As before, if $R_0 > 1$, the disease will spread; if $R_0 < 1$, it will not.

As will be seen in what follows, endemic rather than epidemic dynamics are a result of infection occurring over long time scales, which allows for the replenishment of susceptibles. This can happen in a number of ways. For example, recent evidence indicates that vampirism is transmitted deliberately and only in a very small fraction of biting events [4, 5]; thus the infection rate of vampires is much lower than that of zombies. Similarly, in the case of HIV, transmission occurs in a fraction of sexual encounters. With relatively slow transmission, two rates that could be ignored in the case of zombies—birth and death in the S class—are now relevant. This allows for the possibility that the S class population is growing (through birth) at a rate that meets or exceeds the rate at which it is being depleted (through natural death and

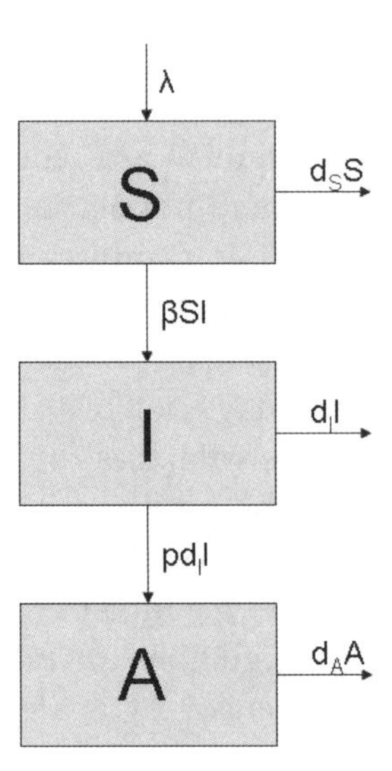

Figure 2.4: A schematic depiction of a model suitable for studying vampires or HIV. Boxes represent classes of individuals: S = susceptible, I = infected, A = abstinent. Arrows indicate rates associated with movement into and out of classes; λ = the baseline birthrate, βSI = infection, $d_S S$ = natural death of susceptibles, $d_I I$ = removal of infecteds either by death or abstinence, $d_R R$ = death of removed individuals and p = fraction of individuals leaving the infectious class that abstain from transmissible activities.

infection). Ultimately, this will lead to stable relative populations of susceptibles and infecteds. There is also a significant death rate for vampires due to their specific weaknesses, such as sunlight, garlic, infighting amongst vampire communities [6] and small yet highly effective groups of vampire hunters [26], which balances population growth due to new infections in the I class. The death rate of HIV-infected patients will also be elevated, ultimately, due to AIDS-related infections.

The similarities between vampires and endemic diseases such as HIV go beyond the appropriate model structure and include some qualities that contribute to the rates. For example, whereas zombies and sufferers of the plague are easily identified by their shambling gait, red eyes, open sores and so on, it is not possible to identify vampires or sufferers of HIV by visual inspection. This may seem an obvious point, but the result is profound: the ability to go unnoticed within a population makes responsive strategies like quarantine or selective vaccination much less effective, ensuring that a low rate of infection can be sustained. The low rate of infection itself is a common feature of endemic diseases, including HIV where transmission is estimated to occur in 1 per 1000 to 1 per 100 sexual contacts between infected and uninfected individuals [27]. Another common feature of vampires and HIV is that simple means of preventing infection are available. In the case of HIV, for example, latex condoms or abstinence from intercourse are highly effective, while, in an interesting analogy, vampire infection can be avoided by simply not inviting the vampire into your home. It's been four hundred years people, how difficult is this to comprehend?

Understanding these similarities, we can now undertake the crucial step of determining realistic rate parameters for our model. It is not feasible to use current birth and death rates for the S class since, because of the zombies, the death rate is inordinately high. However, using life expectancy statistics from a few years ago, we will assume that the average lifetime of a healthy individual in the population is 60 years ($1/d_S = 60$) [28]. For our purposes, we will assume that the size of a totally healthy population is constant such that $\lambda = d_S S_0$, where S_0 is the number of susceptibles at the time of infection. There are ample records of biting by vampires, but accounts are conflicting as to infection. Early accounts often suggest transmission rates of close to 100% [25], but more recent studies are almost unanimous in indicating low rates of transmission, with the majority of victims either being killed (without becoming vampires) or becoming 'blood doll' vampire followers [4, 5, 29, 30]. Taking the more modern view, we estimate the infection rate of vampire bites at 1 in 500,000 contacts between humans and a vampire per year ($\beta = 2 \times 10^{-6}$). For HIV, we also assume that the infection rate is very low; however, we assume that this is only slightly elevated to 1 in 250,000 contacts between susceptibles and an infected individual per year ($\beta = 4 \times 10^{-6}$) since HIV patients can presumably have contact during the day (when the sun is out) as well as at night.

Modern weaponry has had little impact on the death rate of vampire, unlike that of zombies. There are reports of guns, missiles or incendiary devices being somewhat effective, but vampires seem more susceptible to older technology such as wooden stakes, fire and religious symbols [26]. Indeed, humans armed with the latter have been known to successfully hunt vampires, resulting in a significant contribution to the vampire death rate (the other main contribution being accidental exposure to sunlight). Death is similarly heightened in the HIV-infected class due to compromised immune systems. HIV patients and vampires may also choose to abstain from activities that facilitate transmission. In the case of HIV, we assume that both AIDS patients and HIV-infected individuals who choose to abstain from these activities make up the A class. It is assumed that this accounts for 86% of individuals leaving the I class in the HIV model [31, 32]. For vampires, we assume that the number of individuals who abstain from transmission activities is reduced since records show that, despite their inner moral battles, they rarely succeed in abstaining from human blood for long [6, 29]. We assume this accounts for only 10% of individuals leaving the I class. Between death and abstinence, we estimate a vampire-infected-class removal rate of 0.1 per year (d_I). For HIV, we assume a removal rate of 0.2 per year (d_I) [31, 32] to reflect a higher death rate, but also increased success at abstinence.

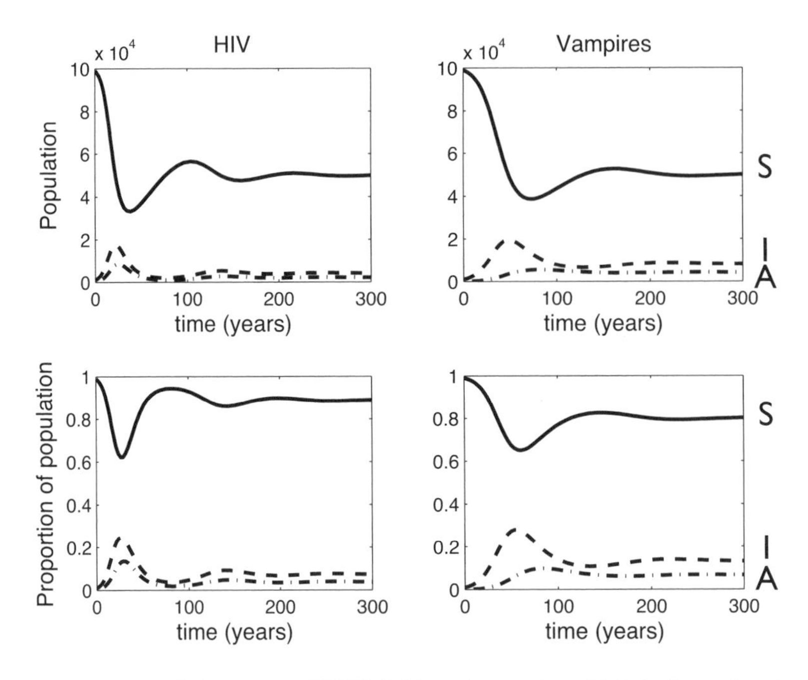

Figure 2.5: Epidemiological dynamics of HIV (left) and vampires (right). In each column, we plot the population dynamics by number (top) and proportion (bottom) of the population in each class. In both cases, the population sizes decrease. The total number of infectious individuals is similar.

Finally, we are left with the removal rates from the A classes. We assume that vampires in this class live longer than those in the I class since they will have less contact with humans (and where they do, they may be protected by vampire slayers or by being virtue of being CEO of an evil law firm). We assume a removal rate of 0.02 per year. For the HIV case, we assume that 30% of the A class is composed of AIDS patients with a high life expectancy of 1 year [32] and 70% is composed of abstinent HIV-infected patients of 20 years. This gives a d_A value of 0.335 per year. With these rates, our model gives the dynamics shown in Figure 2.5.

2.5 Conclusions

In this work, we have shown that the undead, while admittedly the harbingers of human destruction, are also good models for both epidemic and endemic disease. Zombies, with their ravenous appetite for human flesh and high rate of infection, are an excellent model for highly virulent, epidemic diseases such as the bubonic plague. Vampires, with their subtle infestation of humanity and low rate of transmission, provide valuable insights into endemic diseases such as HIV. The team at CRUD have found the generation of the models to be a highly engrossing exercise, to the point of almost forgetting about the groans and window pounding immediately outside our facility. We can therefore recommend this line of thought in future work, not only for the revelation of powerful new epidemiological insights, but also for comforting distraction. For example, one may wish to apply or extend the above models to investigate specific questions, such as the most cost-effective way to fight zombies and vampires, given the available countermeasures [33]. Or perhaps it might be informative to probe the obvious correlation between the periodic appearance of werewolves and endemic diseases with outbreaks such as herpes. We have initiated a study on this subject but, based on the increased rate of gunfire emanating from our security perimeter and our decreasing supply of ammunition (which is an interesting modelling question in itself), we may be forced to leave these important questions to others.

Appendix

A Glossary

Apocalyptic. Pertaining to the end of the world and/or physical universe.

Basic reproductive ratio. The number of secondary infections produced by an infected individual in a totally susceptible population.

Bubonic plague. A plague that repeatedly . . . plagued . . . the world starting in ca. 550 AD. The causative agent was most likely the bacterium *Yersinia pestis.*

Epidemic. A severe, but transient and localized outbreak of disease.

Epidemiology. The study of disease spread at the population level.

Endemic. A disease is endemic if it reaches an equilibrium of infection within a population and remains present at constant levels indefinitely (or until the equilibrium is strongly perturbed).

Herpes. A latent infection that can reactivate itself periodically causing symptomatic and asymptomatic episodes of transmission.

HIV. Human immunodeficiency virus, a virus that primarily attacks the immune system. HIV is a chronic infection that can progress to AIDS (acquired immune deficiency syndrome).

Outbreak. A period characterized by a rapid rise in infection.

Pandemic. A transient but severe and nonlocalized outbreak of disease.

Parameter. A varying number that represents a component of the system under study.

SIR model. A type of mathematical model for making epidemiological predictions of disease. Composed of susceptible (S), infected (I) and removed (R) classes.

Spanish flu. An unsually deadly pandemic of influenza virus starting in 1918.

Umbrella Corporation. An evil multinational corporation that is actually responsible for most zombie outbreaks. Also the primary sponsor for the present study pointing out that zombies can be useful models for disease.

Undead. A human (or occasionally animal) that has died and, for one reason or another, has subsequently un-died.

Vampires. Sophisticated undead who generally reside in urban centres in order to more easily feed on the blood of the living. Vampires generally try to keep their presence a secret.

Werewolves. The undead werewolf is a rare creature that, for the most part, lives as a human. The human, however, periodically turns into a wolf in the light of the full moon.

World Health Organization (WHO). A body of the United Nations responsible for monitoring global human health.

Zombies. Shambling undead with a propensity for eating the flesh of the living, particularly brains.

References

[1] A. Garland, *28 Days Later*, directed by D.A. Boyle. 20th Century Fox, 2002.

[2] J. Gunn. *Dawn of the Dead*, directed by Z. Synder. Universal Pictures, 2004.

[3] M. Protosevich, A. Goldsman. *I Am Legend*, directed by F. Lawrence. Warner Bros. Pictures, 2007.

[4] D.S. Goyer. *Blade*, directed by S.S. Norrington. New Line Cinema, 1998.

[5] A. Ball. *True Blood*. HBO, 2008–2014.

[6] M. Rosenberg. *Twilight*, directed by C. Hardwicke. Summit Entertainment, 2008.

[7] P.W.S. Anderson. *Resident Evil*, directed by P.W.S. Anderson. Constantin Film, 2002.

[8] R. Kent, L. Kimble. *Zombies on Broadway*, directed by G. Dines and G.M. Douglas. RKO Radio Pictures, 1945.

[9] D. Tenney. *I Eat Your Skin*, directed by D. Tenney. Cinemation Industries, 1964.

[10] G.A. Romero, J.A. Russo. *Night of the Living Dead*, directed by G.A. Romero. The Walter Reade Organization, 1968.

[11] R.W. Titball. An Elusive Serial Killer. *Nature* 430:145–146, 2004.

[12] CDC Plague Homepage, http://www.cdc.gov/ncidod/dvbid/plague/

[13] P.W.S. Anderson. *Resident Evil: Extinction*, directed by R. Mulcahy. Constantin Film, 2007.

[14] R. Joffé, J.C. Fresnadillo, E.L. Lavigne, J. Olmo. *28 Weeks Later*, directed by J.C. Fresnadillo. 20th Century Fox, 2007.

[15] E. Wright, S. Pegg. *Shawn of the Dead*, directed by E. Wright. Universal Pictures, 2004.

[16] S. Scott, C.J. Duncan, S.R. Duncan. The plague in Penrith, Cumbria, 1597/8: Its Causes, Biology and Consequences. *Ann. Human Biol.* 23(1):1–21, 1996.

[17] J.V. Nobel. Geographic and Temporal Development of Plague. *Nature* 250:726–728, 1974.

[18] G.F. Raggett. Modelling the Eyam Plague. *Inst. Math. Appl.* 18:221–226, 1982.

[19] M. Keeling, C. Gilligan. Metapopulation Dynamics of Bubonic Plague, *Nature* 407:903–906, 2000.

[20] F. Brauer, L.J.S. Allen, P. van den Driessche, J. Wu. *Mathematical Epidemiology*, Springer Scientific, New York, NY, 2008.

[21] L. Walløe. Medieval and Modern Bubonic Plague: Some Clinical Continuities, *Med. Hist. Suppl.* 27:59–73, 2008.

[22] J.M. Heffernan, R.J. Smith, L.M. Wahl, Perspectives on the basic reproductive ratio, *J. R. Soc. Interface* 2:281–293, 2005.

[23] J.M. Heffernan, L.M. Wahl. Improving estimates of the basic reproductive ratio: Using both the mean and the dispersal of transition times. *Theor. Pop. Biol.* 70:135–145, 2006.

[24] B.J. Frost. *The Monster with a Thousand Faces: Guises of the Vampire in Myth and Literature*. University of Wisconsin Press, Madison, WI, 1989.

[25] B. Stoker. *Dracula*. Archibald Constable and Company, London, UK, 1897.

[26] J. Steakley. *Vampire$*. Roc Books, New York, 1992.

[27] K.A. Powers, C. Poole, A.E. Pettifor, C. Myrenm. Rethinking the heterosexual infectivity of HIV-1: A systematic review and meta-analysis. *Lancet* 8(9):525–582, 2008.

[28] CIA World Factbook, https://www.cia.gov/library/publications/the-world-factbook/

[29] A. Rice. *Interview With the Vampire: The Vampire Chronicles*, directed by N. Jordan. Warner Bros., 1994.

[30] M. Rein-Haggen. *Vampire: the Masquerade.* White Wolf Publishing, Stone Mountain, GA, 1991.

[31] R.M. Anderson, G.F. Medley, R.M. May A.M. Johnson. A preliminary study of the transmission of the human immunodeficiency virus (HIV), the causative agent of AIDS. *IMA J. Math. Appl. Med. Biol.* 3:229–63, 1986.

[32] M.J. Keeling, P. Rohani. *Modeling infectious diseases in humans and animals.* Princeton University Press, Princeton, 2008.

[33] C.T. Bauch et al. Cost-utility of universal hepatitis A vaccination in Canada. *Vaccine* 25(51):8536–48, 2007.

WHEN ZOMBIES ATTACK! ALTERNATE ENDING

Phil Munz

Movie Poster

Authors of an earlier paper on mathematical modelling of zombies, *When Zombies Attack!: Mathematical Modelling of an Outbreak of Zombie Infection* used impulsive differential equations to model the eradication of zombies, leaving none undead. While the paper presented an effective conclusion to the zombie catastrophe, it omitted an important idea from zombie entertainment: that many zombie works provoke sequels. This suggests that a more effective model should include cyclic tendencies between human and zombie populations in order to describe fluctuations in the various populations (that is, nearly all zombies are destroyed in the conclusion of the original work; however, they find a way to rise again at the commencement of the sequel). An alternate ending to the original paper must therefore be considered to keep it true to the canon of such great works as *Resident Evil*, *Return of the Living Dead* and George A. Romero's *Dead* series. This chapter will first recap the pertinent ideas from the original paper. It will introduce the new model—a variation of Lotka–Volterra's predator–prey model—determine whether this cyclic model would be useful for measuring human–zombie dynamics and back up the model assumptions using popular zombie references. These ideas will be presented in the context of an alternate ending: an epidemic in its own right currently wreaking havoc on the DVD industry.

3.1 Title Sequence

THE original paper, *When Zombies Attack!: Mathematical Modelling of an Outbreak of Zombie Infection* by Munz et al. [1], concluded with the eradication of zombies using impulsive differential equations (the spontaneous younger sibling of ordinary differential equations). The reasons for this were twofold: i) the authors tried to provide a happy ending to the paper, one in which the humans were victorious; and ii) the paper was greatly influenced by the movie *Shaun of the Dead* [2], which similarly concluded with the purging of the zombie hordes.

Some discussion has arisen on how many humans are really left once all the zombies are destroyed. Is it still a happy ending if only a small fraction of humans remain? There is speculation that, due to the grim content of the paper, the number of humans that survive is probably not very high. Concerns have also been brought to the attention of this author regarding the realism of the zombie eradication scenario. Can enough resources be assembled to eliminate all zombies in a way that is reflective of the original paper? How can so many zombies be instantaneously eradicated as was modelled? Another solution is needed, one that allows audiences and fans of zombie media to applaud humanity's victory over the zombie hordes and go home happy, confident that some realistic outcome has been achieved, rather than simply a hand-waving solution. The original paper therefore needs a new ending; an alternate ending, if you will.

The movie industry has created a plethora of special features that are now included in their new film and DVD releases. The zombie film sub-industry has been subject to this epidemic. Some special features include: scenes added to the end of or during the credits (*Dawn of the Dead* [3]), director's special, or unrated cuts of the film on DVD (George A. Romero's *Land of the Dead* [4]), 3-D films (*Resident Evil: Afterlife* [5]) and, of course, alternate endings (*Resident Evil* [6]).

The author posits that directors, producers and writers must come up with many different ideas on how to conclude a movie during its production.[1] While ultimately only one ending is released as the 'official' theatrical ending, many variations are still filmed. These unused ending scenes are usually included in the DVD release as a special feature to give the viewer a more complete experience (or are they just another shameless marketing tactic used to raise the selling price of the DVD?).

3.2 The Story So Far . . .

Munz et al. [1] chose to model the slow-moving,[2] flesh-eating, undead variety of zombie, introduced by George A. Romero in *Night of the Living Dead* [7] and popularized by media such as *Shaun of the Dead* [2], *Return of the Living Dead* [8], *Resident Evil* [6, 9] and *Dead Rising* [10]. As a quick aside, it is interesting to note that, of all the films and games mentioned above, only *Shaun of the Dead* does not have a sequel.

First, a basic model was developed, slightly tweaking the ideas developed by the SIR epidemic model of mass-action dynamics [11]. As seen in the previous chapter, such models are useful for understanding a vast array of epidemics, both

[1]Or the idea comes to one of the writers as an afterthought after original production was completed . . . sound familiar?

[2]While the authors used the term "slow-moving," no spatial dynamics were considered in the paper. This chapter also does not consider any spatial dynamics, except that the references only feature the slow-moving variety of zombie.

living and (un)dead. Early on, it was clear that this basic model required revision to bring it closer to reality. As the basic model dealt with instantaneous change from human to zombie, the next step was to develop the idea of latency, inspired by the work of Max Brooks [12]. This model did not produce any positive results, and it was determined that, even with longer latency periods, humanity would still be eradicated quickly by zombies (upwards of seven days [1]).

An important concept of the classical SIR analysis (and consequently of the somewhat less classical SZR analysis) is measuring the basic reproductive ratio (R_0). R_0 is used as a threshold value to predict whether a disease will persist ($R_0 > 1$) or die out ($R_0 < 1$). Diseases that are endemic are modelled under various control strategies in an attempt to slow down the rate of infection (i.e., lower R_0) with the ultimate goal of reaching an $R_0 = 1$ threshold, essentially breaking the chain of transmission. Re-evaluation of R_0 after each intervention can inform us of the efficiency of such measures. The dismal results of the latent model included determining that the disease was endemic. Considering interventions in the latent model became necessary since, in any prototypical zombie tale, a few bands of humans are always left after an apocalypse, so clearly something was done to save humanity (if only just barely).

One such control strategy was the containment and quarantine of zombies. New problems arose and were analyzed using this model. It was realized that quarantine would work as long as large numbers of those infected by a zombie but not yet zombified (due to keeping the idea of latency in this model) were discovered. The problem was the impossibility of knowing who was infected and who was not, as the physical attributes of an infected person would be no different from those of a healthy human, except for a bite wound of some kind. The other problems were the infrastructure and time limitations of trying to coordinate an effective quarantine. Thus the proposed solution would only delay the inevitable zombie invasion.

A further intervention was hypothesized: would it be possible to quickly produce a cure for zombie-ism and, if so, what impact would it have? It was proven that there can be cohabitation between zombies and humans who are given a highly effective treatment with no immunity to reinfection. The downside only a small fraction of humans would be left fighting against the massive hordes of zombies. This idea also assumed the unlimited resource of the treatment. Thus the outcome was still grim.

At this point in the analysis of a zombie outbreak, the authors turned to the impulsive eradication model, which is where our alternate ending begins. The scenario we are left with is this: it is assumed that the analysis of the effectiveness of the previous intervention strategies has caused the zombies to gain more of a foothold in our closed system. Recall that the original paper started with a large number of susceptibles (the population of a metropolitan city). Here, due to the time lag from earlier analysis, the situation has become more dire and the proportions have changed to be 50-50: only half of the population in this city is human.

3.3 A Behind-the-Scenes Look at the Making of the Alternate Ending

In a blatant case of *deus ex machina*,[3] the author has brought in new characters to solve the zombie problem. Alfred Lotka and Vito Volterra, two mathematicians of the late nineteenth and early twentieth centuries, who both independently developed predator–prey equations.[4] Lotka–Volterra models have been widely used to describe the dynamics of various organic and biological systems such as predator–prey interactions, competition and disease, and are also used extensively in economic theory.

Instead of considering zombies as a disease and the interactions between humans and zombies as transmission dynamics, suppose we view the current scenario under a different light. It is not much of a jump to consider that, while zombies are the carriers of the zombie disease, they are also preying on humans in order to transmit said disease. If we implement a model considering predator–prey interactions, then this model should also be a good approximation to human–zombie interactions during a zombie outbreak.

From a mathematical perspective, very little has changed. Instead of using an epidemic model—a system of first-order (impulsive) differential equations—we will use a variation of Lotka–Volterra's model, which is also a system of first-order differential equations.

The equations used in the mathematical analysis are from the traditional Lotka–Volterra predator–prey model, where the prey in this case are susceptible humans and the predators are zombies:

$$\begin{aligned}\frac{dS}{dt} &= \alpha S - \beta SZ \\ \frac{dZ}{dt} &= \delta SZ - \gamma SZ - \zeta Z.\end{aligned} \tag{3.1}$$

Here, S represents the number of humans and Z represents the number of zombies. α is the rate at which the human population grows, β is the rate at which zombies and humans meet, δ is the rate at which the zombie population grows, γ is the rate at which zombies are destroyed by humans and ζ is the rate at which zombies expire from other means. (These rates are discussed further, later on.) S and Z represent the number of humans and zombies, respectively. As a caution, the rates used here are not the same as those in the original paper, even though the equations look similar to those represented in the basic model.

There are no big twists here, as the SZR model from [1] was to the classic SIR model. The only difference is the distinction between zombie deaths by humans and other zombie removal from the system. In order to be considered pertinent to the

[3]A plot device whereby a seemingly insoluble problem is suddenly solved with the unexpected intervention of some new character or previously unmentioned device.

[4]Whether they came to the author in a dream or as zombies is not integral to this scenario.

current situation, the model assumptions and physical meanings will be brought to light and backed up by actual events observed in zombie media.

The first few assumptions come from Hastings [13], slightly tweaked, and form the basis of the equations given above:

- In the absence of zombies, the human population grows exponentially.
  The human population since the 1920s has grown at a near-exponential rate [14]. The focus of this chapter is still on a short time scale and only devoted to a metropolis. Little of the short-term growth is attributed to births. Rather, it could be argued that the growth in this city mostly comes from the immigration of survivors. This can be seen in *28 Weeks Later*[5] [15] and in George A. Romero's *Land of the Dead* [4].

- In the absence of humans, zombies die off exponentially.
  Zombie films where humanity is eradicated typically end on this note, without exposing the viewer to the long-term effects on zombies. Other films never reach this point; either the zombies do not survive or they cohabitate with the remaining humans in some way. Being organic creatures, it is likely that zombies would expire without brains to sustain them, as seen in *28 Days Later* [16].

- The per-zombie rate at which humans are killed is a linear function of the number of humans.
  This assumption gives the nonlinear term βSZ from the above equations. It goes back to the original paper, which says that humans and zombies must interact in order for humans to be killed and become zombified. This can be seen in all traditional zombie media. This assumption also suggests that the per zombie rate at which humans are killed is independent of the number of zombies (i.e., the zombies act independently). This is also prevalent in zombie works. While the zombies all have the same objective in mind, and many will attack a human at once, the concept of teamwork does not exist among the zombie ranks.

- Each human death contributes to the growth of the zombie population.
  This comes straight from zombie canon, since those preyed on by zombies usually wind up becoming zombies themselves. All works referenced by this chapter can serve as an example.

A number of additional assumptions regarding the environment and evolution of the human and zombie populations [17] have also been considered:

- The human population can find ample food at all times.
  As this chapter is only looking at events in the short term, this assumption

[5]While the movie does not contain slow-moving zombies, the canon of this film series is very similar to the canon of the classic zombie movies.

is reasonable. In films that have looked at human survival in the long term, such as *Resident Evil: Afterlife* [5], where many years have elapsed since the initial zombie outbreak, the human characters still have enough food.

- The food supply of zombies depends entirely on the human populations. This was also an assumption of the original paper. In nearly all zombie media, humans serve as prey. There are few examples where zombies have been seen consuming something else (in [7], for example) and these instances are not considered in the model.

- The environment does not change during the process in favour of either species. Since we are only attending to the short term, this assumption is realistic.

Given these assumptions and various physical interpretations, the following considerations were given to the variables in the Lotka–Volterra equations:

- α: The rate at which the human population grows will not consist only of births; these were ignored in the original paper that was modelled on a short time scale. It will include immigration into the system (survivors, reserves, etc.), which will make up nearly the entirety of the rate. The context of the αS term in comparing it to zombie entertainment makes sense, as more survivors will gather together when the human number is large already, while lower numbers of humans will have more difficulty finding each other. α should be a relatively small rate to stay consistent with the original paper.

- β: This was discussed earlier; to prey on humans, zombies must interact with humans. This rate will be a number relatively close to the β rate of the original paper: the rate at which a human succumbs to an encounter with a zombie and subsequently becomes one.

- δ: The rate at which the zombie population grows depends on an encounter with a human. This could be considered the success rate at which a human is zombified through an encounter with a zombie. Since zombies are mindless, we assume that they do not realize that their numbers increase through attempts at human consumption. Thus zombie efficiency in adding to their ranks is low, and the value for δ will be less than the value for β. However, due to the way in which the zombie population increases (dead humans becoming undead so long as they are not completely destroyed), the value for δ will be greater than in conventional Lotka–Volterra models (e.g., lynx and snowshoe hare dynamics [13]).

- γ: The rate at which zombies are destroyed in an encounter with humans. This is similar to the second nonlinear term in the original paper and will have values close to those given to the α value used in [1].

- ζ: The rate at which zombies die by means other than through an encounter with humans. This is used to be consistent with the assumption that, in the absence of humans, zombies cannot survive. Early on, this value could be used to describe zombies that, as humans, had been mutilated to such a degree that they (now zombified) are unable to prey on humans and are therefore removed. This number is quite small.

3.4 The End?

Now that you have suffered through the writer's commentary,[6] it is time to get to the climax of the feature presentation. To be consistent, the analysis follows the same format as in the original paper. One major difference that should be clarified here is that the original model presented zombies as a disease. In the following analysis, using the Lotka–Volterra system, the focus is not so much on transmission of a disease, but rather on the rates of change in two populations, one predators (zombies) and the other its prey (humans).

Setting the model equations (3.1) equal to zero provides our equilibrium points. From the first equation, one obtains $S = 0$ or $Z = \frac{\alpha}{\beta}$, and from the second, $Z = 0$ or $S = \frac{\zeta}{\delta-\gamma}$. Using the Jacobian matrix, it can be shown that the equilibrium obtained when both humans and zombies are absent, $(S, Z) = (0, 0)$, is unstable.[7] Therefore the rest of the analysis deals only with the second equilibrium point, $(S, Z) = \left(\frac{\gamma}{\delta-\gamma}, \frac{\alpha}{\beta}\right)$.

The Jacobian of the system is

$$J = \begin{bmatrix} \alpha - \beta Z & -\beta S \\ Z(\delta - \gamma) & S(\delta - \gamma) - \zeta \end{bmatrix}.$$

Evaluating the Jacobian at the equilibrium point gives:

$$J\left(\frac{\zeta}{\delta - \gamma}, \frac{\alpha}{\beta}\right) = \begin{bmatrix} 0 & -\frac{\beta\zeta}{\delta-\gamma} \\ \frac{\alpha(\delta-\gamma)}{\beta} & 0 \end{bmatrix}.$$

Thus

$$\det(J - \lambda I) = \lambda^2 + \alpha\zeta > 0.$$

The trace of the Jacobian matrix at the equilibrium point here is 0. This results in closed solution cycles (see Figure 3.1A). Further analysis of other basic models [13] has suggested that the current model of this system is inadequate for determining whether this equilibrium is stable or unstable.[8] At this point, it is still possible to do some biological analysis and make conclusions.

---

[6] A special feature that is usually optional on DVDs, but, due to the writer's ego, has been forced upon the reader in this case.

[7] This equilibrium would also make for a very uninteresting piece of zombie fiction.

[8] For the non-mathematically inclined, stable could be read as likely, unstable as not likely.

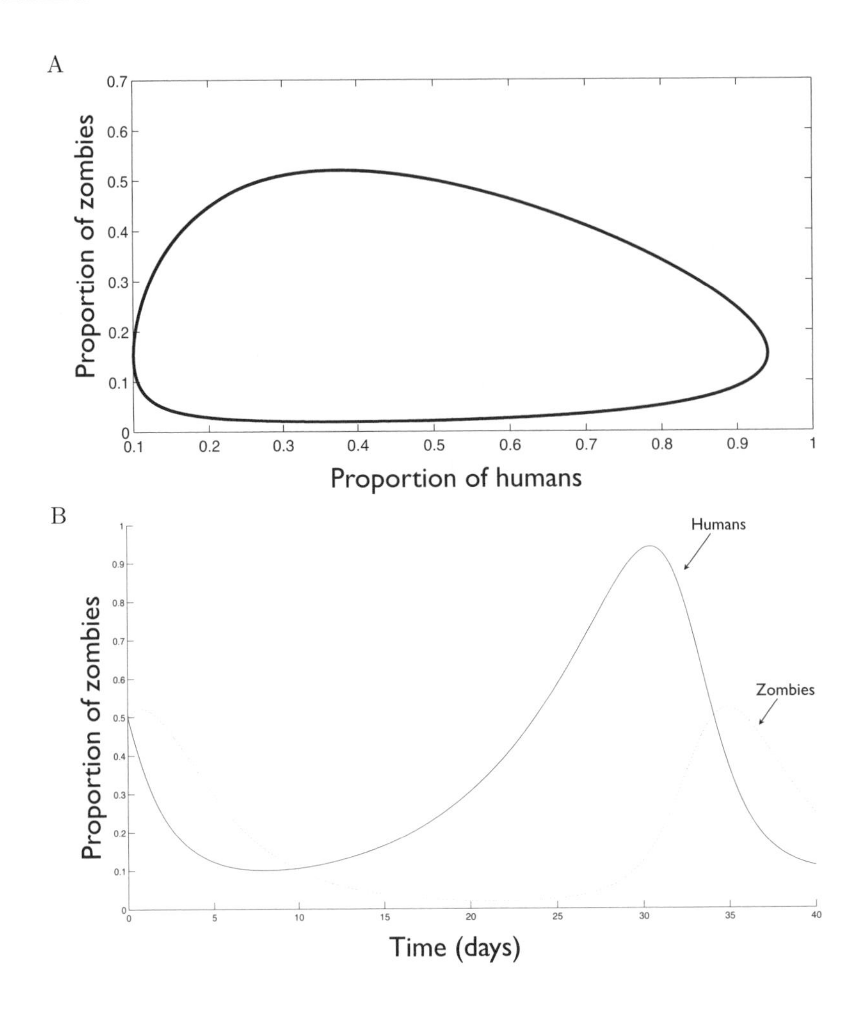

Figure 3.1: A: Human–zombie phase portrait. B: Time series. The initial population was 50% human and 50% zombie. Parameters were $\alpha = 0.0015$, $\beta = 0.0098$, $\gamma = 0.003$, $\delta = 0.008$ and $\zeta = 0.0001$. The phase portrait demonstrates the existence of cycles and therefore hope for humanity.

The existence of cycles for the time being gives hope to humanity that they can endure a zombie outbreak (at least, longer than they could based on the other models). This may have been influenced somewhat by the assumption in this model that other humans may enter the system leading to an increase in humanity's numbers. However, this could also have allowed the zombie population to grow as well due to a greater food supply. Clearly, humanity appears to have benefited more than the zombies from the assumption on α.

In the context of zombie media, limit cycles are also useful. Cycles demonstrate the existence of sequels.[9] When zombie population numbers are low enough, it could signify the ending of one movie. Humans figure that the threat of zombies is over and they go back to their ordinary lives. Then something catastrophic occurs that allows the zombie hordes to increase in number once again, and the sequel has begun (this is essentially the plot idea of the movies in the *Return of the Living Dead* series [8]). From the general theory of basic Lotka–Volterra models [13], the model oscillates with whatever amplitude is determined by the initial conditions; in this case, these conditions would be based on the criteria of the writers, creators and directors of zombie media.[10]

3.5 DVD Extras

This chapter only looked at a very basic model, but it has at least laid the groundwork for more rigorous mathematical analysis of using predator–prey models to explain human–zombie dynamics. Nonetheless, the model remained effective in demonstrating the variations in population levels that usually come from zombie media that are part of a series. It was left unproven whether these cycles are stable or unstable. Perhaps this was done on purpose, leaving the audience with a cliffhanger ending to take to the water cooler and discuss *ad nauseam* with their peers (or perhaps the author is trying to milk this series of publications further; not even he knows for sure).[11]

The focus of this chapter was different from that of the original, which may have influenced the change in both mathematical and biological conclusions. While every effort was made to keep some form of continuity between the two, a more developed model would help to smooth out the differences, or possibly show which type of model better describes the interactions between humans and zombies. One could argue that, since this model was more basic and less rigorous than the original, the conclusions above are even less reasonable. Does that mean zombie sequels are not realistic?

Whether these mathematical models and their conclusions are realistic or not, one thing is certain: when the dead rise and zombies do in fact attack, humanity will be prepared, thanks to the efforts of the many writers, directors and producers (and mathematicians) involved in zombie media.

---

[9]Actually, they demonstrate the necessary conditions for sequels. These limit cycles don't imply that sequels *have* to happen. Indeed, some sequels should never happen, mathematically acceptable or not.

[10]I am sure a rigorous mathematical analysis is done by most writers and creators before beginning a zombie series, in the hopes of producing multiple sequels.

[11]This manuscript was meant to be an entertaining application of an already-established theory. Due to the author's basic understanding of the theory, the outcome was expected. This is not necessarily what is done in practice.

Awards

A Oscar Speech

The author wishes to thank George Romero, Simon Pegg and all the other references for their zombie influences. Special thanks go to Robert Smith? for his invitation to submit to this volume. The author is extremely grateful for the help from those he knows who have better grammar and communication skills than he does, especially Phillip Gallant, Doug Woolford and Craig Olinski, whose efforts are evident in this document.

B IMDB

Impulsive differential equations. A subset of differential equations containing impulsive effects. The solutions of impulsive differential equations are piecewise continuous in their domain, which is due to the nature of impulsive system.

Jacobian. A matrix of partial derivatives, created by taking the partial derivative of each equation with respect to each variable separately.

Latency. The time between exposure to a pathogen or disease and the appearance of symptoms or signs of infection.

Limit cycles. Isolated closed trajectories in two-dimensional phase space (the space in which all possible states of the system are represented). Any neighbouring trajectories of limit cycles are not closed, and spiral either towards or away from a limit cycle. Limit cycles are generally found in nonlinear dynamical systems.

R_0. The basic reproductive ratio. It is a threshold value that indicates the expected number of secondary infections produced by a single infectious individual throughout its lifetime.

SIR model. One basic type of compartmental model found in epidemiology. Compartmental models illustrate the progress of a disease in a large population over time. The compartments of this model are Susceptible, Infected and Removed, which are all functions of time: $S(t), I(t), R(t)$.

Trace. The sum of the elements of the main diagonal of a square matrix.

End Credits

[1] P. Munz, I. Hudea, J. Imad, R.J. Smith?. When zombies attack!: Mathematical modelling of an outbreak of zombie infection. In *Infectious Disease Modelling Research Progress* J.M. Tchuenche and C. Chiyaka, eds., Nova Science, Happuage, NY 2009, pp. 133–156.

[2] E. Wright, S. Pegg. *Shawn of the Dead*, directed by E. Wright. Universal Pictures, 2004.

[3] J. Gunn. *Dawn of the Dead*, directed by Z. Synder. Universal Pictures, 2004.

[4] G.A. Romero. *George A. Romero's Land of the Dead*, directed by G.A. Romero. Universal Pictures, 2005.

[5] P.W.S. Anderson. *Resident Evil: Afterlife 3D*, directed by P.W.S. Anderson. Constantin Film, 2010.

[6] P.W.S. Anderson. *Resident Evil*, directed by P.W.S. Anderson. Constantin Film, 2002.

[7] G.A. Romero, J.A. Russo. *Night of the Living Dead*, directed by G.A. Romero. The Walter Reade Organization, 1968.

[8] D. O'Bannon. *The Return of the Living Dead*, directed by D. O'Bannon. Orion Pictures, 1985.

[9] S. Makimi. *Resident Evil*, directed by S. Makimi. Capcom, 1996–2014.

[10] M. Ikehara. *Dead Rising*, directed by Y. Kawano. Capcom, 2006.

[11] F. Brauer. Compartmental Models in Epidemiology, In *Mathematical Epidemiology*, F. Brauer, P. van den Driessche, J. Wu, eds., Springer, Berlin, Germany, 2008, pp. 19–79.

[12] M. Brooks *The Zombie Survival Guide: Complete Protection from the Living Dead.* Three Rivers Press, New York, NY, 2003.

[13] A. Hastings. *Population Biology: Concepts and Models.* Springer, New York, NY, 1997.

[14] F. Krausmann, S. Gingrich, N. Eisenmenger, K.H. Erb, H. Haberl, M. Fischer-Kowalski. Growth in global materials use, GDP and population during the 20th century. *Ecol. Econ.*, 68(10):2696–2705, 2009.

[15] R. Joffé, J.C. Fresnadillo, E.L. Lavigne, J. Olmo. *28 Weeks Later*, directed by J.C. Fresnadillo. 20th Century Fox, 2007.

[16] A. Garland, *28 Days Later*, directed by D.A. Boyle. 20th Century Fox, 2002.

[17] R.W. Hiorns, D. Cooke, eds. *The Mathematical Theory of the Dynamics of Biological Populations II.* Academic Press, London, UK, 1981.

WHEN HUMANS STRIKE BACK! ADAPTIVE STRATEGIES FOR ZOMBIE ATTACKS

Bard Ermentrout and Kyle Ermentrout

Abstract

We describe a novel model for interactions between zombies and humans such that humans can have an advantage over zombies when the population density of the zombies is low; however, with greater numbers of zombies, the advantage is lost. This endows the system with bistability and oscillatory behaviour. We then extend the model to allow for the humans to become complacent when zombie attacks are infrequent. This provides a mechanism for periodic zombie outbreaks.

4.1 Introduction

IN a recent article, Munz et al. [1] suggested a number of models for zombie attacks on human populations, likening these to disease outbreaks. In addition to the trends of a standard disease outbreak, the zombie apocalypse may show some more interesting dynamics such as multistability and oscillations, as Munz himself has shown in the zombie paper's sequel (Chapter 3). To see this in more detail, we allow for some reaction of humans to their zombie enemies. Rather than playing a defensive role against disease, humans are capable of becoming the aggressors against the undead. The zombie makes for a relatively vulnerable victim: it seems to lose any agility and speed that it may have had while living and it is mindless, meaning that it is incapable of any strategic action. The average person should easily be able to dispatch a lone walking corpse, provided they are equipped with the proper knowledge and minimal armament. The true danger of the zombie becomes apparent when he is in the company of his peers. Through involuntary vocalizations (those despair-inducing moans), the zombie unwittingly attracts other hungry fellows when he becomes aware of his next potential meal [2]. Acting without fear, zombies will unhesitatingly swarm any source of warm flesh; if a human is not careful, she can easily become surrounded and overwhelmed. In this chapter, we consider a number of mechanisms through which humans strike back at the zombie invaders over the long run. We introduce a new model for zombie–human interactions which allows for multiple stable end states where either zombies or humans are dominant. We then introduce a model in which humans can slowly adapt to their zombie attackers and find oscillations and other dynamics.

4.2 Multiple Zombies

To motivate this first simple model, we consider the scene at the house early in *Night of the Living Dead* [3]. Ben and others are able to fend off attacks by single zombies and successfully kill them (using the basic principle "Kill the brain and you kill the ghoul"). This is because the zombies of this particular strain are slow and rather uncoordinated. Thus in one-on-one encounters with alert and prepared susceptibles, the single zombie will generally be destroyed. What allows success in zombies is the sheer number of living dead. Groups of two or more zombies can successfully attack and defeat all but the most well-armed human. For example, a human with a chain gun is likely to survive a concerted attack by many zombies. However, a single person with a screwdriver, such as in the *Dawn of the Dead* [4], would probably be killed in an attack by multiple zombies. With these empirical observations, we consider our first simple model. Let $Z(t)$ denote the population of zombies and $S(t)$, the susceptible population. We describe interactions using the laws of mass action:

$$\star \xrightarrow{a_1} S \tag{4.1}$$

$$S \xrightarrow{a_2} \star \tag{4.2}$$

$$S + Z \xrightarrow{a_3} S \tag{4.3}$$

$$S + 2Z \xrightarrow{a_4} 3Z \tag{4.4}$$

$$Z \xrightarrow{a_5} \star \tag{4.5}$$

Here (4.1) represents the migration of new humans into the area of zombie infestation while (4.2) is the natural death rate of humans leaving the infested area. (4.3) depicts the ability of the susceptible population to pick off zombies when they occur individually. (4.4) represents the idea that packs of zombies (here represented by two for simplicity) are able to kill susceptibles. To keep this initial model simple, we have assumed that the killed humans are instantly transformed into zombies. (4.5) represents the death of zombies by natural or other means, such as the accidental beheading by the airplane in an early scene in *Dawn of the Dead* [4]. Let (s, z) denote the population of susceptible humans and zombies, respectively.

Then

$$\frac{ds}{dt} = a_1 s_0 - a_2 s - a_4 s z^2$$
$$\frac{dz}{dt} = -a_3 s z + a_4 s z^2 - a_5 z.$$

Here s_0 is the number of people outside of infested areas who are able to enter the region where the zombie outbreak has occurred. One of the first steps in modelling is to reduce the number of parameters and to make the equations dimensionless [5]. If we divide the first equation by a_2, and let $\hat{t} = a_2 t$ be a dimensionless time, let $s = \hat{S}S$ and $z = \hat{S}Z$ where $\hat{S} = a_1 s_0 / a_2$, then we can eliminate several parameters from the model and obtain:

$$\frac{dS}{d\hat{t}} = 1 - S - \hat{a}_4 S Z^2 \tag{4.6}$$
$$\frac{dZ}{d\hat{t}} = -\hat{a}_3 S Z - \hat{a}_5 Z + \hat{a}_4 S Z^2. \tag{4.7}$$

There are now just three dimensionless parameters, $\hat{a}_3 = a_3 a_1 s_0 / a_2^2$, $\hat{a}_4 = a_4 s_0^2 a_1^2 / a_2^3$ and $\hat{a}_5 = a_5 / a_2$. For notational simplicity, we will drop the hats for the remainder of the chapter. The reader should keep in mind that the three rate constants, $a_{3,4,5}$ can all be expressed in terms of the rate of migration of fresh human meat and the steady-state levels of humans in the absence of attacks. For example, $S = 0.25, Z = 0.75$ means that the humans are reduced to 25% of their population before the attack and zombies represent 75% of the pre-attack population of humans.

There can be up to three equilibrium points to this model and we will see that two of these can be stable. Recall that an equilibrium point satisfies $dS/dt = dZ/dt = 0$ [6]. The point $(S, Z) = (1, 0)$ is always an equilibrium and it is always asymptotically stable. This means that for low-density zombie attacks, humans always survive. Add equations (4.6) and (4.7) together to get

$$1 - S - a_3 S Z - a_5 Z = 0$$

and solve this to get

$$S = \frac{1 - a_5 Z}{1 + a_3 Z}.$$

Substituting this expression into equation (4.7), we get

$$Z(a_3 + a_5 - a_4 Z + a_4 a_5 Z^2) = 0.$$

As expected, we find $Z = 0$, but, additionally, two other roots:

$$Z = \frac{a_4 \pm \sqrt{a_4^2 - 4 a_4 a_5 (a_3 + a_5)}}{2 a_4 a_5}. \tag{4.8}$$

If the term inside the radical is positive, we have two positive roots and thus there are three equilibria. If the natural death rate of the zombies is small and the rate at which humans successfully kill zombies is small, then the zombies will be able to mount an attack that overwhelms the humans and wipe out a substantial portion of them.

We still must show that the new equilibrium is stable. To determine stability, we linearize the right-hand sides of equations (4.6) and (4.7) about the equilibrium point. The eigenvalues of the resulting matrix (called the Jacobian) tell us whether or not the equilibrium is stable. If the eigenvalues have negative real parts then we have stability and if there is at least one eigenvalue with positive real parts, then the equilibrium is unstable [6]. For a two-dimensional system, as we have here, stability is assured if the trace of the Jacobian matrix is negative and the determinant is positive [5]. The Jacobian matrix for our model is

$$J = \begin{pmatrix} -1 - a_4 z^2 & -2a_4 zs \\ -a_3 z + a_4 z^2 & -a_5 - a_3 s + 2a_4 zs \end{pmatrix}.$$

For $Z = 0$, it is clear that the trace is negative and the determinant positive. The nontrivial roots are a bit tougher to study. We note that for a_5 small, the middle root is approximately a_3/a_4 and the large Z root is roughly $1/a_5$. Plugging these approximations in, we find that the trace is negative for both roots and that the determinant is positive for the large root and negative for the middle root. Thus, as suspected, the large root is stable and the middle root is unstable.

It is convenient to sketch the phase plane for this system in order to understand the qualitative dynamics. Figure 4.1A shows the (S, Z) phase plane along with the nullclines $dS/dt = 0$ and $dZ/dt = 0$. Intersections show the three equilibrium points: two stable (circles) and one unstable (square). The unstable equilibrium is a saddle point and thus has a stable manifold (labelled SM) [6]. This means there is a pair of trajectories that go into the saddle point as $t \to \infty$. They form a separatrix between the two stable equilibria: all initial conditions above the curve go to a persistent high zombie state while all those below go to a zero zombie state. To see this, we also show two trajectories starting at $S = 1$ and Z above (b) and below (a) the stable manifold. Thus if $Z(0)$ is less than about 1.75 zombies, then the zombie population will collapse while if $Z(0)$ is larger than 1.75, the zombie population explodes and humans are nearly wiped out. This is not good news.

The key parameter that determines the threshold is a_3. This is the ability of a human to beat a zombie in a one-on-one interaction. If the humans are totally unprepared, then this term could, in fact, be negative and we would have to adjust the equations to reflect the fact that the susceptible human involved in the battle was lost to the population. Thus, to incorporate humans completely unready to do battle with zombies, we add the term

$$+\min(a_3 SZ, 0)$$

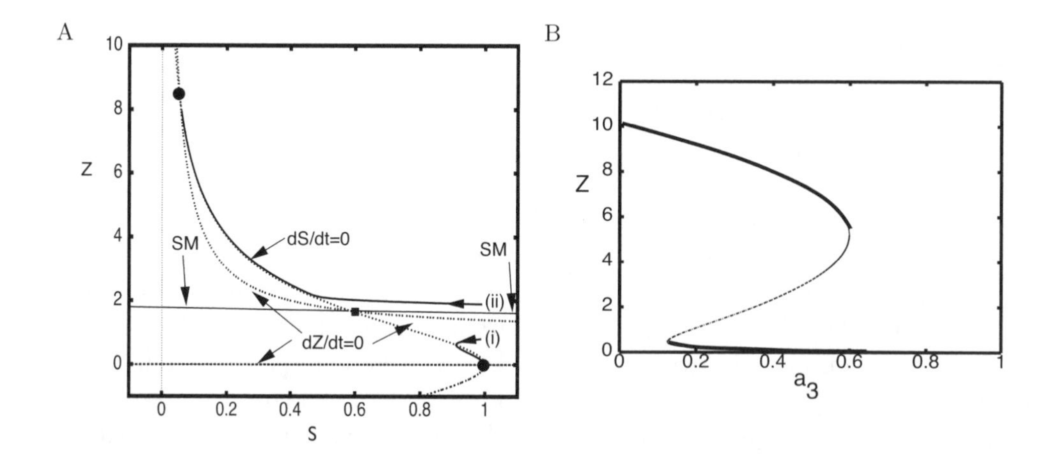

Figure 4.1: Analysis of bistable zombie model. A: Phase plane of simple zombie model. Z-nullclines ($dZ/dt = 0$) and S-nullclines ($dS/dt = 0$) are shown. Stable equilibria are shown with filled circles ($a_3 = a_4 = 0.25$, $a_5 = 0.1$). The black square is a saddle point. B: Bifurcation diagram as a_3 varies showing the equilibrium value of Z ($a_4 = 0.25, a_5 = 0.1$). Solid curves are stable and dashed are unstable. This and all other diagrams in this chapter are computed using XPPAUT [7].

to the dS/dt equation. If $a_3 < 0$, then we subtract from the population. We remark that when $a_4 = 0$ and $a_3 < 0$, we recover a simplified version of the original Munz model [1].

Let us add one more small alteration to our present model. We allow for the possibility of zombie migration. That is, zombies from somewhere else are allowed to enter our local population. To this end, we add a small source term to the zombie equation and obtain the general two-dimensional model that includes group attacks by zombies:

$$\frac{dS}{dt} = 1 - S - a_4 S Z^2 + \min(a_3 S Z, 0) \tag{4.9}$$

$$\frac{dZ}{dt} = -a_3 S Z - a_5(Z - Z_0) + a_4 S Z^2. \tag{4.10}$$

It is no longer so simple to find the equilibria, so instead we compute them numerically as the parameter a_3 varies. Figure 4.1B shows the equilibrium zombie population as the parameter a_3 changes. Because of the small migration term, there is a positive number such that if a_3 is smaller than this value, the lower (nearly zombie-free) state does not exist and zombies rule. Similarly, for large enough values of a_3, the zombies can never take hold and the population is largely zombie-free. (A few will always be wandering around due to migration; we suspect you may have encountered them occasionally in your life.)

In conclusion, in this section we have shown that preparedness for zombie attacks can be sufficient to maintain a low or zero population of zombies. However, if groups of them form, and there are enough, the zombies can overcome the defences and rise to reign supreme.

4.3 Adaptive Strategies for Humans

It would seem to take a good deal of effort to maintain a high preparedness for zombie attacks. Indeed, carrying around an ice pick or club whenever we want to go shopping is an inconvenience that most of us would rather avoid. Thus we can suppose that if the zombie attacks remain infrequent, we might allow the parameter a_3 to begin to fall, perhaps even to fall below zero, where an isolated zombie could attack and successfully kill such a citizen. For example, in one of the early scenes in *Night of the Living Dead*, Johnny is easily killed by a lone zombie due to his naïve attitude toward the threat. We can then imagine that eventually the attacks would be very common as the isolated attacks increase the population of zombies and this is then amplified by the group attacks. As more and more zombies are created, the populace might be compelled to step up their readiness and thus increase the parameter a_3. Hence, in this section, we study the effects of adapting the 'readiness' parameter, a_3, to the zombie population.

We amend equations (4.9) and (4.10) by now allowing a_3 to evolve according to the zombie population:

$$\frac{da_3}{dt} = F(Z, a_3).$$

What would be a good choice for the function F? We want a_3 to increase if the zombie population is large and to decrease if it is small. How small or how large depends on our tolerance to the presence of zombies. Obviously, some people find them so utterly abhorrent that they will choose to set a minimum of zero. But, as we noted earlier, such preparedness comes at a cost (at the least, in inconvenience and, more seriously, in collateral damage due to accidents with anti-zombie devices). A simple linear equation would seem to suffice:

$$\tau \frac{da_3}{dt} = Z - \bar{Z} - ca_3. \tag{4.11}$$

The parameter $\bar{Z}$ sets the level of zombies that you are willing to tolerate. The parameter c is just a decay of a_3 and could optionally be set to zero. A nonzero value of c means that there is some natural decay of readiness to a neutral value ($a_3 = 0$). Finally, the parameter τ sets the time scale for the reaction to the zombies. If humans are slow to react to the increasing zombie attacks, then we should make τ large. On the other hand, if the humans are acutely aware of the zombies and react quickly, then τ should be small.

Figure 4.2 repeats Figure 4.1B and thus shows the steady state population of zombies as a function of a_3. Equilibria of equation (4.11) satisfy $Z = \bar{Z} + ca_3$. This is just a straight line. If we plot this line along with the diagram in Figure 4.1B, intersections tell us the equilibrium values. For example, if $c = 0$, then the

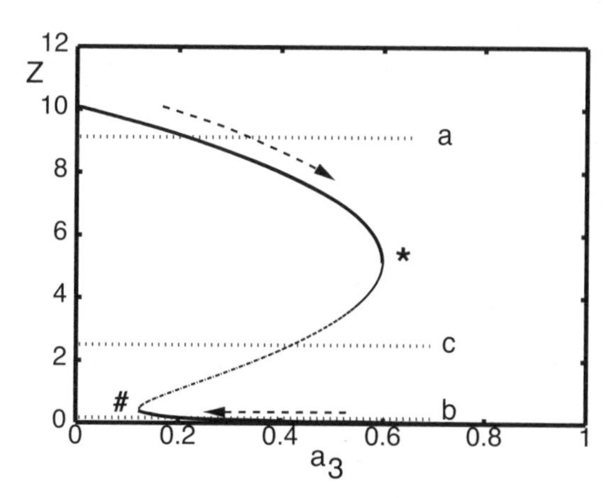

Figure 4.2: Different levels of zombie tolerance: (a) high, (b) low or (c) intermediate. Parameters as in Figure 4.1B.

equilibrium values are found by drawing horizontal lines at $Z = \bar{Z}$. Suppose you have a high tolerance of zombies, such as line (a) in the figure. Then you can choose a_3 quite low. Of course you and most of your neighbours will be nearly exterminated, but, hey, you didn't spend any energy, so if you are one of the lucky ones to survive, you can leave your icepick at home. On the other hand, if you choose to tolerate very few zombies, then you might want to set your tolerance to be very low, such as line (b) in the figure. In fact, if you want to avoid any zombie attacks at all, then set the line low enough so that a_3 is larger than about 0.6 in this particular example. In that case, there can never be a dominant zombie presence. Suppose that you hedge your bets and pick an intermediate tolerance of zombies, like line (c) in the figure. Then the intersection is on the unstable part of the zombie equilibrium curve and, in reasonable circumstances, you can expect to see periodic fluctuations in the zombie and human populations as well as in the readiness parameter, a_3.

Figure 4.3 shows an example simulation of such an oscillation. The zombie population rises and wipes out a substantial fraction of the people. The remainder arm themselves to the teeth and cut down the zombies to a low level. They then become complacent, allowing the zombie population to once more rise. We can understand this oscillation by looking again at Figure 4.2. Suppose that τ is very large so that the people adapt really slowly. At a high zombie population, since $Z > \bar{Z}$, da_3/dt is positive and a_3 starts to increase (dashed arrow on the top of Figure 4.2). This increase in weaponry causes the zombie population to slowly decrease until the point marked with (*) is reached. At this point, the zombie population crashes to the nearly zero level. Since Z is now below $\bar{Z}$, the tolerance level, a_3 begins to decrease (complacency sets in, shown by the dashed line at the

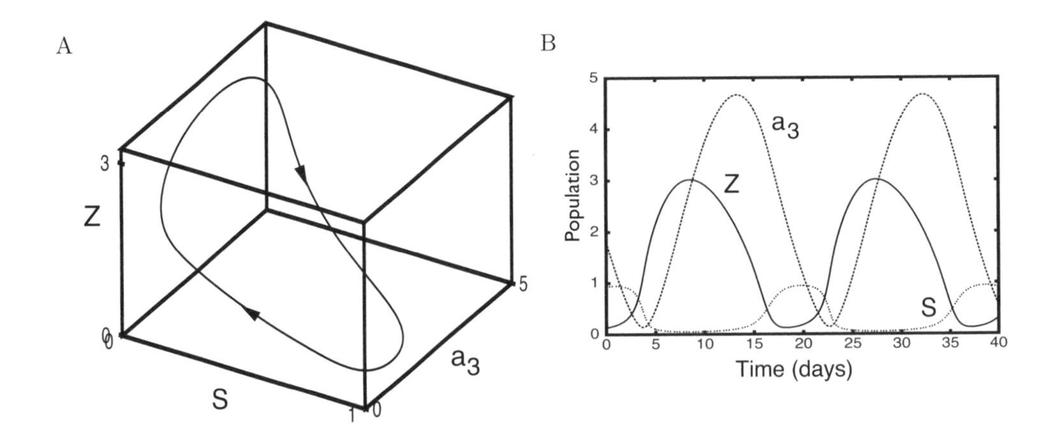

Figure 4.3: Behaviour of the adaptive strategy model. A: Solution in phase space of a limit cycle behaviour. B: Time series over several cycles. Parameters are $a_4 = 2, a_5 = 0.25, Z_0 = 1, \bar{Z} = 1.1, c = 0.2, \tau = 2$.

bottom). The zombie level rises slightly but almost imperceptibly until the point (#) is reached where there is a sudden explosion in the zombie attacks and the zombie population rises to dominance once again.

If we treat tolerance as a parameter, we can numerically determine how the dynamics of the model change with $\bar{Z}$. Figure 4.4A shows the behaviour of equations (4.9)-(4.11) as we change the tolerance from low to higher values (e.g., we move the dashed line (c) in Figure 4.2). The equilibrium value of a_3 is stable for near-zero tolerance, but, at a fairly low value, it loses stability (first arrow) as a Hopf bifurcation [6]. That is, the equilibrium switches from having a damped oscillation to having a growing oscillation. A periodic orbit emerges as the only stable behaviour (the curve labelled SPO) and a_3, Z, and S all oscillate. The oscillation grows in amplitude until it is abruptly lost (at (*) in the figure) and there is a return to a stable equilibrium point. Thus if an intermediate tolerance is chosen, then the zombie attacks wax and wane in a rhythmic manner. Interestingly, there is a region of $\bar{Z}$ that expresses both rhythmicity and stable equilibrium. We will explore this shortly.

Three-dimensional dynamics are more difficult to understand than two-dimensional ones, so we might ask if there is a way to reduce our three-variable system to a simpler one. For a moment, consider Figure 4.1A, the (Z, S)-model where a_3 is fixed. Two trajectories are drawn (black arrows) and both of them appear to move horizontally until they hit the S-nullcline where they essentially follow it nearly perfectly to the equilibrium. This suggests that the dynamics of S are much faster than Z, a reasonable assumption given that classic zombies are slow compared to humans.

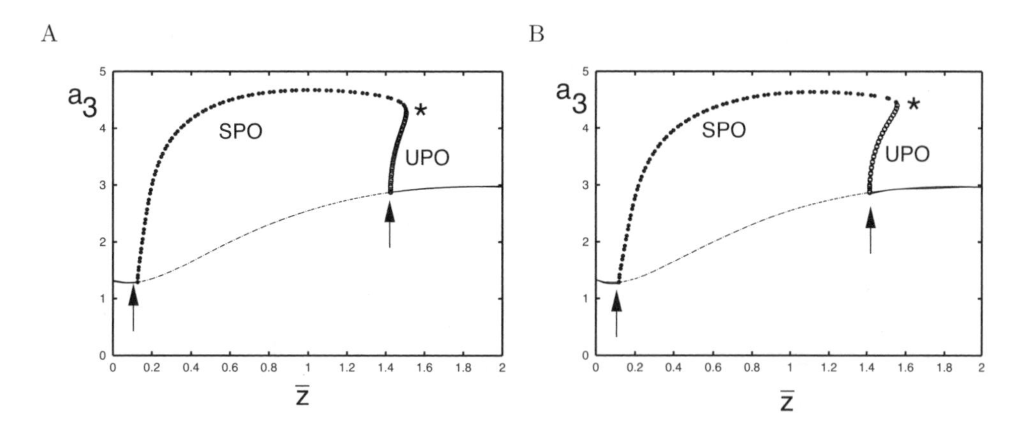

Figure 4.4: Bifurcation diagrams for the adaptive zombie model. A: Full three-dimensional model. Arrows denote Hopf bifurcations and (*) denotes the collision of unstable (UPO) with stable (SPO) periodic orbits. B: Same as A with the reduced model. Parameters as in Figure 4.3.

Thus we could let S reach its equilibrium value found by setting equation (4.9) to zero:

$$S = S_{eq}(Z) \equiv \frac{1}{1 + a_4 Z^2 - \min(a_3 Z, 0)}.$$

If we make this substitution, then the three-dimensional model becomes a two-dimensional model:

$$\frac{dZ}{dt} = -a_3 S_{eq}(Z) Z + a_4 Z^2 S_{eq}(Z) - a_5 Z \tag{4.12}$$

$$\tau \frac{da_3}{dt} = \bar{Z} - Z - c a_3. \tag{4.13}$$

Very little is lost in making this simplification. Indeed, Figure 4.4B is almost identical both quantitatively and qualitatively to Figure 4.4A. Figure 4.5A shows the phase plane and nullclines for the same parameters as Figure 4.3. Z- and a_3-nullclines are shown as well as the limit cycle. Comparing the time series for Figure 4.5B to that of Figure 4.3B shows little difference. As $\bar{Z}$ changes, we can get the a_3-nullcline to intersect in different parts of the Z-nullcline and thus vary the qualitative dynamics. Intersections away from the middle part of the Z-nullcline will lead to stable equilibria and thus a stable adaptation to zombie attacks.

We close our discussion of the adaptive model by returning to the region in Figure 4.4 where there was both a stable equilibrium solution and a stable limit cycle solution (the values of $\bar{Z}$ between the right arrow and the asterisk). Figure 4.6 shows the phase plane behaviour for the reduced (Z, a_3) dynamics when $\bar{Z} = 1.4$. The closed circle is an unstable periodic orbit (UPO) and divides the plane into two regions. Any initial values of a_3, Z starting inside the UPO are attracted to the stable equilibrium point at the intersection of the nullclines. On the other hand, initial values outside the UPO will go to the limit cycle where the zombie population

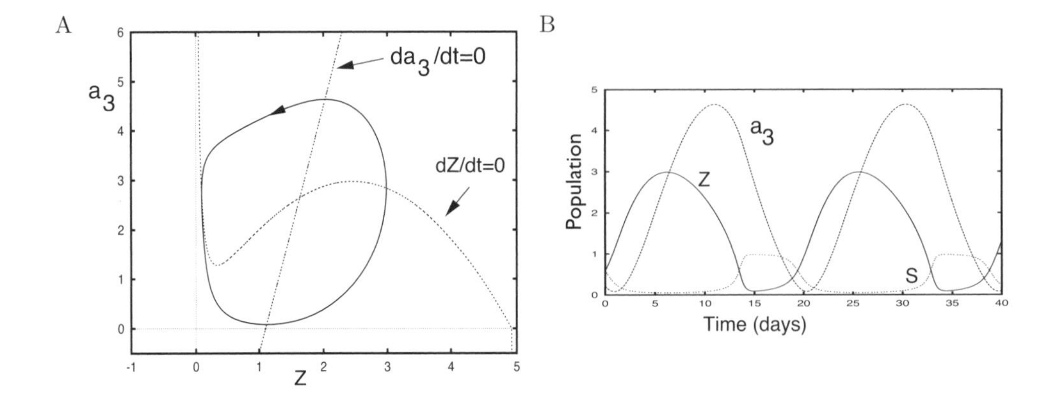

Figure 4.5: Reduced adaptive model where $S = 1/(1 + a_4 Z^2 - \min(a_3 Z, 0)$. A: Phase plane showing the Z-nullclines and the a_3-nullclines as well as the limit cycle. The single equilibrium point is unstable. B: Time series for the cycle shown in A. Parameters as in Figure 4.3.

oscillates. This suggests an interesting phenomenon. Suppose that we are at the stable equilibrium. A brief influx of new zombies could push the initial conditions to the right, past the UPO, and result in a massive decrease in zombies, subsequent complacency, and a rebound to an oscillation. Only a carefully timed culling of the zombies could take you back to the stable equilibrium.

In conclusion, we see that having a zombie-dependent behaviour in readiness to confront the undead hordes can lead to instabilities that result in waxing and waning of the total zombie population. This is reminiscent of many other disease and population cycles seen in nature [8].

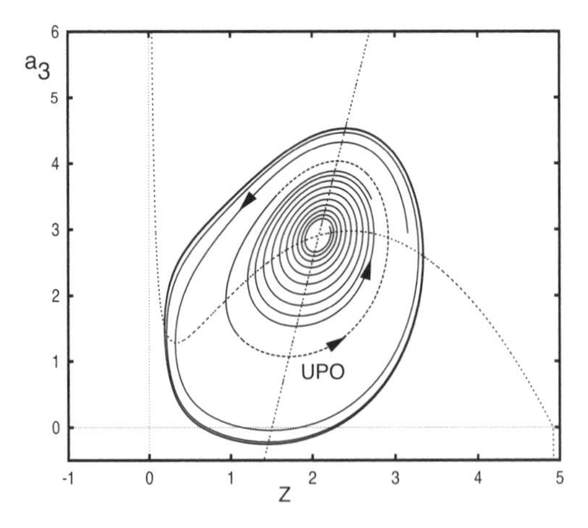

Figure 4.6: Reduced adaptive model with $\bar{Z} = 1.4$, in the bistable region. The unstable periodic orbit is labelled UPO. Parameters as in Figure 4.3.

4.4 Discussion

In this chapter, we have introduced a new type of interaction between zombies and humans that is based on some empirical observations from classic films. Specifically, we hypothesize that low populations of zombies are easily overcome by humans that are suitably armed. These defences can be overcome when zombies attack in groups. It is well known [2] that zombies like to congregate, so this latter part of the hypothesis also seems well-founded. By introducing this novel interaction, we are able to endow zombie–human interactions with various interesting nonlinear behaviours. In particular, by considering two types of interactions (zombies lose at low density and win at high density), we can find model systems where there can be two qualitatively distinct outcomes for exactly the same parameters: humans dominate or zombies dominate. This *winner-take-all* behaviour is common in population models where there is competition. Here, there is not direct competition; rather, the prey can become the predators if they are well-armed. We also introduced adaptive strategy for humans. That is, we suggested that when zombie levels were low, then the alertness and readiness of humans became low. This results in a massive amplification of zombie attacks and a near decimation of humans until they slowly return to their vigilant and armed condition. Then comes the inevitable return to complacency and the cycle begins anew.

We can imagine several ways to extend and generalize the present model. In Munz et al. [1], humans killed by zombies did not immediately turn into zombies. Rather there was a waiting period to become living dead. This addition could be applied to the present model. Another generalization could be to break the human population into two groups, those who are well-armed and those who are not. As the zombie population drifts downward, the well-armed drop their weapons and become members of the passive crowd until the zombies attack again. We could also include a penalty for increasing the vigilance as this could be viewed as taking resources away from other useful tasks such as working, building roads, etc. Indeed, too many complacent humans (free riders) depending on a few well-armed individuals could lead to decimation of the humans, an example of the so-called *tragedy of the commons* [9]. As humans actually are able to cooperate and communicate, we could also introduce a situation where multiple humans could defeat multiple zombies. Recent evidence has accumulated indicating that zombies, themselves, may cooperate. In *Zombie Strippers* [10], the female zombies cooperated for the most part in attacking male humans and on occasion fought among themselves. The male zombies in this film, however, behaved like classic zombies. Thus, one might be tempted to add some degree of zombie cooperation into the model. Finally, it is known that zombies in temperate latitudes slow down as winter arrives [2]. They are technically dead, so their blood must get colder and thus they might approach a torpid state during the winter. Hence one could envision adding a seasonal fluctuation to the parameter a_4 such that it is quite small during the winter but rises to a large value in the spring as thousands of hungry zombies awaken and search for fresh brains. Periodic forcing of oscillatory and nearly oscillatory systems can often lead to complex dynamics including chaos. And what could be more chaotic than a horde of flesh-eating undead descending unpredictably on their human hosts!

Appendix

A Glossary

Bifurcation diagram. A plot of some aspect of a differential equation (such as the number of zombies) as a function of some parameter of interest. A *bifurcation* occurs when the qualitative behaviour changes suddenly as the parameter varies. For example, the number of equilibrium points might change, or their stability might.

Bistability. A system is bistable if it has two stable equilibria.

Equilibrium point. A solution to a differential equation that does not change in time. Also called a steady state or fixed point.

Limit cycle. An isolated periodic solution; that is, there are no other periodic solutions near it.

Nullclines. Curves in the phase plane along which one of the variables remains constant. When nullclines intersect, we have an equilibrium.

Periodic solution. A trajectory of a differential equation that oscillates.

Phase plane. A plot of the solutions to a differential equation with the variables of interest on the axes.

SPO. Stable periodic orbit.

Stability. An equilibrium solution is stable if it returns to equilibrium after being perturbed away from it.

UPO. Unstable periodic orbit.

References

[1] P. Munz, I. Hudea, J. Imad, R.J. Smith?. When zombies attack!: Mathematical modelling of an outbreak of zombie infection. In *Infectious Disease Modelling Research Progress* J.M. Tchuenche and C. Chiyaka, eds., Nova Science, Happuage, NY 2009, pp. 133–156.

[2] M. Brooks *The Zombie Survival Guide: Complete Protection from the Living Dead.* Three Rivers Press, New York, NY, 2003.

[3] G.A. Romero, J.A. Russo. *Night of the Living Dead*, directed by G.A. Romero. The Walter Reade Organization, 1968.

[4] G.A. Romero. *Dawn of the Dead*, directed by G.A. Romero. United Film Distribution Company, 1978.

[5] L. Edelstein-Keshet. *Mathematical Models in Biology.* SIAM, Philadelphia, PA, 2005.

[6] S. Strogatz. *Nonlinear Dynamics and Chaos.* Perseus Books, Cambridge, MA, 1994.

[7] B. Ermentrout. *Simulating, Analyzing, and Animating Dynamical Systems.* SIAM, Philadelphia, PA, 2002.

[8] H. Hethcote and S.A. Levin. Periodicity in epidemiological models. In *Applied Mathematical Ecology Biomathematics 18*, S.A. Levin and T. Hallam, editors, Springer-Verlag, Berlin, 1989, pp. 193–211.

[9] Hardin, G. The Tragedy of the Commons. *Science*, 162:1243–1248, 1968.

[10] J. Lee. *Zombie Strippers*, directed by J. Lee. Triumph Films, 2008.

INCREASING SURVIVABILITY IN A ZOMBIE EPIDEMIC

Ben Tippett

Abstract

Once the zombie epidemic has begun, it is not clear which strategy a person or community should adopt to confront the threat. Is it better to run and hide, confront the ghouls gun in hand or keep on with our regular workaday lives? Is the probability of survival higher in rural communities or urban ones? Here, epidemiological models are constructed to explore these possibilities. The living humans in the models are divided into three classes: workers, who accumulate food and supplies; militia, who hunt the undead; and moles, who hide and wait for rescue. These populations change according to the number of zombies, and also the quantity of food and supplies available to the humans. Numerical simulations suggest that humans have a better chance of surviving in cities than in places with low population densities. We also see that survivability is increased if zombie-ism comes about as a result of a transmissible disease, as opposed to being the result of the reanimation of the recently deceased.

5.1 Introduction

It is said that the world is only 70 days away from starvation [1]. By this, we mean that, given our rates of consumption and global stockpiles of food, if workers were to stop farming, fishing or gathering food, the human population would not survive beyond two months.

In the midst of a zombie pandemic, where national and municipal quarantines would grind trade to a halt, it is safe to assume that famine's skeletal spectre would pose as much of a threat as the literal skeletal anthropophagite ghouls. For this reason, survivors would face a morbid calculus as they are forced to choose between gathering food (exposing themselves to slimy teeth and clammy grasping hands), hunting down the zombies in militias (on an empty stomach) or laying low and hiding until the horrific scenario resolves itself (one way or another). As seen in the previous chapter, complacency can be deadly—or at least result in periodicity. Given the variety of outcomes seen in the preceding chapters when different aspects of our model are tweaked, our aim here is to build a comprehensive model of this scenario to compare the effectiveness of different strategies at ensuring the survival of humanity.

Our model will allow survivors to choose one of three different survival strategies: working, fighting or hiding. Workers who find and produce the food required to keep the community from starving will make themselves more vulnerable to zombie attack. Militia members who spend all their time hunting and eliminating zombies may eat through the community's entire food reserves before starving to death. Cowards who go into hiding will find themselves exposed to the least risk, but the community's problems will grow worse if no one strives to fix them. Survivors face a choice, not unlike those found in game theory, wherein the survivability of an individual depends on both the choices the individual makes and the choices everyone else in the group makes. Going into hiding may be the best strategy for an individual's short-term survival, but an individual in hiding will not survive in the long term if other members of the community are not killing zombies and procuring food. Clearly, the community's survival depends on the the use of a blend of different strategies. Our aim is to investigate how the adoption of these different strategies affects the survivability of the human population.

Our model will adopt a deterministic compartmental model, where survivors are sorted according to one of three survival strategies. The survivors are allowed to switch their survival strategies according to the size of the zombie population and the risk of starvation. The survivors can also turn into zombies. It is similar to other epidemic models [2, 3], and models of immune system cells responding to viral infection [4, 5].

5.2 The Model

5.2.1 Interaction Rates Assuming a Uniform Population Density

Define the Undead Wandering Area (UWA) to be the area through which the average zombie can actively hunt for food over the course of a day. We assume that the urban or rural landscape for our model is divided into N squares, where $N =$ Geographic Area/UWA. Realistically, the UWA should depend on the geography of the region, as well as the mobility and awareness of the undead. We will, for simplicity's sake, assume that the UWA is uniform for all zombies in the geographic region under consideration.

If we assume a uniform population density, we can quantify the number of people who will, on average, encounter a zombie in the course of a day. If there are S_L

living people in the city and S_Z undead, then over the course of the day there will be $\frac{S_L S_Z}{N}$ encounters between members of the two populations.

Our model keeps track of five different quantities: the undead population, the worker population, the militia population, the mole population and the total quantity of supplies available to the community. We divide the human population into different classes because we expect that different individual survival strategies will be differentially vulnerable to zombie attack, will differentially increase or consume the supply stockpile and will differentially cull the zombie population.

Living individuals should be able to change their class according to their circumstance. In the case where a worker encounters a zombie and lives, there will be a chance that this worker joins the militia. Let this probability be given by the 'bravado coefficient.' There is also a chance that he or she will become a 'mole,' hiding from the zombies; we define this probability by the 'cowardice coefficient.' We also assume that an individual who is hunting or in hiding will act as a drain on the overall stockpile of supplies. When the quantity of supplies is reduced below some threshold, we will force some percentage of the militia and the mole populations to rejoin the workforce in order to prevent general starvation. In addition, we note that if the total number of supplies ever reaches zero, the entire living human population will be completely exterminated.

5.2.2 Zombie Population Dynamics

Let S_Z denote the zombie population and let S_W, S_M, S_H denote the worker, militia and mole population, respectively. Then, on average, over a set amount of time Δt (assumed to be a day), the zombie population will change by:

$$\frac{\Delta S_Z}{\Delta t} = Z_1 \frac{S_W S_Z}{N} + Z_2 \frac{S_H S_Z}{N} + (Z_3 - M_1) \frac{S_M S_Z}{N}.$$

The constant coefficient Z_1 is the probability that an average zombie will convert a worker, while Z_2 is the probability that an average zombie will convert a mole (we would expect that $Z_2 < Z_1$). Z_3 is the probability that an average zombie will convert a militia member (we would expect that $Z_3 > Z_1$, since the militia are actively seeking confrontations with zombies), and M_1 is the probability that a militia member will cull a zombie.

5.2.3 Worker Population Dynamics

A variety of factors will drain and replenish the worker population. A worker who has been scared by a zombie might react by joining the militia or the mole population. There are also natural death rates from disease and old age, as well as deaths by accidents. For example, an untrained militia member with a hair-trigger reflex might easily mistake a worker with his head down for a ghoul. In addition, the worker population can be replenished through natural birth and by militia members and moles who are forced back to work out of hunger.

An important quantity, the Days of Food Left (DFL), is the number of units of time Δt (assumed to be days for illustrative purposes) of food left. In our model, the worker population changes according to the following equation:

$$\begin{aligned}\frac{\Delta S_W}{\Delta t} = & -(Z_1 + \alpha(1 - Z_1) + \beta(1 - Z_1 - \alpha(1 - Z_1)))\frac{S_Z S_W}{N} \\ & + f_1(DFL)S_M + f_2(DFL)S_H + F_4 S_W - F_5\frac{S_W S_M}{N}.\end{aligned}$$

The parameter α, called the coefficient of cowardice, is the probability ($0 \leq \alpha \leq 1$) that a worker will join the mole community after seeing a zombie. β, called the bravado coefficient, dictates the portion of the remaining survivors who will join the militia. DFL denotes the estimated number of time units Δt of supplies remaining as defined in equation (5.1). $f_1(DFL)$ and $f_2(DFL)$ are equations describing the community's response to a food shortage. If supplies run short, new workers will be recruited from the militia and mole populations respectively. In our models, we will set

$$\begin{aligned}f_1(DFL) &= \frac{1}{2}H(4 - DFL) \\ f_2(DFL) &= \frac{1}{2}H(2 - DFL)\end{aligned}$$

where $H(x - DFL)$ is a Heaviside function centred around x units of time. Thus if there are fewer than 4 days of food left, half of the militia will join the workforce whereas if there are fewer than 2 time units of food remaining, half the moles will come out of hiding and join the workforce. This feature is similar to the discontinuities considered by Guo [6]. F_4 will be the population increase-rate per Δt per worker, and F_5 will be the probability ($0 \leq F_5 \leq 1$) that a worker will be shot accidentally by a militia member. F_4 will be small considering how long the gestation rate is for humans compared to the reaction rates of the epidemic and also how the collapse of public infrastructure will increase the mortality rate.

5.2.4 Militia Population Dynamics

The militia population feels stresses similar to those of the worker population. The population will feel a drain from the loss of some members to the zombie populace, or through accidental homicide. We can also expect that some portion of the militia populace will be spooked after surviving a zombie encounter and join the mole population in hiding. Daily, we would expect the militia to change according to the equation:

$$\begin{aligned}\frac{\Delta S_M}{\Delta t} = & -Z_3\frac{S_Z S_M}{N} + \beta(1 - Z_1 - \alpha(1 - Z_1))\frac{S_Z S_W}{N} - M_3\frac{S_M S_M}{N} \\ & - f_1(DFL)S_M - \alpha(1 - Z_3)\frac{S_Z S_M}{N} + \beta(1 - Z_2)\frac{S_H S_Z}{N},\end{aligned}$$

where M_3 $(0 \leq M_3 \leq 1)$ is the probability that a militia member will accidentally shoot and kill another member of the militia.

5.2.5 Mole Population Dynamics

Finally, the population in hiding feels the same population pressures as the other living human populations. It will follow the equation:

$$\begin{aligned}\frac{\Delta S_H}{\Delta t} = &- Z_2 \frac{S_Z S_H}{N} + \alpha(1 - Z_1)\frac{S_Z S_W}{N} - H_3 \frac{S_H S_M}{N} \\ &- f_2(df)S_H + \alpha(1 - Z_3)\frac{S_Z S_M}{N} - \beta(1 - Z_2)\frac{S_H S_Z}{N},\end{aligned}$$

where H_3 $(0 \leq H_3 \leq 1)$ is the probability that one militia member will accidentally shoot and kill a member of the mole population.

5.2.6 Supply Stockpile Dynamics

In our model, only workers will be able to increase the quantity of supplies. Let R denote the total quantity of supplies held by the human community. Let us normalize the units of R in terms of the average quantity required by a worker over the course of one unit of time (assumed to be a day). Thus $R = 1$ units of supplies will keep a worker alive for Δt amount of time.

The different population classes will withdraw supplies from the stockpile at different rates due to differences in their activity levels over the course of the day. Let us suppose that, on average, the supply stockpile will change over Δt according to the following formula:

$$\frac{\Delta R}{\Delta t} = (R_1 - 1)S_W - (R_2 + 1)S_M - (1 - R_3)S_H.$$

R_1 is the average number of normalized supply units found or produced by a worker over a unit of time Δt. R_2 is the number of normalized supply units used by a militia member over a unit of time Δt above the consumption rate of a worker $(R_2 > 0)$. For example, if the increased activity of a militia member demands 3 normalized supply units per Δt, we would set $R_2 = 2$. R_3 is the number of normalized supply units used by a mole over Δt below the consumption rate of a worker $(0 < R_3 < 1)$. For example, if the sedentary lifestyle of a mole requires only 1/5 the calories of a worker over Δt, we would set $R_3 = \frac{4}{5}$. Parameters R_1, R_2 and R_3 can be altered to describe models where workers are more or less efficient at acquiring resources, or where the militia or mole populations are a larger or smaller drain on resources.

Supposing that workers stopped accumulating supplies, how many days worth of supplies would remain (if population levels remained fixed)? We define this quantity to be:

$$\text{days of food left } (DFL\ ) = \frac{R}{S_W + (R_2 + 1)S_M + (1 - R_3)S_H}. \tag{5.1}$$

In our model, we will assume that this quantity is carefully monitored by the community. If the number of days of food left falls below a threshold, then a percentage of moles and militia members will respond by returning to the workforce.

5.2.7 Comments

The time resolution (or length of time step) Δt of our simulations is one day, which we feel encompasses the rate at which decisions on a municipal level can be made and announced. One day also exceeds the amount of time it takes for a living victim to be turned into an undead ghoul [7], and so the minutiae of any specific zombie attack can be blurred out and discussed in terms of averages. For instance, if two workers are bitten (one in the neck and the other in the foot) it might take minutes for one worker to turn into a zombie, and the other will take hours as the infection spreads through her. However, these details will not matter to our model since, by the end of the day, both have turned into zombies.

The parameter N deserves a comment, since a large N can ambiguously mean either a large geographic area, or a species of slow-moving zombies. For the duration of this chapter, we will interpret a large N as being a rural region (where a zombie can shamble all day and only cover a small portion of the total area); and a small N as being an urban region (where a shambling zombie might still be able to cross a fair segment of the city in a day). The results drawn below could equivalently be described as being scenarios with slow (shambling) zombies, or fast (running) zombies respectively. An alternative way of describing rural/urban scenarios would be to increase the human population in these regions, where a large S_W would indicate a city, and a small S_W would indicate the countryside.

5.3 Modelling Specific Scenarios

A variety of different scenarios can be described using the model as we have just outlined in Section 5.2, depending on the assigned parameter values (see Table 5.1). Since the purpose of this chapter is to explore the effect that choices made in facing a zombie epidemic will have upon the overall survivability, we will consider and vary only a few of the parameters.

The most interesting of the parameters are: N, which can be used in describing the population density (or the mobility of zombies); M_1, which describes the militia's effectiveness at culling zombies; α, the coefficient of cowardice; and β, the coefficient of bravado. We feel that these coefficients are the only ones over which an individual or a government might have control during a zombie epidemic. An individual might choose to move to a rural area (with a larger N) over an urban one. A government might encourage the citizens to run and hide (a large α), or to organize themselves and fight (a large β), or it might choose to train its militia to be more effective (a large M_1). All other parameters—the population increase rate, the rate of accidental homicide or the virulence of the zombie epidemic—are determined externally.

Table 5.1: Parameter values.

Name	Value	Description
N	> 0	The ratio of the total geographic area to the area patrolled by a zombie over the course of one time unit Δt.
S_Z	≥ 0	Total zombie population, calculated for each time step.
S_W	≥ 0	Total worker population, calculated for each time step.
S_M	≥ 0	Total militia population, calculated for each time step.
S_H	≥ 0	Total mole population, calculated for each time step.
Δt	1 day	The time resolution of the model.
R	≥ 0	The community's stockpile of supplies in units normalized to worker population's demands per time unit Δt.
DFL	≥ 0	The estimated number of time units' worth of supplies remaining in stockpile, calculated for each time step.
α	$[0, 1]$	The coefficient of cowardice, or the probability that a human who survives a zombie encounter will go into hiding.
β	$[0, 1]$	The bravado coefficient, or the probability that a human who does not run away after a zombie encounter joins the militia.
M_1	$[0, 1]$	The probability that a militia member who encounters a zombie will exterminate the zombie.
Z_1	0.1	The probability that a zombie who encounters a worker will convert the worker.
Z_2	0.05	The probability that a zombie who encounters a mole will convert the mole.
Z_3	0.15	The probability that a zombie who encounters a militia member will convert the militia member.
R_1	3	Average number of supply units acquired per individual worker per time unit Δt.
R_2	1	Average number of supply units used by militia member per time unit Δt *above* the daily use required by a worker.
R_3	0.4	Average number of supply units used by a mole per time unit Δt *below* the daily use required by a worker.
F_4	6×10^{-5}	Average population increase rate per Δt per worker.
F_5	0.01	Probability that a militia member who encounters a worker will accidentally kill the worker.
M_3	0.01	Probability that a militia member who encounters a militia member will accidently kill the militia member.
H_3	0.01	Probability that a militia member who encounters a mole will accidentally kill the mole.

Let us fix all alternative parameters initially. Let the zombie conversion rate of a worker be $Z_1 = 0.1$ (an average individual worker has little chance of surviving more than 10 zombie encounters in the course of a day). The zombie conversion rate for a mole will then be $Z_2 = 0.05$, and the conversion rate for a militia member will be $Z_3 = 0.15$. Let the natural population increase rate be $F_4 = 0.00006$, while the probability of accidental homicide in these dangerous times will be set to $F_5 = M_3 = H_3 = 0.01$. We will suppose that workers, over the course of the day, will add $R_1 = 3$ supply units to the stockpile, while militia members will use an additional $R_2 = 1$ units per day (since they burn more calories over the course of a day, and also require ammunition, fuel and armour) and a mole will decrease his or her personal consumption by $R_3 = 0.4$ (supposing that they are rationing their food).

There are four different outcomes for zombie plague scenarios: humans could be overwhelmed by the undead hoard and die out; humans could successfully fight off the zombie plague, reducing the number of zombies to zero; humans could fall victim to their reactionary management system as they move between fending off starvation and the zombies (as the zombie plague relentlessly eats away at their population); humans and zombies could coexist (this requires a very slow zombie plague, and will not be considered). Through the rest of this section, we will consider the first three of these scenarios. The issue of which values of α, β, M_1 and N result in the extinction of the human race will be dealt with in Section 5.4.

5.3.1 Humanity Is Exterminated Outright

In specific scenarios in our phase space, humanity will be effectively extinguished. Consider the following example, occurring in a rural community (N=1000) where the initial population consists of $S_W = 3000$ workers and $S_M = 100$ militia, and one zombie $S_Z = 1$. Let our population's response to the zombie epidemic follow: $\alpha = 0.1$, $\beta = 0.1$ (where 10% of people prove cowards once they have encountered a zombie, and 10% of the remaining people take up arms in the militia) and $M_1 = 0.1$ (where an inexperienced and unskilled militia member will need to shoot at 10 zombies to cull one). The population over time is graphed in Figure 5.1. In this model, most conflicts between militia and zombies end with the zombies winning, so it is not surprising that the zombies will exterminate the humans outright.

Note that in this scenario, a survivor will be better off hiding in a closet (the dot-dashed curve denotes the moles), though everyone will die before three months are through.

5.3.2 The Zombie Population Is Culled

In other scenarios, the zombies are efficiently and effectively exterminated. The following example is a suburban setting $N = 1000$ where $S_W = 3000$ and $S_Z = 1$. The parameters $\alpha = 0.1$, $\beta = 0.5$ indicate that the population is very eager to fight

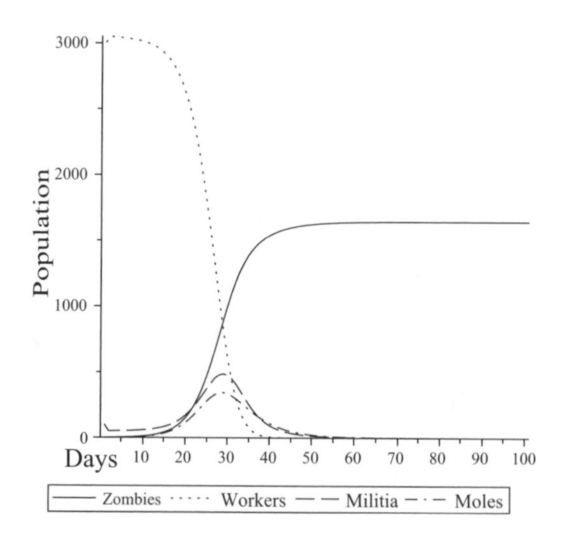

Figure 5.1: The three human populations and the zombie population as functions of time. Once zombies become sufficiently common, the militia and mole populations increase. The militia is not effective at culling the zombie population. Initial values: $R = 8000$, $N = 1000$, $S_W = 3000$, $S_M = 100$, $S_H = 0$ and $S_Z = 1$. Parameters: $\alpha = 0.1$, $\beta = 0.1$ and $M_1 = 0.1$.

the zombie threat, and setting $M_1 = 0.9$ indicates that the militia is very competent at fighting the zombie threat. The degree to which the humans win in this scenario is so outstanding that we needed to graph it on a log scale (Figure 5.2)!

Note that in this scenario, once the zombie threat has been dealt with, a large militia population remains. Longer time scale simulations show that the continuing activity of the militia (and the military accidents that result) will cause a gradual decline in the overall population.

5.3.3 Humans Nearly Starve

A third possible outcome is one where the initial reaction to the zombie threat is overenthusiastic, although ultimately ineffective. The overconsumption of resources by an oversized militia population will lead to a drop in resources and, though the zombie population is partially extinguished, too many people are forced to return to work before the zombie population can be completely culled. Eventually, the zombie population re-establishes itself and the events of the cycle repeat themselves: zombies slowly eating away into the total human population until eventual extinction. Graphs in Figure 5.3 depict the simulation of a rural setting.

Traditional anti-zombie strategies have relied on quick, universal responses to a zombie epidemic, where a good percentage of the populace is called to arms. We note here that, since a zombie epidemic will inevitably cause a breakdown in trade, a comprehensive response to a zombie threat should account for and respond to the issue of dwindling resources. There is clearly a tradeoff between the bravado of a

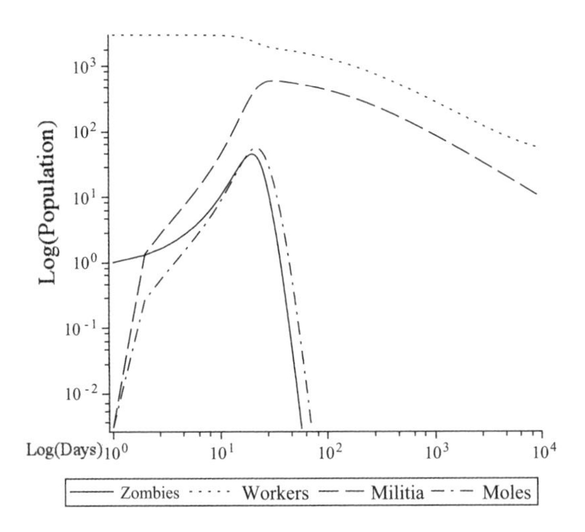

Figure 5.2: The four populations, on a log scale, showing that, once zombies become sufficiently common, the militia population will increase and then cull the undead population. Initial values: $R = 8000$, $N = 1000$, $S_W = 3000$, $S_M = 0$, $S_H = 0$ and $S_Z = 1$. Parameters: $\alpha = 0.1$, $\beta = 0.5$ and $M_1 = 0.9$.

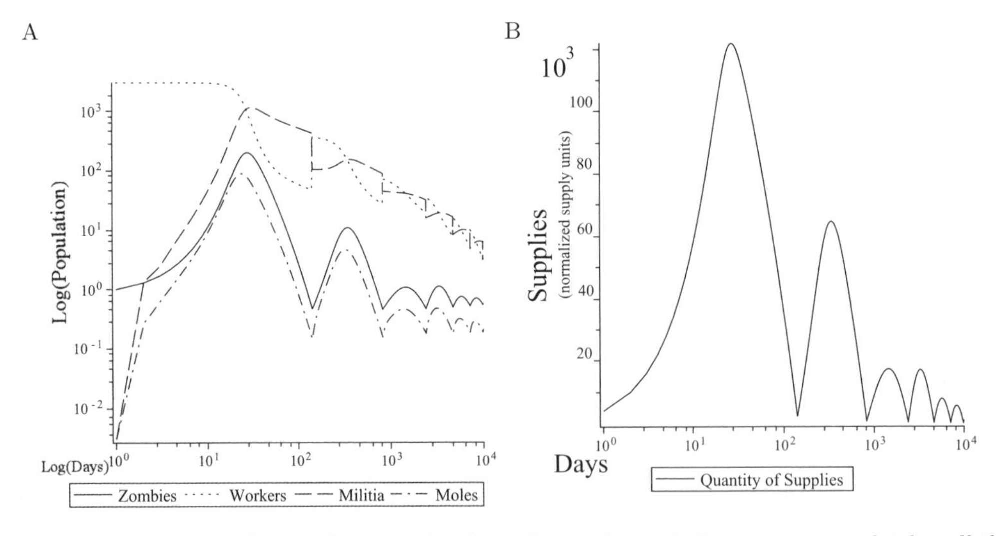

Figure 5.3: The time evolution of a scenario where the rural population cannot completely cull the zombie population before it begins to starve to death. The human population repeatedly switches between dealing with the zombie threat and the threat of starvation, as its numbers dwindle. A: The population graph shows clearly the periodic nature of the behaviour. B: The supplies graph demonstrates that the human population never manages to break the cycle, or successfully kill off the zombie population. Initial values: $R = 4000$, $N = 1000$, $S_W = 3000$, $S_M = 0$, $S_H = 0$ and $S_Z = 1$. Parameters: $\alpha = 0.1$, $\beta = 0.5$ and $M_1 = 0.25$.

populace and its productivity; with this in mind, an effectively trained militia is more important than a very large (and ineffective) one.

5.4 How Public Policy Affects Survivability During a Zombie Epidemic

What percentage of zombie survivors should we ask to join the militia? How much training should they be given? Will a rural setting require a different strategy from an urban one? What difference will it make if the population is cowardly? What difference will it make if, rather than panicking at the sight of the grasping undead, the public retains a deliberately unflappable mindset and continues to diligently work at their assigned jobs?

We have repeatedly run simulations of the type seen in Section 5.3, describing a community starting with 1000 workers and 5000 units of food with 1 initial zombie for 2000 days (6 years) in a rural scenario and 1000 days (3 years) in an urban scenario. Over the course of these simulations, we vary the bravado coefficient β and the militia's culling efficiency M_1. We then judge the effectiveness of the zombie cull by plotting the average number of zombies over the last year as a function of the bravado β and the killing efficiency M_1. By looking only at the cases where the final number of zombies is nearly zero, we can determine how brave and well-trained a population must be in order to completely cull the zombies. We then filter these results to differentiate between the scenarios where the final human population is zero, and the scenarios where some humans will survive.

We repeat this process for cases where $\alpha = 0$ (very little cowardice) and $\alpha = 0.3$ (one-third of people who see a zombie take to hiding); and for $N = 30$ (corresponding to an urban environment where you will see 33 people per area unit) and $N = 3000$ (corresponding to a rural environment where you will see only one person every three area units).

5.4.1 Managing a Zombie Epidemic in an Urban Scenario

Cowardly city

In this case, the parameters have been set for $\alpha = 0.8$, $N = 30$, corresponding to a situation where 80% of the people will panic and become moles after surviving a zombie encounter. The shaded region in Figure 5.4B depicts the region of the parameter space where no humans are left alive at the end of the simulation. If we focus on the parts of the parameter space where humanity has survived and all the zombies were eliminated, we end up with Figure 5.4A. This demonstrates that, even if most of the populace proves cowardly, so long as the militia is well trained, humanity can survive. Note that this result holds even if less than a quarter of the survivors of a zombie attack who do not run away enlist in the militia.

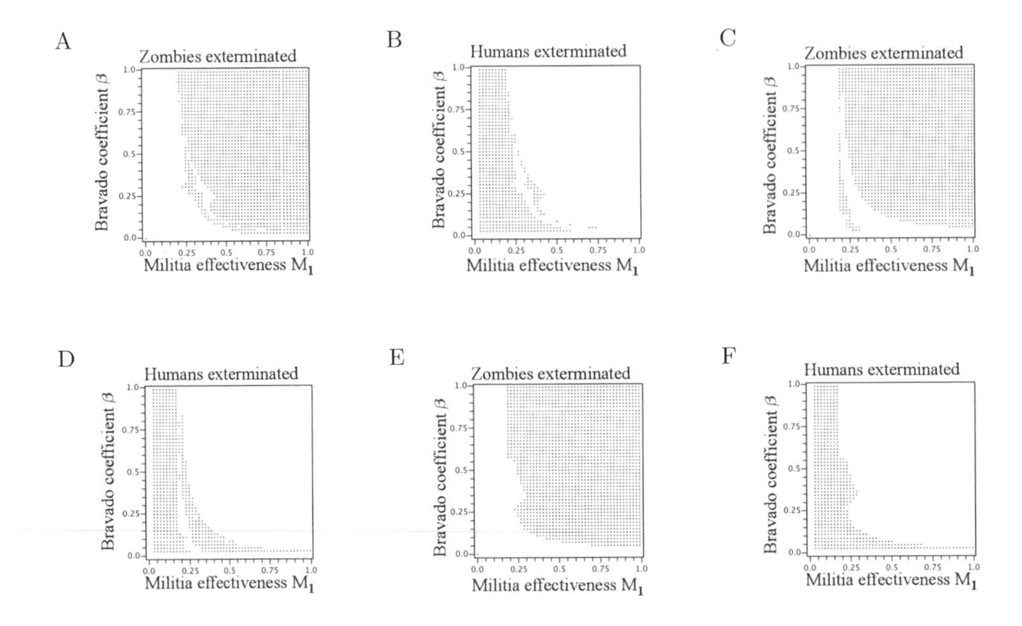

Figure 5.4: Culled populations after 1000 days in an urban setting. Initial values: $N = 30$, $R = 1000$, $S_W = 1000$, $S_M = 0$, $S_H = 0$ and $S_Z = 1$. Final time $= 1000$ days. 2500 evenly spaced points are chosen from $M_1 \in [0.01, 0.99]$, $\beta \in [0.01, 0.99]$. A: Zombie population exterminated in cowardly urban setting ($\alpha = 0.8$). B: Human population exterminated in cowardly urban setting ($\alpha = 0.8$). C: Zombie population exterminated in average urban setting ($\alpha = 0.3$). D: Human population exterminated in average urban setting ($\alpha = 0.3$). E: Zombie population exterminated in unflappable urban setting ($\alpha = 0$). F: Human population exterminated in unflappable urban setting ($\alpha = 0$).

Average city

In this set of simulations, the parameters have been set for $\alpha = 0.3$, $N = 30$, corresponding to a situation where one-third of people panic and go into hiding after meeting a zombie. Figure 5.4D clearly shows the parameter space where the human race has been exterminated. If we focus only on the parameter space where humanity has survived and has also completely culled the zombie population, we end up with Figure 5.4C.

Unflappable city

In this case, the parameters have been set for $\alpha = 0$, $N = 30$, corresponding to a situation where no one in the city panics and hides after seeing a zombie. The shaded region in Figure 5.4F corresponds to the parameter space where humanity has died out. The parameter space where humanity has survived and has culled the zombie population is depicted in Figure 5.4E.

Comparing the graph in Figure 5.4E to the graph in Figure 5.4A (where the city was full of cowards) suggests some interesting conclusions. A cowardly city requires the use of a (much) smaller militia (their numbers will be smaller since α is so much larger), so long as the militia is well trained. Conversely, if the militia is poorly trained (an effectiveness of about $M_1 = 0.2$), a city will have a better chance at surviving if its citizens continue working, rather than running away and hiding. This distinction becomes even more pronounced in rural situations.

5.4.2 The Countryside

The dynamics of a zombie epidemic in a rural setting are different from those in the city, since the rate at which zombies are randomly encountered is relatively low compared to the rate at which supplies are gathered and consumed. Whereas the urban zombie scenarios resolve themselves relatively quickly, it will take a long time for any conclusion to be reached in the rural scenarios.

In our simulated scenarios, we shall take the presence of any zombies above a threshold after six years as a sign that the community has failed to cull the undead from their streets and fields.

Cowardly countryside

In this scenario, the parameters have been set to $\alpha = 0.8$ and $N = 3000$, corresponding to a very rural environment (one person can wander for three days at random before seeing another person) where 80% of the workers who see a zombie will go into hiding. Figure 5.5B shows the region of the parameter space where all humans have been exterminated after six years.

Figure 5.6A demonstrates that, even after six years, there is a large area of the parameter space where both humans and zombies remain (with the zombie population numbering in the hundreds).

Figure 5.5A plots only the area of the parameter space where humanity has succeeded in culling the undead population, and the results are depressingly small when compared to the urban scenario.

We conclude that if humanity is to survive a zombie epidemic in a rural setting, and if most of the human population are cowardly, then most of those who are not will have to fight skillfully.

Normal countryside

In this set of simulations, the parameters have been set to be $\alpha = 0.3$ and $N = 3000$, corresponding to a very rural environment (you can wander for three days at random before seeing a person) where one-third of the workers who survive a zombie encounter will go into hiding. Figure 5.5D plots the regions of the parameter space where all of humanity has vanished after six years.

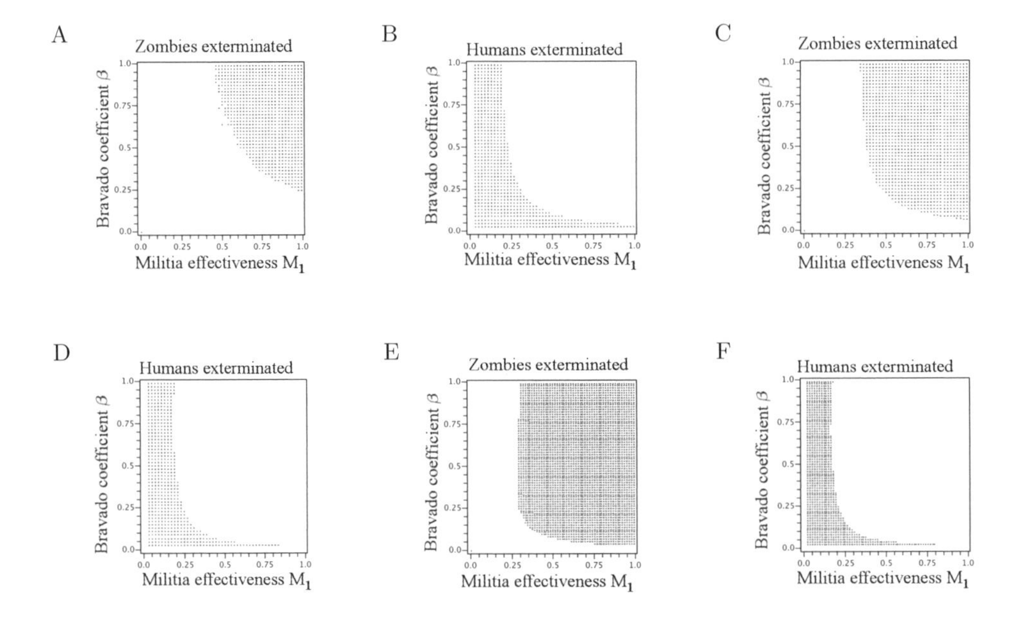

Figure 5.5: Culled populations after 2000 days in a rural setting. Initial values: $N = 3000$, $R = 1000$, $S_W = 1000$, $S_M = 0$, $S_H = 0$ and $S_Z = 1$. Final time = 2000 days. 2500 evenly spaced points are chosen from $M_1 \in [0.01, 0.99]$, $\beta \in [0.01, 0.99]$. A: Zombie population exterminated in cowardly rural setting ($S_Z < 0.02$, $\alpha = 0.8$). B: Human population exterminated in cowardly rural setting ($\alpha = 0.8$). C: Zombie population exterminated in average rural setting ($S_Z < 0.02$, $\alpha = 0.3$). D: Human population exterminated in average rural setting ($\alpha = 0.3$). E: Zombie population exterminated in unflappable rural setting ($S_Z < 0.02$, $\alpha = 0$). F: Human population exterminated in unflappable rural setting ($\alpha = 0$).

Figure 5.6B plots the area of the parameter space where some population of humans will remain after six years.

Figure 5.5C plots the area of the parameter space wherein there are fewer than 0.02 zombies remaining in the last year on average.

Unflappable countryside

In this set of simulations, the parameters have been set to be $\alpha = 0$ and $N = 3000$, corresponding to a very rural environment, where no one who survives a zombie encounter will go into hiding. As before, Figure 5.5F demonstrates the area of the parameter space where all humanity has died within six years.

Figure 5.6C plots the average number of zombies remaining over the last year as a function of bravado and the effectiveness of the militia. Figure 5.5E shows the area of the parameter space where the undead population has been completely culled.

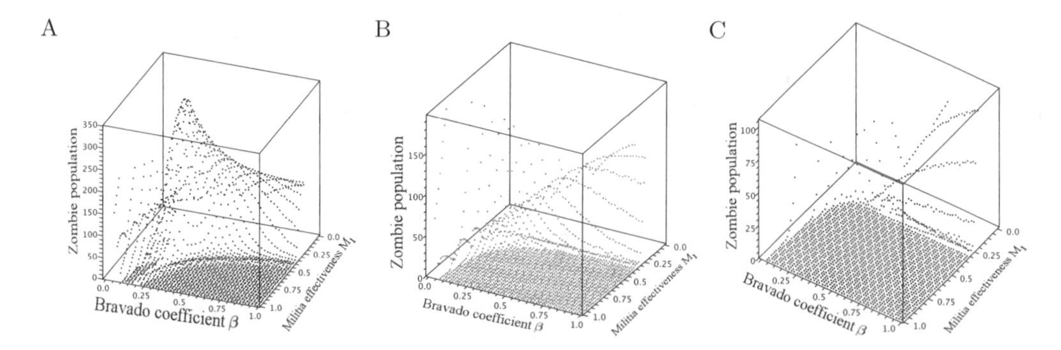

Figure 5.6: Average number of zombies remaining during the last year as a function of bravado and the effectiveness of the militia in a cowardly rural scenario after 2000 days in a rural setting. Initial values: $N = 3000$, $R = 1000$, $S_W = 1000$, $S_M = 0$, $S_H = 0$ and $S_Z = 1$. Final time = 2000 days. 2500 evenly spaced points are chosen from $M_1 \in [0.01, 0.99]$, $\beta \in [0.01, 0.99]$. A: Average zombie population in cowardly rural setting ($\alpha = 0.8$). B: Average zombie population in average rural setting ($\alpha = 0.3$). C: Average zombie population in unflappable rural setting ($\alpha = 0$).

Remarks

Given the slower rate at which zombies appear in rural regions, people in these regions need to be, on average, much braver than people in urban regions. As in the urban scenario, total eradication of the zombies depends more strongly on the efficiency of the militia than on the bravado of the general populace. On the other hand, cowardice seems to play a very important role in eradicating the zombies.

Comparing Figure 5.5A with Figure 5.5E demonstrates that a 'stiff upper lip' is requisite in exterminating the zombie presence. The conflict with the zombie population burns much more slowly in the rural setting than in cities. This explains the dramatic contrast between Figure 5.4B and 5.4E. The human population cannot afford to waste time and resources hiding, lest they starve.

5.5 The Romero Scenario

In the previous model, it is understood that the only cause of zombification of an individual is through a direct physical encounter with a member of the undead population. In these situations, zombification is due to a transmissible agent, and it travels much like a disease [7, 3]. This is not necessarily the only scenario where zombification can occur: there is a well-known and far grimmer scenario where zombification is due to the reanimation of the dead [8].

5.5.1 Altered Transmission Equations

In this scenario, every human who has died but whose brain has not been destroyed will become a ghoul. We refer to this as the Romero scenario and we can model it by using a slightly altered set of equations. We note that, in our previous model, there was a good deal of mortality due to the dangers of having a militia wander about the community shooting at everything that moves. In the prior model, victims of accidental death were not added to the zombie population but they are in the Romero scenario, which is given by the following equation:

$$\begin{aligned}\frac{\Delta S_Z}{\Delta t} &= Z_1\frac{S_W S_Z}{N} + Z_2\frac{S_H S_Z}{N} + (Z_3 - M_1)\frac{S_M S_Z}{N} \\ &+ F_5\frac{S_W S_M}{N} + M_3\frac{S_M S_M}{N} + H_3\frac{S_H S_M}{N}.\end{aligned}$$

In the prior model, accidental homicide was not a dominant effect on the outcomes of the simulations, and so the relevant parameters were fixed by ansatz. In the Romero scenario, accidental homicide can act as a mechanism for generating undead. We feel that the values of the coefficients should depend on how much training the militia has had: a well-trained and efficient militia will cause fewer accidental homicides, and vice versa. Thus, as we model the Romero scenario, we will relate the coefficients of accidental homicide H_3, F_5, M_3 to the coefficient of zombie-killing efficiency M_1:

$$F_5 = M_3 = H_3 = 3\frac{1 - M_1}{100}\ .$$

Let us reconstruct the plots of Section 5.4 for the Romero scenario.

5.5.2 Simulation Results

Unflappable city ($\alpha = 0$, $N = 30$)

The parameter space where humanity is exterminated after eleven years is plotted in Figure 5.7B. The parameter space showing the scenarios where humanity has successfully culled the zombies is plotted in Figure 5.7A. Even in an unflappable city, a good deal of bravado and skill is required by the populace to survive a Romero scenario.

Cowardly city ($\alpha = 0.8$, $N = 30$)

The parameter space where humanity is exterminated after eleven years is plotted in Figure 5.7D, while the parameter space showing the scenarios where the humans have successfully culled the zombie population after eleven years is plotted in Figure 5.7C.

Note that, while a high efficiency is required from the militia, only a small amount of bravado is required. In this scenario, a bravado of 0.5 corresponds to one out of every ten people joining the militia after meeting a zombie. Indeed, the lower region of the phase space corresponds to only one out of every 100 people who see a zombie joining the militia. Also note that in the urban Romero scenario, a highly unflappable attitude in a city offers lower odds of survivability than a cowardly one,

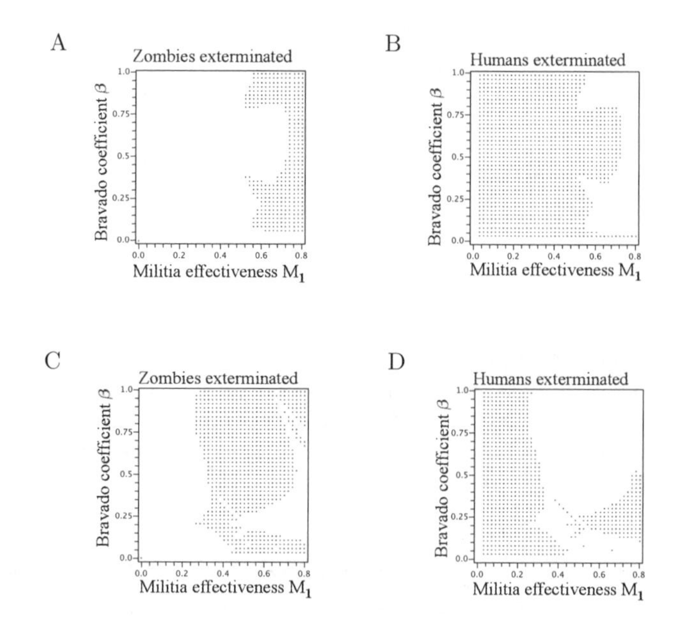

Figure 5.7: Culled populations after 4000 days in an urban setting with 'Romero zombies.' Initial values: $N = 30$, $R = 10000$, $S_W = 1000$, $S_M = 0$, $S_H = 0$ and $S_Z = 1$. Final time $= 4000$ days. 1600 evenly spaced points are chosen from $M_1 \in [0.01, 0.8]$, $\beta \in [0.01, 0.99]$. A: Zombie population exterminated in unflappable rural setting ($S_Z < 0.02$, $\alpha = 0$). B: Human population exterminated in unflappable urban setting ($\alpha = 0$). C: Zombie population exterminated in cowardly urban setting ($S_Z < 0.02$, $\alpha = 0.8$). D: Human population exterminated in cowardly urban setting ($\alpha = 0.8$).

since the survivable regions of the parameter space in Figure 5.7C are more easily attainable than the survivable regions of the parameter space in Figure 5.7A (where a complete cull of the zombie population requires a very enthusiastic population of expert marksmen).

Unflappable countryside ($\alpha = 0$, $N = 3000$)

In Section 5.4.2, we encouraged people in the countryside to adopt a diligently unflappable attitude in order to fix a larger parameter space where humanity would succeed. It is not clear, however, that this advice will still hold in the Romero scenario, given the results in the urban setting.

In a rural Romero scenario, where the populace maintains a stiff upper lip in the face of a zombie epidemic, the plot of the parameter space where humanity has not been exterminated after 4000 days is shown in Figure 5.8A. In this graph, the unplotted data points represent the parameter space where humanity is exterminated. This region is shown explicitly in Figure 5.8C.

Unlike the previous simulations, in this scenario there will be no cases where the zombie population has been completely extinguished (according to the standards

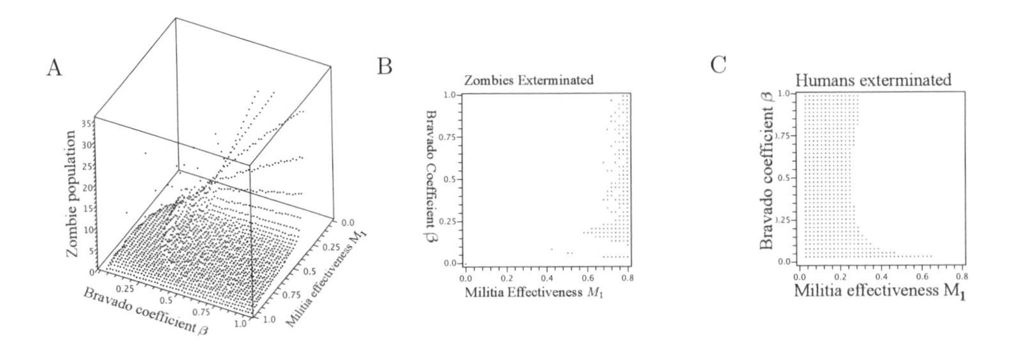

Figure 5.8: Populations after 4000 days in a rural unflappable ($\alpha = 0$) setting with Romero zombies. Initial values: $N = 3000$, $R = 10000$, $S_W = 1000$, $S_M = 0$, $S_H = 0$ and $S_Z = 1$. Parameter: $\alpha = 0$. Final time = 4000 days. 2500 evenly spaced points are chosen from $M_1 \in [0.01, 0.99]$, $\beta \in [0.01, 0.99]$. A: Zombie population remaining after 4000 days. B: Zombie population suppressed ($S_Z < 2$). C: Human population exterminated.

we have been using up until now) within six years. The parameter space where an average of fewer than two zombies remain over the course of the last year (at the end of the eleven years) is displayed in Figure 5.8B.

In the Romero scenario, the strategy recommended in Section 5.4.2 (a stiff upper lip) would lead to an endemic zombie population in rural areas, if not outright extermination of the human population. Clearly, understanding the nature of the zombies you are dealing with must be fundamental to the strategy you adopt.

Cowardly countryside ($\alpha = 0.8$, $N = 3000$)

The results of the rural simulation where the workers are very cowardly are similar to the unflappable rural simulation: there is no region of the parameter space where zombies have been successfully culled. Consider the parameter space where humanity has been completely exterminated in Figure 5.9A, and compare it with the equivalent plot for an unflappable countryside (Figure 5.8C). It is clear that, even in the Romero scenario, the rural workers are better off diligently maintaining an unflappable disposition than giving in to panicked horror and cowardice (though they are still fighting a battle they cannot win).

The results of the cowardly rural scenario are so pessimistic that one wonders if tweaking the initial conditions could increase the survival of the human population. Perhaps adding an initial militia population could increase the survival odds by culling the zombie population before it can take hold. In the next two sections, we consider cases where a standing militia is added as part of the initial conditions.

Standing military in a cowardly countryside ($\alpha = 0.8$, $N = 3000$, $S_M = 100$)

Suppose that there is an army base amid the countryside where a Romero zombie epidemic is under way. How will having a trained militia population at the start

A

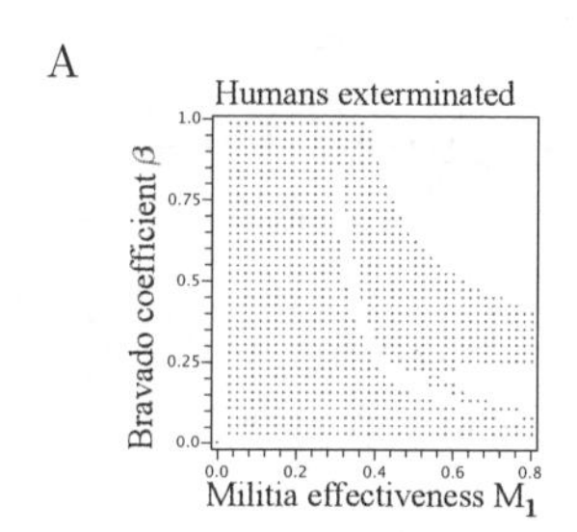

B

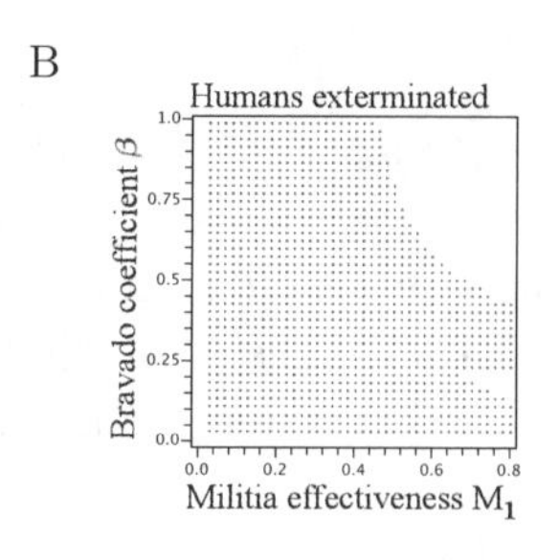

C

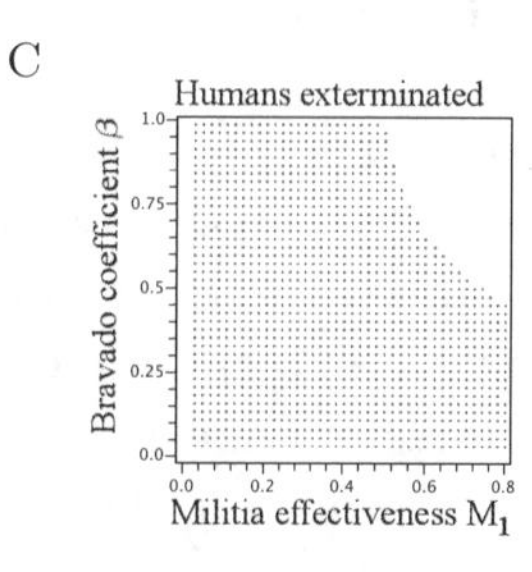

Figure 5.9: Human extinction after 4000 days in a cowardly rural setting with Romero zombies. Initial values: $N = 3000$, $R = 10000$ $S_W = 1000$, $S_H = 0$, $S_Z = 1$. Parameters: $\alpha = 0.8$ and 1600. Final time = 4000 days. Evenly spaced points are chosen from $M_1 \in [0.01, 0.8]$, $\beta \in [0.01, 0.99]$. A: Human population exterminated with no initial militia ($S_M = 0$). B: Human population exterminated with moderate initial militia ($S_M = 100$). C: Human population exterminated under military dictatorship ($S_M = S_W = 1000$).

of the zombie plague affect the outcome we saw in Section 5.5.2? We modelled this scenario by adding an extra 100 militia members at the start of the simulation we performed in Section 5.5.2. Again, there was no region of the parameter space where the zombie population was culled within six years. The parameter space where all of humanity had died out (Figure 5.9B) seemed quite pessimistic. In the Romero scenario, having a standing army initially present in a rural environment will decrease the worker community's odds of survival.

Military dictatorship

Could a military dictatorship, which possesses a very large standing army, survive the Romero scenario? Here we assume that, prior to the zombie epidemic, the country will have the maximum stable military population:

$$\frac{\Delta R}{\Delta t} = 0 \rightarrow S_W \frac{R_1 - 1}{R_2 + 1} = S_M.$$

For our assumed values of R_1 and R_2, $\frac{R_1 - 1}{R_2 + 1} = 1$. Thus, given the production levels of the workers, $S_W = S_M$ (initially).

We repeated our simulation for the cowardly rural scenario with $S_M = 1000$. The parameter space where humankind was completely exterminated is displayed in Figure 5.9C.

Again, there were no solutions where the zombie population had been successfully culled, and we see that the existence of a very large standing military will decrease the survivability of the scenario.

5.6 Discussion

Our model described the complexity of resource management during a zombie epidemic. We showed that the populace's lack of productivity (if workers become consumed with fighting rather than producing) can lead to starvation, which can result in a resurgent zombie population. It is clear that, when it comes to successfully culling an undead population, a very efficient and small militia is more effective than a large but poorly trained mob.

In addition, we explored the effect that the cowardice or bravado of the population had on the overall survivability in a zombie epidemic. In the urban scenario, where the population density was high, it did not matter whether the population was cowardly or unflappable. Provided that the militia members are on average at least twice as deadly to the zombies as the zombies are to the militia members, a militia of almost any size would serve to clear the city of zombies over the course of a few years.

The model oulined in this chapter might be overly simple in many ways. A uniform population density does not account for humans intelligently using geometry and geography to their advantage when confronting the mindless zombies. Future simulations could include human populations in adjoining regions: an urban landscape and a rural landscape with a bridge between them, for instance.

In a rural scenario, the personal attitudes in the populace played a much more important role in overall survivability. Since the zombie hunt will burn at a much slower rate than in the city, it is important that the population not quit working to hide or fight each time they see a zombie. It seems that unflappability is a requirement of living in the countryside.

We repeated these simulations for the Romero scenario, where the zombie population is made up of the reanimated dead, rather than infected insane anthropophagites. These models differ from the prior model since victims of accidental homicide will also join the zombie population. Thus an untrained militia will act as a double-edged sword. In the Romero scenarios, we showed that a zombie population will become endemic to rural environments independent of how society reacts to the epidemic. If the dead rise, we would advise individuals to move to the city.

Appendices

A Acknowledgements

The author would like to thank Phil Ledwith and James Watmough for their helpful advice on this paper.

B Glossary

Ansatz. An assumption about the system made at the outset.

Anthropophagite. A cannibal. Someone who eats people.

Heaviside function. A discontinuous function $H(x)$ that has two values: $H(x) = 0$ when $x < 0$ and $H(x) = 1$ when $x > 0$. This function works really well as a switch, which is zero until a certain criterion is met. (In this case, when humanity is about to starve to death.)

Initial population. The original population on the morning of day one.

Log scale. A rescaling of an axis to highlight variations. There are occasions where the details in a graph vary in scale too dramatically: a large drop in size might wash out little wiggles, or it might get very close to zero. In these circumstances, it is useful to take a logarithm of the coordinates of the points on a graph.

Normalization. The procedure by which we measure the relative increase and decrease of a variable by setting its value at some point in time to equal 1.

Parameter space. The representation of two non-variable coordinates in a plane. This is particularly useful when we are interested in the outcome of a system over a variety of different possible values of the parameters.

Romero zombie. A dead human who has become reanimated.

Uniform population density. The assumption that the population density is constant throughout the city.

References

[1] L. Brown. *Full Planet, Empty Plates: The New Geopolitics of Food Scarcity.* W.W. Norton Co., New York, NY, 2012.

[2] F. Brauer. Epidemic Models with Heterogeneous Mixing and Treatment. *Bull. Math. Biol.* 70:1869–1885, 2008.

[3] P. Munz, I. Hudea, J. Imad, R.J. Smith?. When zombies attack!: Mathematical modelling of an outbreak of zombie infection. In *Infectious Disease Modelling Research Progress* J.M. Tchuenche and C. Chiyaka, eds., Nova Science, Happuage, NY 2009, pp. 133–156.

[4] R.V. Culshaw, S.G. Ruan. A delay-differential equation model of HIV infection of CD4(+) T-cells. *Math. Biosci.* 165:27–39, 2000.

[5] N.I. Stilianakis, K. Dietz, D. Schenzle. Analysis of a model for the parthogenesis of AIDS. *Math. Biosci.* 145:27–46, 1997.

[6] Z. Guo, L. Huang. Impact of discontinuous treatments on disease dynamics in an SIR epidemic model. *Math. Biosci. Engin.*, 9(1):97–110, 2012.

[7] M. Brooks *The Zombie Survival Guide: Complete Protection from the Living Dead.* Three Rivers Press, New York, NY, 2003.

[8] G.A. Romero, J.A. Russo. *Night of the Living Dead,* directed by G.A. Romero. The Walter Reade Organization, 1968.

HOW LONG CAN WE SURVIVE?

Thomas E. Woolley, Ruth E. Baker, Eamonn A. Gaffney and Philip K. Maini

Abstract

Knowing how long we have before we face off with a zombie could mean the difference between life, death and zombification. Here, we use diffusion to model the zombie population shuffling randomly over a one-dimensional domain. This mathematical formulation allows us to derive exact and approximate interaction times, leading to conclusions on how best to delay the inevitable meeting. Interaction kinetics are added to the system and we consider under what conditions the system displays an infection wave. Using these conditions, we then develop strategies that allow the human race to survive its impending doom.

6.1 Introduction

HUMANS would not survive a zombie holocaust, or so current work suggests [1]. The fact that reanimated corpses do not stop unless their brain is destroyed, coupled with an insatiable appetite for human flesh, has proven to be a deadly combination [2]. However, the work produced by Munz et al. [1] assumes that zombies and humans are well mixed, meaning that zombies can be found everywhere there are humans. Realistically, the initial horde of zombies will be localized to areas containing dead humans, such as cemeteries and hospitals. In addition, because humans and zombies are not initially separated, humans are not able to run and hide in order to try and preserve themselves. It is a well-documented fact that zombies are deadly but slow-moving [3, 4, 5]. Due to their slow movements, it is quite possible that, given sufficient warning, we would be able to outrun the zombies and produce a defensible blockade where humans could live safely [4]. In order to do so, it would be useful to know just how long an infestation of zombies would take to reach our defences; this would give us an estimate of how long we would have to scavenge for supplies and weaponry in order that we may protect ourselves from these oncoming, undead predators.

Another useful question to ask would be: 'If a zombie were to infect a group of humans, could we slow down the rate of infection?' This gives a chance for potential survivors to split from the infected group and move to another defensible area.

In order to answer these and other important questions, we add a spatial dimension to a simple infection model. Previous models in this book have looked at time- and population-dependent interactions between humans and the undead. We now add an obvious extension: that of geography, because zombies do not move homogeneously. This allows zombies to shuffle around, giving a much more realistic picture of an invasion.

If the reader currently does not have the time to understand the justification of the mathematics due to their impending doom and would rather skip to useful results, we recommend skipping Section 6.2, where we justify our model; Section 6.3, where we introduce the mathematical formalism; and Section 6.4, in which analytic solutions are provided. Instead, we suggest that the reader jump straight to Section 6.5, which details a quick way of approximating the time left before zombies will be nibbling on their brains. Section 6.6 then looks at adding interaction rules between the populations of the humans and zombies, and can probably be skipped by those looking for immediate answers. The conclusion in Section 6.7 is absolutely vital as we consider all the results we have achieved and interpret them in the context of a strategy for survival.

6.2 Random Walks and Diffusion

To simplify the mathematics, we assume we are working in only one space dimension, denoted x. This is not too restrictive as most results hold in two or even three dimensions. Realistically speaking, this assumption means that there is only one possible access point to our defences and thus the shortest distance between the group of zombies and us is a straight line.

According to Munz et al. [1], "the 'undead' move in small, irregular steps." This makes their individual movement a perfect model of a random walk [6]. The 'drunkard's' or random walk, as you would expect from the name, is a mathematical description of movement in which no direction is favoured. To get an intuitive idea of the motion, consider inebriated people. They stagger backwards and forwards, lurching from one resting place to another.

In Figure 6.1, we illustrate this motion with zombies; since they will move around randomly bumping into one another, they too will spread over the domain. Diffusion has two primary properties: it is random in nature and movement is from regions of high densities to low densities.

We could model the motion of each zombie individually but, since their movement is random, their positions will not be certain unless we track them all exactly [7]. Probabilistic descriptions can be used but, due to the assumed large population

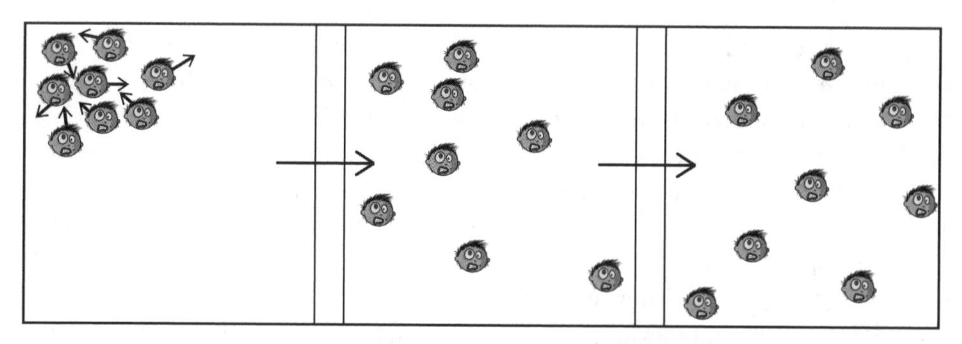

Figure 6.1: An example of two-dimensional diffusion. On the left, an initial group of zombies is placed in the top left corner, close together, their initial directions denoted by the small black arrows. After a time, their random motion will cause them to spread themselves out over the domain.

of zombies, the computational power needed by such simulations would soon get out of hand, even when we are not running for our lives. Instead, we consider the case where the number of zombies is initially large. This produces a continuous model of diffusion that is much easier to use, as it needs less computational power and analytical solutions are available (see Section 6.4). However, what we have gained in simplicity we have lost in accuracy. Realistically the density of zombies in a region will be a discrete integer value: $1, 2, 3, \ldots$ zombies/metre (counting even a dismembered zombie as a whole one). Since we are assuming a large number of zombies, we are forced to allow the density to take any value, including non-integers. Thus the density, originally a discrete function, has been smoothed out to form a continuous function.[1]

6.3 A Mathematical Description of Diffusion

Since time may be running out, we introduce the diffusion equation in an intuitive rather than rigorous way. Further, although word equations are used, a working knowledge of partial derivatives will be useful for the following section.

If we let the density of zombies at a point x and at a time t be $Z(x, t)$, then the density has to satisfy the diffusion equation

$$\frac{\partial Z}{\partial t}(x,t) = D\frac{\partial^2 Z}{\partial x^2}(x,t). \tag{6.1}$$

Explicitly,

$$\frac{\partial Z}{\partial t}(x,t) = \text{rate of change of } Z \text{ over time at a point } x.$$

[1]The interested reader who would like to learn about the rigorous definitions of continuous functions is first advised to wipe out the walking dead, then rebuild civilization and finally look for any real-analysis textbooks that may have survived intact, such as Bartle and Sherbert [8].

This means that if $\partial Z/\partial t$ is positive, then Z is increasing at that point in time and space. This term allows us to consider how $Z(x,t)$ evolves over time.

The factor D is a positive constant that controls the speed of movement. The units of D are chosen to make the diffusion equation consistent, since the units on each side of the equality must match. The term $\partial Z/\partial t$ has units of [density]/[time] and $\partial^2 Z/\partial x^2$ has units of [density]/[space]^2; thus D must have units of the form [space]^2/[time]. The units could be of the form of kilometres^2/hour or nanometres^2/second. However, for the equation to be most informative we choose scales that are relevant to a zombie invasion; we thus fix the units to be metres^2/minute.

The term on the right-hand side is a little more complex than the left, but essentially it encapsulates the idea that the zombies are moving from high to low densities. This is illustrated with the help of Figure 6.2. Initially, there are many more zombies on the left than the right. Just before the peak in density the arrow (which is the tangent to the curve, or $\partial Z/\partial x$ at this point) is pointing upwards. This means that as x increases, so does the zombie density, Z. At this point,

$$\frac{\partial Z}{\partial x} = \text{rate of change of } Z \text{ as } x \text{ increases} > 0.$$

Just after the peak the arrow is pointing down; thus, at this point,

$$\frac{\partial Z}{\partial x} = \text{rate of change of } Z \text{ as } x \text{ increases} < 0.$$

Hence, at the peak, $\partial Z/\partial x$ is decreasing as x increases. Since $\partial^2 Z/\partial x^2$=rate of change of $\partial Z/\partial x$ as x increases and we have just deduced that $\partial Z/\partial x$ is decreasing at the peak,

$$\frac{\partial^2 Z}{\partial x^2} < 0.$$

From the diffusion equation, we see that, at the peak,

$$\frac{\partial Z}{\partial t} = D\frac{\partial^2 Z}{\partial x^2} < 0.$$

This means that the peak of zombie density is decreasing over time. Similarly, $\partial Z/\partial x < 0$ just before the trough and $\partial Z/\partial x > 0$ after the trough, as shown in Figure 6.2. Thus, at the trough, $\partial^2 Z/\partial x^2 > 0$, implying that the density of zombies is increasing. Overall we see that diffusion causes zombies to move from regions of high density to low density.

It is impossible to overstate the importance of equation (6.1). Whenever the movement of a modelled species can be considered random and directionless, the diffusion equation will be found. This means that, by understanding the diffusion equation, we are able to describe a host of different systems such as heat conduction through solids, gases spreading out through a room, proteins moving around the body, molecule transportation in chemical reactions, rainwater seeping through soil and predator–prey interaction, to name but a few of the great numbers of applications [9, 10, 11].

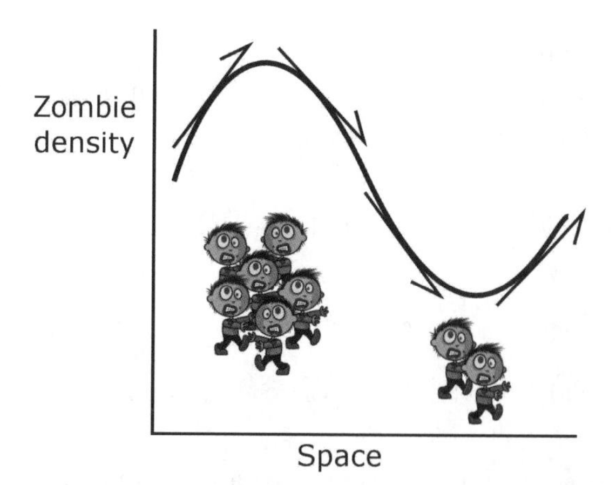

Figure 6.2: An example of how diffusion smoothes out peaks and troughs. At the peak of the density, there are more zombies than the surrounding area. Since the rate of change of number of zombies over space (the tangent shown by the arrows on the curve) is positive before the peak and negative after the peak, the rate of change of number of zombies over space is decreasing. Using this knowledge and equation (6.1), we see that the density at the peak is decreasing. Similarly, at the trough, the density will be increasing.

6.4 Solution to the Diffusion Equation

The diffusion equation gives us a mathematical description of zombie movement. By solving it, we produce the function $Z(x,t)$, the density of the zombies at each position x for all time $t \geq 0$. However, we must add some additional information to the system before we can solve the problem uniquely.[2]

Initially, we have a density of Z_0 zombies/metre. All of the zombies are assumed to start in the region $0 \leq x \leq 1$ (see equation (6.3)). The implication of this is that all of the undead will originate from one place, such as a cemetery or mortuary in a hospital.

The final assumption we make is that the zombies cannot move out of the region $0 \leq x \leq L$. That is, we define L as the length of the domain. This creates a theoretical boundary at $x = 0$ that the population cannot cross; the zombies will simply bounce off this boundary and be reflected back into the domain. We place our defences at $x = L$, thus creating another theoretical boundary here. Thus, at $x = 0$ and $x = L$, the 'flux of zombies' is zero. Various other boundary conditions are possible but for simplicity we chose 'zero flux' conditions so that the boundaries may prevent the zombies from leaving the region but do not alter their population.

The flux of zombies at $x = 0$ and $x = L$ is simply the rate of change of the numbers of zombies through these boundaries. Mathematically, this is the spatial

[2]Uniqueness is an important concept in mathematics, particularly when solving differential equations. If your solution is not unique, then you cannot be sure which one of the possible solutions is the one needed.

derivative at these points and, since the flux is assumed to be zero, then $\partial Z/\partial x = 0$ at $x = 0, L$. These initial and boundary conditions can be seen in Figure 6.3A.

The system is then fully described as

$$\frac{\partial Z}{\partial t}(x,t) = D\frac{\partial^2 Z}{\partial x^2}(x,t) \qquad \text{(the partial differential equation)} \tag{6.2}$$

$$Z(x,0) = \begin{cases} Z_0 & \text{for } 0 \le x \le 1 \\ 0 & \text{for } x > 1 \end{cases} \qquad \text{(the initial condition)} \tag{6.3}$$

$$\frac{\partial Z}{\partial x}(0,t) = 0 = \frac{\partial Z}{\partial x}(L,t) \qquad \text{(zero-flux boundary conditions).} \tag{6.4}$$

Equation (6.2) with initial condition (6.3) and boundary condition (6.4) can be solved exactly and has the form

$$Z(x,t) = \frac{Z_0}{L} + \sum_{n=1}^{\infty} \frac{2Z_0}{n\pi} \sin\left(\frac{n\pi}{L}\right) \cos\left(\frac{n\pi}{L}x\right) \exp\left(-\left(\frac{n\pi}{L}\right)^2 Dt\right). \tag{6.5}$$

The interested reader can find more details in Appendix A or in [9].

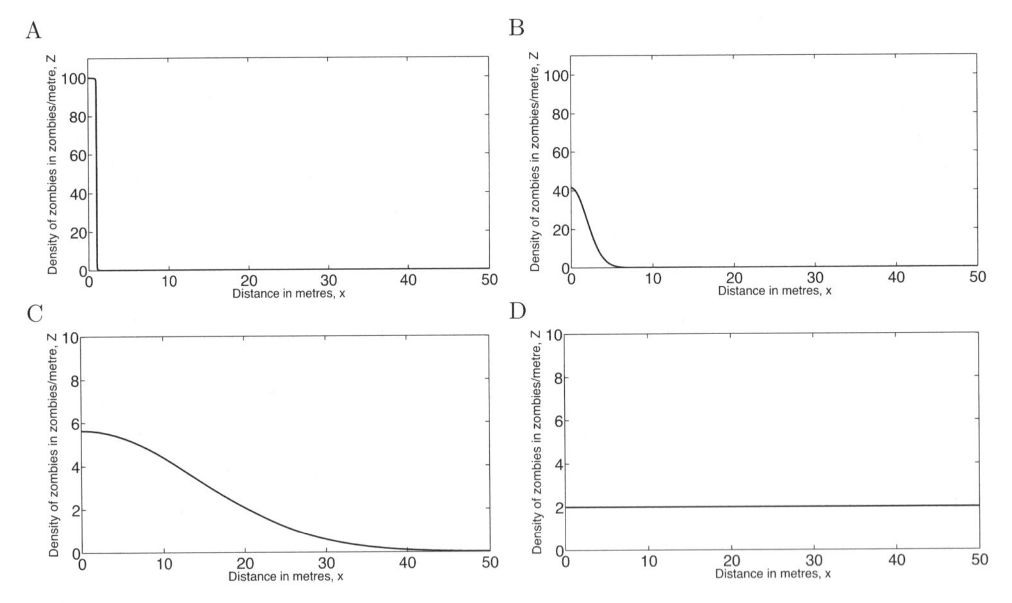

Figure 6.3: Evolution of a system of zombies over 100 minutes. The zombies are assumed to be confined within a domain of length 50 metres and have a diffusion coefficient of $D = 100$ metres^2/minute. A: $t = 0$ minutes. B: $t = 1$ second. C: $t = 1$ minute. C: $t = 100$ minutes.

The sine and cosine functions are those that the reader may remember from their trigonometry courses. The exponential function, 'exp,' is one of the fundamental operators of mathematics, but for now the only property that we are going to make use of is that if $a > 0$ then $\exp(-at) \to 0$ as $t \to \infty$ [8]. Using this fact and equation

(6.5), as t becomes large, most of the right-hand side of the equation becomes very small, approximately zero. Hence, for large values of t, we can approximate

$$Z(x, t) \approx \frac{Z_0}{L}. \tag{6.6}$$

This means that as time increases, zombies spread out evenly across the available space, with average density Z_0/L everywhere.

In Figure 6.3, we have plotted zombie density for a number of times. Figure 6.3A shows the initial condition of a large density of zombies between $0 \leq x \leq 1$. Figure 6.3B and Figure 6.3C then illustrate how diffusion causes the initial peak to spread out, filling the whole domain. Note that we rescaled the axes in Figures 6.3C and 6.3D in order to show the solution more clearly. Figure 6.3D verifies equation (6.6) since, after 100 minutes, the density of zombies has become uniform throughout the domain.

6.5 Time of First Interaction

Equation (6.5) gives the density of zombies at all places, x, and for all time, t. There are many questions we could answer with this solution; however, the most pressing question to any survivors is, "How long do we have before the first zombie arrives?" If we had proceeded with a stochastic description of the system, we could use the theory of 'first passage time' [7]. However, deterministically, the mathematical formulation of this question is, "For what time, t_z, does $Z(L, t_z) = 1$?" The time t_z is then the time taken (on average) for a zombie to have reached $x = L$.

Unfortunately, this does not have a nice solution that can be plotted like equation (6.5). In Appendix B, we introduce the bisection search technique [12], which enables us to find this solution, approximated to any degree of accuracy we desire.

Using this technique, we can now vary the distance and speed of the zombies to calculate the various interaction times. In Figure 6.4, we plot the interaction time of an initial density of 100 zombies against a diffusion rate ranging from $100\text{m}^2/\text{min}$ to $150\text{m}^2/\text{min}$, which (from empirical evidence) is realistic for a slow shuffling motion and for distances between 50 and 90 metres. Notice that, for distances greater than $L = 100\text{m}$, the average density of zombies is less than 1 zombie/metre and so there will be no solution, t_z, to $Z(L, t_z) = 1$ since $Z(L, t) < 1$ for all $t \geq 0$. In this situation, the assumption that we have a large number of zombies has failed, making the continuous model an inadequate description.

6.5.1 Diffusive Time Scale

When the apocalypse does happen, we have to ask ourselves the question, "Do we want to waste time computing solutions when we could be out scavenging?" We

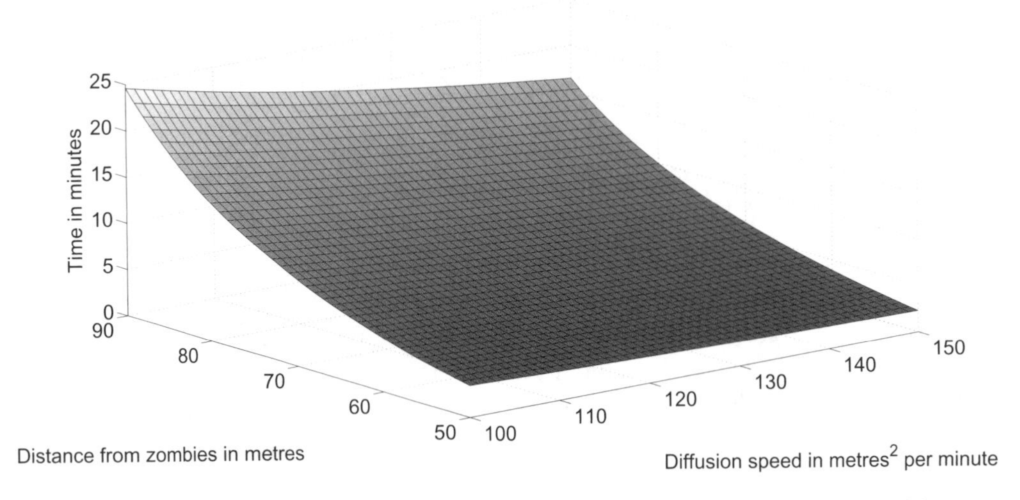

Figure 6.4: Time in minutes until the first zombie arrives for various rates of diffusion and distances.

thus introduce the diffusive time scale

$$t = \frac{1}{D}\left(\frac{L}{\pi}\right)^2 \approx 0.32\frac{L^2}{D}. \tag{6.7}$$

From equation (6.5), we can see that this is the time it takes for the first term of the infinite sum to fall to $\exp(-1)$ of its original value. The factor of $\exp(-1)$ is used due to its convenience.

Only the first term of the expansion is considered, as the exponential function is monotonically increasing. This means that if $a > b$ then $\exp(a) > \exp(b)$ and so as n increases, the contribution from the term

$$\exp\left(-\left(\frac{n\pi}{L}\right)^2 Dt\right)$$

rapidly decreases. Thus the first term gives an approximation to the total solution.

Hence equation (6.7) gives a rough estimate of how quickly the zombies will reach us. For example, if zombies are 90 metres away and have a diffusion rate of $100\text{m}^2/\text{min}$, then $t \approx 26$ minutes, comparable to the solution in Figure 6.4. It also implies a very important result about delaying the human–zombie interaction.

There are two possible ways we could increase the time taken for the zombies to reach us. We could run further away or we could try and slow the zombies down, since the time taken is proportional to the length squared, L^2, and inversely proportional to the diffusion speed, D. This means that if we were to double the distance between ourselves and the zombies, then the time for the zombies to reach us would approximately quadruple. However, if we were to slow the zombies down by half, then the time taken would only double. Since we want to delay interaction with the zombies for as long as possible, then from the above reasoning, we see that it is much better to expend energy travelling away from the zombies than it is to try and slow them down. Without some form of projectile weaponry or chainsaw, killing zombies is particularly difficult as they do not stop until their brains are destroyed. These conclusions are confirmed in Figure 6.4.

It should be noted that the time derived here is a lower bound. In reality, the zombies would be spreading out in two dimensions and would be distracted by obstacles and victims along the way, so the time taken for the zombies to reach us may be longer. This conservative estimate will keep us safe, especially since the authors would prefer to be long gone rather than chance a few more minutes of scavenging!

6.6 Slowing the Infection

In the previous sections, we assumed that the main source of zombies were the primary infected; that is, the dead who have somehow been brought back to life. However, the primary infected are able to produce a secondary generation of infected individuals through biting. These are victims who, although bitten, survive the initial encounter with a zombie and will eventually turn into zombies themselves [1].

Note that this hijacking of a host species is not restricted to fiction. The natural world has a number of parasites that infect and control their prey [13]. One particularly well-known example is *Cordyceps unilateralis*, a fungus that infects ants and alters their behaviour. These infected ants are then recruited in the effort to distribute fungus spores as widely as possible to other ants.

6.6.1 Interaction Kinetics

Initially, we do not consider a spatial model. This simplification allows us to investigate the interactions of the two species more easily. However, in Section 6.6.3, we will use diffusion once again to model movement and thus see how this can affect the dynamics.

To model human–zombie interactions, we suppose that a meeting between the two populations can have three possible outcomes.

1. The human kills the zombie.

2. The zombie kills the human.

3. The zombie infects the human and so the human becomes a zombie.

See Figure 6.5.

Allowing H to stand for the human population and Z to stand for the zombie population, these three rules can be written as though they were chemical reactions:

$$H + Z \xrightarrow{a} H \qquad \text{(humans kill zombies)}$$
$$H + Z \xrightarrow{b} Z \qquad \text{(zombies kill humans)}$$
$$H + Z \xrightarrow{c} Z + Z \qquad \text{(humans become zombies).}$$

The letters above the arrows indicate the rate at which the transformation happens and are always positive. If one of the rates is much larger than the other two, then this 'reaction' would most likely happen.

Figure 6.5: The possible outcomes of a human–zombie interaction: (a) humans kill zombies, (b) zombies kill humans, or (c) zombies convert humans.

To transform these reactions into a mathematical equation, we use the Law of Mass Action [10]. This law states that the rate of reaction is proportional to the product of the active populations. Simply put, this means that the above reactions are more likely to occur if we increase the number of humans and/or zombies. Thus we can produce the following equations, which govern the population dynamics:[3]

$$\frac{dH}{dt} = -bHZ - cHZ = -(b + c)HZ \tag{6.8}$$
$$\frac{dZ}{dt} = cHZ - aHZ = (c - a)HZ. \tag{6.9}$$

6.6.2 Is It Possible to Survive?

Equation (6.8) shows that the population of humans will only ever decrease over time, because $b, c > 0$ and $H, Z \geq 0$. We could add a birth term into this equation, which would allow the population to also increase in the absence of zombies.

[3]Our use of the symbol d in this case rather than ∂ implies that we are working with one variable dimension—in this case time, t. When dealing with diffusion, we consider both space and time and thus we use ∂.

However, as we have seen from Section 6.5.1, the time scale we are working on is extremely short, much shorter than the nine months it takes for humans to reproduce! Thus we ignore the births since they are not likely to alter the populations a great deal during this period.

Equation (6.9) is not so clear, as the term '$(c - a)$' may either be positive or negative. If $(c-a) > 0$ then the creation rate of zombies, c, must be greater than the rate at which we can destroy them, a. In this case, the humans will be wiped out as our model predicts that the zombie population will grow and the human population will die out. However, there is a small hope for us. If the rate at which humans can kill zombies is greater than the rate at which zombies can infect humans, then $(c - a) < 0$. In this case, both populations are decreasing and thus our survival will come down to a matter of which species becomes extinct first.

If we now let $b + c = \alpha$ then α is the net removal rate of humans; similarly, $c - a = \beta$ is the net creation rate of zombies. We can now write equations (6.8) and (6.9) as

$$\frac{d(\beta H + \alpha Z)}{dt} = \beta\frac{dH}{dt} + \alpha\frac{dZ}{dt} = -\beta\alpha HZ + \alpha\beta HZ \equiv 0.$$

This means that $(\beta H + \alpha Z)$ does not change over time so, although $H(t)$ and $Z(t)$ do change over time (as mentioned above, $H(t)$ is always decreasing), they must change in such a way to keep the value of $(\beta H(t) + \alpha Z(t))$ a constant. Since we can estimate the initial populations of zombies and humans as Z_0 and H_0, respectively, we can define this constant exactly:

$$\beta H(t) + \alpha Z(t) = \beta H_0 + \alpha Z_0. \tag{6.10}$$

As mentioned above, the only way humans can survive is if we are more deadly than zombies[4] and thus $\beta < 0$. To make the following result easier to see, let $\beta = -\gamma$, with $\gamma > 0$. If the humans are to survive, then all of the zombies must be wiped out so $Z(\infty) = 0$ (here $Z(\infty)$ is taken to mean the population of the zombies after a long time has passed; similarly, $H(\infty)$ is the human population after a long time). Upon rearranging equation (6.10), we find that

$$\gamma H(\infty) = \gamma H_0 - \alpha Z_0,$$

and since we desire the human population to still exist, this implies $H(\infty) > 0$. Thus, for our species to survive the war of attrition against the zombies, the initial populations must satisfy

$$\gamma H_0 > \alpha Z_0. \tag{6.11}$$

[4]If the zombie apocalypse should occur and you are in fact a zombie reading this, please accept the authors' apologies. There is nothing personal against you or your kind. Humans simply do not take kindly to being thought of as lunch. Indeed, if you are reading this, we must assume that you are an intelligent species, in which case you may like to work through these results and see how they might favour zombies instead of humans.

Essentially, the inequality echoes the sentiments we would expect. To survive extinction, the humans need a large enough initial population which is capable of being more deadly than the zombies. If this condition is not satisfied, then the zombie revolution is certain. See Figure 6.6 for an example of this.

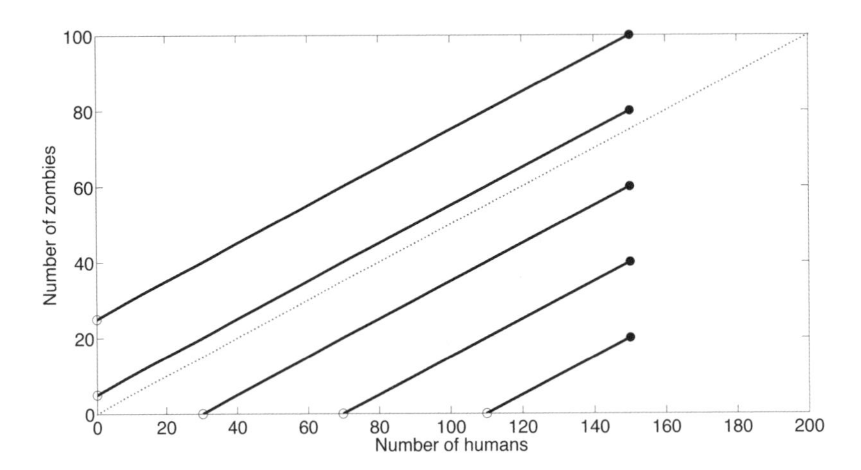

Figure 6.6: The solid lines show the evolution of the populations for a number of different initial zombies. The initial condition is shown as a filled in circle and the final state of the system is shown as a hollow circle. The dotted line, $\gamma H = \alpha Z$, separates the two possible cases, where γ is the net destruction rate of zombies and α is the net human removal rate. Above the dotted line, zombies survive and kill all humans; below the dotted line, humans survive and kill all zombies. If γ were increased (zombies are removed more quickly than they are created) the line would become steeper, meaning that it would become easier for the population to be beneath the line and thus survive. If α were to increase, the line would become shallower, implying that it would be easier to be above the line and thus the zombies would wipe us out. Parameters used are $\gamma = 0.5$ per human per minute and $\alpha = 1$ per zombie per minute.

6.6.3 Infection Wave

The above analysis does not take into account spatial effects. However, the zombies will not be everywhere, but will rather be localized. Thus, in Section 6.5, we considered only the time before the zombies came into contact with the human population. Once this occurs, we then have to consider not only how fast the zombies are moving but also how fast they can infect people. Intuitively, if the infection rate was very small, we would have little to worry about because, even if the zombies managed to reach us, it would be very hard for them to infect us while very easy for us to pick them off one by one. To consider the spatial situation, we now add the diffusion term to each of the equations

$$\frac{\partial H}{\partial t} = D_H \frac{\partial^2 H}{\partial x^2} - \alpha H Z \tag{6.12}$$

$$\frac{\partial Z}{\partial t} = D_Z \frac{\partial^2 Z}{\partial x^2} + \beta H Z. \tag{6.13}$$

Note that the diffusion coefficient of humans (D_H) is likely to be much smaller than the zombies' (D_Z) as, being in control of their mental faculties, humans do not move randomly. Since we are considering the initial stages of the invasion by the undead, we are assuming that humans have not figured out what is going on yet and are not yet running for their lives.

To see how quickly the infection moves through the human population, we look for a wave solution: a particular solution form of a differential equation that does not change its shape as it moves in space and time. This infection wave will move at a certain speed, v, which we hope to find. For this wave solution that we are looking for, we know that space, x, and time, t, will be linked through this speed; we can thus reduce the dimensions of the system from the two coordinates (x, t) to the single coordinate $u = x - vt$. See Appendix C for more justification and the implication of this coordinate change.

The system simplifies to

$$\begin{aligned} 0 &= D_H H'' + vH' - \alpha HZ \\ 0 &= D_Z Z'' + vZ' + \beta HZ, \end{aligned}$$

where, for the sake of brevity, we have used the notation

$$' = \frac{d}{du}.$$

In Appendix C, we are able to show that the speed has a minimum value of

$$v_{\text{min}}^2 = 4D_Z \beta H_0. \tag{6.14}$$

In order to slow the infection, we should try to reduce the right-hand side of equation (6.14). This in turn reduces the minimum speed for v, causing the infection to move more slowly [14]. To do this, we can reduce either D_Z, β or H_0. We discuss the reality of these decisions in the next section. In Figure 6.7, we show the effect of reducing the zombie diffusion rate from 100 m^2/min to 50 m^2/min. Note that the infection wave is further back in the slower case, as expected.

6.7 Conclusion

In Section 6.5.1, we showed that running away from zombies increases the time before the initial interaction far more than trying to slow the zombies down. Thus fleeing for your life should be the first action of any human wishing to survive. However, we cannot run forever; interaction with zombies is inevitable.

Section 6.6.2 then comes into play. Here we showed that the best long-term strategy is to create a fortified society that can sustain the population, while allowing us to remove the zombies as they approach. This, in effect, reduces α to zero which means that equation (6.11) is easy to satisfy and we will survive.

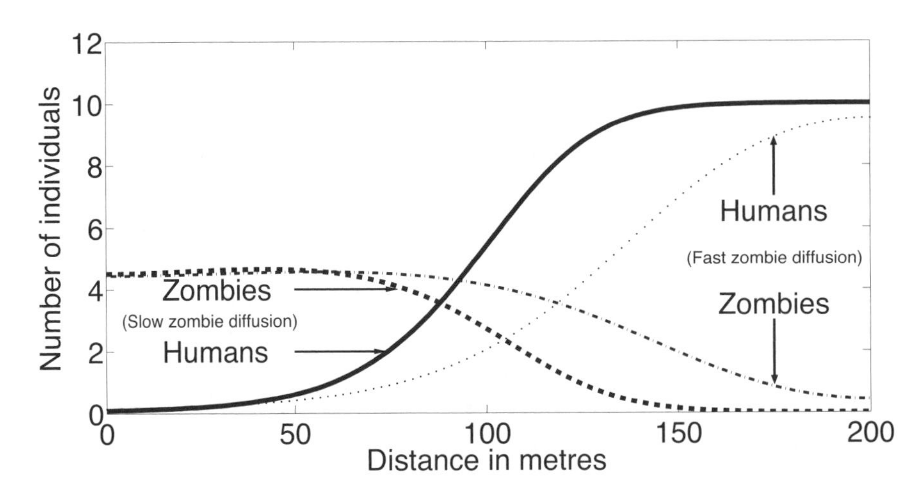

Figure 6.7: The infective wave solution for two different values of D_Z. The solid and dashed curves are the human and zombie populations respectively, for $D_Z = 100$ m^2 per minute. The dotted and dash-dotted curves are the human and zombie populations respectively, for $D_Z = 50$ m^2 per minute. Other parameters are $D_H = 0.1$ m^2 per minute, $\alpha = 0.1$ per zombie per minute and $\beta = 0.05$ per human per minute.

In the event of the apocalypse, it is unlikely that we would be able to support such a commune without raiding parties scavenging for medicinal, food and fuel supplies [4]. Thus, in this case, we fall back on the maxim of being more deadly than the zombies which increases the left-hand side of equation (6.11).

What if the barricade fails? In Section 6.6.3, we consider this possibility and show that our only hope is to try and reduce the speed of infection. To do this, equation (6.14) implies we must reduce either D_Z, β or H_0. Reducing D_Z amounts to slowing the zombies down; thus an effective fortification should have plenty of obstructions that a human could navigate but a decaying zombie, who may not be so athletic, would find challenging [15]. We have already discussed the case of reducing β; ideally, we want $\beta < 0$. If this is true and we are able to kill the zombies more quickly than they can infect, then the system cannot support a wave solution, thus greatly reducing the speed of zombification. Finally, we consider the possibility of reducing H_0, the population of the commune. This leads to the controversial tactic of removing your fellow survivors: if the zombies are unable to infect them, then their population cannot increase. The authors do not recommend this course of action, as reducing the human population also reduces the number of people able to fight the zombies and humans sacrificing other humans would only speed the extinction of their own species. The population will have enough trouble trying to survive the hordes of undead, without worrying about an attack from their own kind!

Our conclusion is grim, not because we want it to be so but because it is so. It had always been the authors' intention to try and save the human race. So to you, the reader, who may be the last survivor of the human race, we say: run. Run as far away as you can get; an island would be a great choice.[5] Only take the chance to fight if you are sure you can win and seek out survivors who will help you stay alive.

Good luck. You are going to need it.

Appendices

A Diffusion Solution Details

In Section 6.4, we claimed

$$Z(x,t) = \frac{Z_0}{L} + \sum_{n=1}^{\infty} \frac{2Z_0}{n\pi} \sin\left(\frac{n\pi}{L}\right) \cos\left(\frac{n\pi}{L}x\right) \exp\left(-\left(\frac{n\pi}{L}\right)^2 Dt\right) \tag{6.15}$$

as the solution to

$$\frac{\partial Z}{\partial t}(x,t) = D\frac{\partial^2 Z}{\partial x^2}(x,t) \tag{6.16}$$

$$Z(x,0) = \begin{cases} Z_0 & \text{for } 0 \leq x \leq 1 \\ 0 & \text{for } x > 1 \end{cases} \tag{6.17}$$

$$\frac{\partial Z}{\partial x}(0,t) = 0 = \frac{\partial Z}{\partial x}(L,t). \tag{6.18}$$

Given the solution, we can easily check by taking the derivatives that it is indeed a solution to the above constraints, but how do we find such a solution?

Separation of variables

Equation (6.16) depends on the variables space, x, and time, t, so we 'guess' the solution has the form $Z(x,t) = f(x)g(t)$. This is known as separating the variables, as we are assuming that Z can be decomposed into functions each depending on only one of the variables.

This guess is now substituted into equation (6.16) which we use to derive constraints on the forms of f and g. The substituted form of the equation, after rearrangement, is

$$\frac{1}{g}\frac{dg}{dt} = \frac{D}{f}\frac{d^2 f}{dx^2}. \tag{6.19}$$

Notice that the left-hand side of this equation does not depend of x in any way. Similarly, the right-hand side does not depend on t. This implies that if we change the variable x, then the right-hand side does not change. If it did, then due to the equality, the left-hand side would do too, but we have already noticed that

[5] Make sure the bay around the island is steep, as zombies can also cross water since they do not need to breathe [4].

the left-hand side does not depend on x. This logic is exactly the same for the variable t. Due to this argument, we deduce that both sides must be a constant, $-\alpha$, say (the negative sign is arbitrary but useful later on). Equation (6.19) is then split to produce

$$\frac{dg}{dt} = -\alpha g \tag{6.20}$$

$$\frac{d^2 f}{dx^2} = -\frac{\alpha}{D} f. \tag{6.21}$$

These are standard differential equations and their solution can be found in numerous text books [9]. First, we consider equation (6.20). This has the well-known solution

$$g(t) = C_1 \exp(-\alpha t),$$

where C_1 is a constant that we find later. This is the same exponential function described in Section 6.4. The value of α has yet to be defined exactly; however, this equation implies the sign of α. We want our solutions to not increase, since the zombies are spreading out and therefore $\alpha \geq 0$. This is useful since equation (6.21) has a different solution depending on whether α is positive, negative or zero. Using this information, equation (6.21) has the general solution

$$f(x) = C_2 \cos\left(\sqrt{\frac{\alpha}{D}} x\right) + C_3 \sin\left(\sqrt{\frac{\alpha}{D}} x\right),$$

where C_2, C_3 are, once again, constants that we need to define. To pin down the constants, we appeal to the boundary conditions, equation (6.18). Since we have a separable equation (i.e., $Z(x, t) = f(x)g(t)$), the boundary conditions simplify to

$$\frac{df}{dx}(0) = 0 = \frac{df}{dx}(L).$$

Since

$$\frac{df}{dx} = -C_2 \sin\left(\sqrt{\frac{\alpha}{D}} x\right) + C_3 \cos\left(\sqrt{\frac{\alpha}{D}} x\right),$$

then

$$\frac{df}{dx}(0) = C_3 = 0$$

$$\frac{df}{dx}(L) = -C_2 \sin\left(\sqrt{\frac{\alpha}{D}} L\right) + C_3 \cos\left(\sqrt{\frac{\alpha}{D}} L\right) = 0$$

$$\implies \frac{df}{dx}(L) = -C_2 \sin\left(\sqrt{\frac{\alpha}{D}} L\right) = 0.$$

To satisfy the last equation, we could let $C_2 = 0$. However, this would mean $Z \equiv 0$. Not only is this a pretty dull solution, but also it does not satisfy the initial condition. In order to satisfy the equation, we use the fact that $\sin(\theta) = 0$ whenever $\theta = n\pi$, where $n = 0, 1, 2, 3, \ldots$, so we define $\alpha = \alpha_n$ through

$$\sqrt{\frac{\alpha_n}{D}} L = n\pi \implies \alpha_n = D\left(\frac{n\pi}{L}\right)^2.$$

Putting f and g back together and defining $C_n = C_1 C_2$ gives

$$Z(x,t) = C_n \cos\left(\frac{n\pi}{L}x\right) \exp\left(-D\left(\frac{n\pi}{L}\right)^2 t\right).$$

Since this is a solution for all values of $n = 0, 1, 2, 3, \ldots$, we can take a linear combination of them and still have a solution. Thus

$$Z(x,t) = \sum_{n=0}^{\infty} C_n \cos\left(\frac{n\pi}{L}x\right) \exp\left(-D\left(\frac{n\pi}{L}\right)^2 t\right). \tag{6.22}$$

A crash course in Fourier series

The form of solution (6.22) is known as a Fourier series expansion. It is a way of representing a function as the sum of different functions. We will not go into the subtleties of Fourier series here; suffice it to say that it is an incredibly powerful technique that allows us to produce solutions to linear differential equations [16, 9].

To solve our specific problem, we must satisfy the initial condition by producing a definite form for the arbitrary constants C_n. To do this, we notice that

$$\int_0^L \cos\left(\frac{m\pi}{L}x\right) dx = \begin{cases} 0 & \text{if } m = 1, 2, 3, \ldots \\ L & \text{if } m = 0 \end{cases}$$

$$\int_0^L \cos\left(\frac{m\pi}{L}x\right) \cos\left(\frac{n\pi}{L}x\right) dx = \begin{cases} 0 & \text{if } m \neq n \\ \frac{L}{2} & \text{if } m = n. \end{cases}$$

Using these, we can deduce that

$$C_0 = \frac{1}{L}\int_0^L Z(x,0)\, dx$$

$$C_n = \frac{2}{L}\int_0^L Z(x,0) \cos\left(\frac{n\pi}{L}\right) dx \qquad \text{for } n > 0$$

and, finally, using the initial condition

$$Z(x,0) = \begin{cases} Z_0 & \text{for } 0 \leq x \leq 1 \\ 0 & \text{for } x > 1, \end{cases}$$

we derive exactly the solution given in equation (6.15).

B Approximating First Interaction Time

As mentioned in Section 6.5, there is no formula we can use to produce the first interaction time exactly. However, the mathematical problem comes down to finding a solution t_z of equation $Z(L, t_z) = 1$. In this case, the probabilistic method has an advantage over the continuous model since the mean first interaction time can be derived exactly [17].

A simple method of solving $Z(L, t_z) = 1$ would simply be to substitute in a value of t. If $Z(L, t) < 1$ then we double t and consider $Z(L, 2t)$. Notice that if $t_1 < t_2$, then $Z(L, t_1) < Z(L, t_2)$ and so $Z(L, t) < Z(L, 2t)$. Thus we keep doubling t until we reach a value such that $Z(L, 2^n t) > 1$. If $t_0 = 2^{n-1}t$ and $t_1 = 2^n t$, then we know that $Z(x, t) = 1$ for some[6] $t \in [t_0, t_1]$. To gain better approximations to the value of t_z, we then start halving this domain and only keep the half that contains the root, as shown in Figure 6.8.

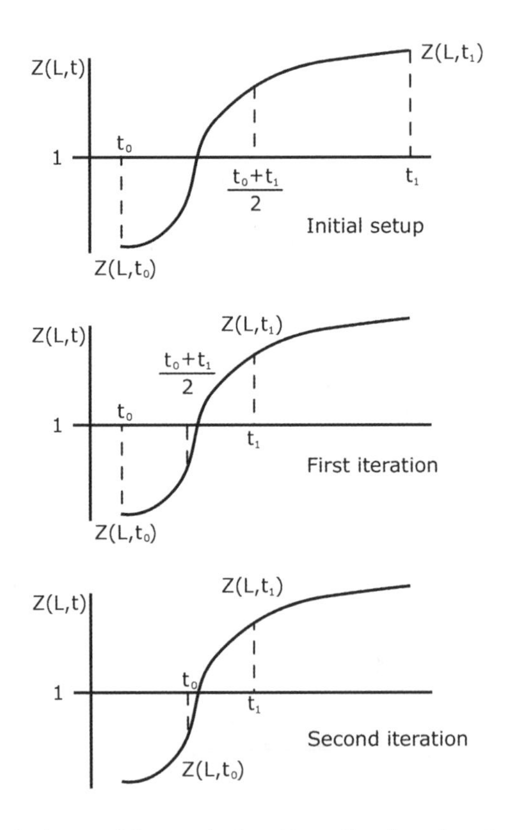

Figure 6.8: Bisection technique. After each iteration, the domain, shown by the dashed lines, becomes half as long as it was before.

[6]The symbol $\in$ means 'contained in.' Thus $t \in [t_0, t_1]$ means t is contained in the interval $[t_0, t_1]$ or, simply, $t_0 \leq t \leq t_1$.

Figure 6.8 illustrates this method of halving the interval. In the initial setup, the solution, t_z, is in the left half, so we know that $t_z \in [t_0, (t_0+t_1)/2]$. We redefine

$$t_1 \equiv \frac{t_0 + t_1}{2},$$

and repeat the process. This time the root is in the right half and so we redefine

$$t_0 \equiv \frac{t_0 + t_1}{2}.$$

After each iteration, we halve the size of the interval and so the interval $[t_0, t_1]$ gets smaller and smaller. Thus, by design, we always have $t_z \in [t_0, t_1]$ and so, by repeating the halving process, we can estimate t_z to any accuracy we like.

The benefit of this method is in its simplicity and reliability; it will always work. However, the cost of this reliability comes at the price of speed. If the initial searching region is very big, it may take a large number of repeats before the process produces an answer to an accuracy with which we are happy. There are quicker methods but these are usually more complex and sometimes they may fail to find a solution altogether. If the zombies are closing in on you and you need to compute their interaction times more quickly, we direct you to consider Newton-Raphson techniques [18] rather than the bisection method; although, if time is really short, we would firmly recommend running away first.

C Changing Coordinates

Justification

In Section 6.6.3, we made the change of coordinates of $(x,t) \mapsto u = x - vt$. This is because we are looking for a specific form of solution to the equations (6.12) and (6.13) that take the form of 'Fisher waves' [19, 10], which look like those in Figure 6.9. Note that ahead of the wave front the human population is high and the zombie population is low. As the wave invades the human species, it causes the zombie population to increase while the human population decreases.

Specifically, we are looking for a wave that moves at a constant speed and does not change in shape as it moves. Fisher waves are widely known to meet these criteria [19]. Thus if we could define the shape of the wave at each space-time point as $F(x,t)$, then, since the wave does not change shape, it can also be described by $F(u = x - vt) \equiv F(x,t)$, where v is the speed of the wave. This can be seen intuitively, since in any interval of time of length t, a point on the curve at a position $u = x$ will move a distance vt (using distance = speed × time) so the position of the point on the curve is now $u + vt = x \implies u = x - vt$. By converting to the coordinate of u, we are able to reduce the dimensions of the system and make it simpler.

By looking for Fisher waves, we implicitly assume that the domain is infinite in size with zero-flux boundary conditions. Since we are using a large, finite domain, the derived results represent a good approximation to those actually seen while the wave is not near either of the boundaries.

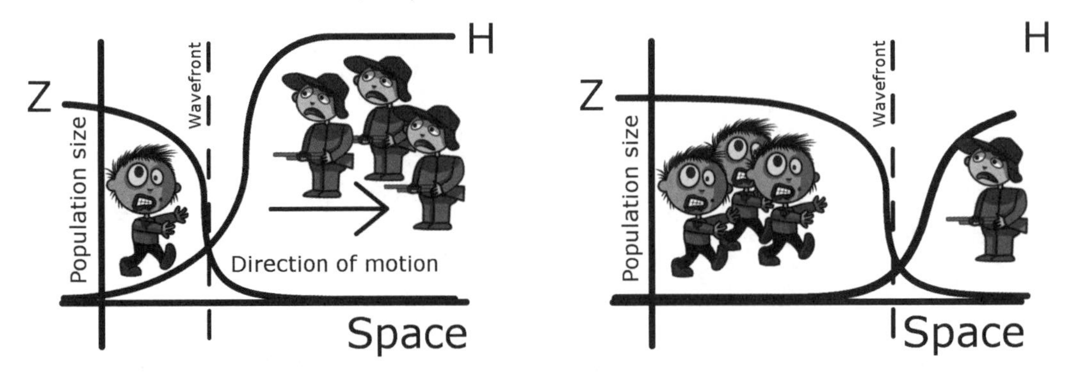

Figure 6.9: Diagram of a 'Fisher wave' form. As time increases, the wave moves to the right with a constant speed, keeping the same shape. The left image evolves to the right image as time increases.

Consequences of changing variables

Since we have changed the dependent variable, we need to change the derivatives as well. This is done using the chain rule [8]:

$$\frac{\partial}{\partial x} = \frac{\partial}{\partial u}\frac{\partial u}{\partial x} = \frac{\partial}{\partial u}$$
$$\frac{\partial^2}{\partial x^2} = \frac{\partial^2}{\partial u^2}$$
$$\frac{\partial}{\partial t} = \frac{\partial}{\partial u}\frac{\partial u}{\partial t} = -v\frac{\partial}{\partial u}.$$

These are then used in equations (6.12) and (6.13) to produce the form

$$0 = D_H H'' + vH' - \alpha HZ \tag{6.23}$$
$$0 = D_Z Z'' + vZ' + \beta HZ, \tag{6.24}$$

where we have used the notation

$$' = \frac{d}{du}.$$

Speed bound

By considering Figure 6.9, we can see that, in front of the wave, $H = H_0$ (the initial population) and $Z = 0$, as the infection has not reached that point yet. We use this information to consider what would happen to a small perturbation of the population ahead of the wave. Either the perturbation will die out or it will increase. To capture the effects of the perturbation, we substitute in the following forms for H and Z:

$$H(u) = H_0 + h\exp(\lambda u) \tag{6.25}$$
$$Z(u) = z\exp(\lambda u), \tag{6.26}$$

where h and z are the small initial perturbations. The outcome of the perturbation is then determined by the sign of λ. If $\lambda < 0$, the perturbations die out and the populations go back to their initial values, but if $\lambda > 0$ then the perturbations grow and so the infection is able to take hold. Most importantly, λ should be a real value. If it were complex (i.e., if it had the form $\lambda = X + iY$, where X and Y are any real numbers and $i = \sqrt{-1}$), then the system would predict that the solution would oscillate like a *sine* or *cosine* function and this would cause the $Z(u)$ function to become negative, which is unrealistic. To stop this situation from occurring, we need λ to be real.

We now substitute equations (6.25) and (6.26) into equation (6.24) and ignore the product hz since h and z are both small factors, so $hz \ll h$ or z. Upon simplification, we are left with $0 = D_z\lambda^2 + v\lambda + \beta H_0$. This can be simply solved with the quadratic formula [20], which means

$$\lambda = \frac{-v \pm \sqrt{v^2 - 4D_z\beta H_0}}{2D_z}.$$

Because λ must be real, we need to make sure the term inside the square root is non-negative and so inequality (6.14) follows.

D Acknowledgements

Thomas Woolley would like to thank Martin Berube for the use of his delightful zombie pictures and Lorraine Broaders for walking like a zombie in order to gain empirical data. Thomas Woolley would also like to thank the EPSRC for a studentship in Mathematics. Philip Maini was partially supported by a Royal Society Wolfson Research Merit Award. Ruth Baker would like to thank Research Councils UK, for an RCUK Fellowship in Mathematical Biology, and St Hugh's College, Oxford, for a Junior Research Fellowship.

E Glossary

In this glossary of terms, we have, at times, sacrificed rigour for readability.

Continuous. A function is continuous if a small change in the input results in a small change in the output.

Diffusion equation. A partial differential equation that describes a population's random movement from regions of high density to low density.

Discrete. A function is discrete if it is defined only on isolated points.

First passage time. The first time an individual of the population reaches a specific location.

Fisher wave. A solution that travels in time, moving with constant speed and shape. The solution connects a low population state to a high population.

Flux. The rate of flow of material at a specific point.

Fourier series. An infinite series representation of a function that uses trigonometric functions sine and cosine.

Law of Mass Action. The rate of a reaction is proportional to the density of reactants multiplied together.

Monotonically decreasing. A function is monotonically decreasing if it never increases as its variable increases.

Random walk. Movement of an individual in which each step is independent of any previous step.

Stochastic. Random.

Wave solution. A particular solution form of a differential equation that does not change its shape as it moves in space and time.

Zero flux boundary condition. A specific form of boundary condition that assumes that the solution domain is insulated and thus individuals can neither leave nor enter.

References

[1] P. Munz, I. Hudea, J. Imad, R.J. Smith?. When zombies attack!: Mathematical modelling of an outbreak of zombie infection. In *Infectious Disease Modelling Research Progress* J.M. Tchuenche and C. Chiyaka, eds., Nova Science, Happuage, NY 2009, pp. 133–156.

[2] M. Brooks *The Zombie Survival Guide: Complete Protection from the Living Dead.* Three Rivers Press, New York, NY, 2003.

[3] G.A. Romero, J.A. Russo. *Night of the Living Dead*, directed by G.A. Romero. The Walter Reade Organization, 1968.

[4] G.A. Romero. *George A. Romero's Land of the Dead*, directed by G.A. Romero. Universal Pictures, 2005.

[5] G.A. Romero. *Diary of the Dead*, directed by G.A. Romero. Dimension Films, 2007.

[6] G. Grimmett, D. Stirzaker. *Probability and Random Processes.* Oxford University Press, Oxford, UK, 2001.

[7] N.G. van Kampen. *Stochastic Processes in Physics and Chemistry.* Elsevier, Amsterdam, North Holland, 2007.

[8] R.G. Bartle, D.R. Sherbert. *Introduction to Real Analysis.* Wiley, New York, NY, 1982.

[9] E. Kreyszig. *Advanced Engineering Mathematics.* Wiley-India, 2007.

[10] J.D. Murray. *Mathematical Biology: An Introduction.* Springer-Verlag, Berlin, Germany, 2002.

[11] J.D. Murray. *Mathematical Biology: Spatial Models and Biomedical Applications.* Springer-Verlag, Berlin, Germany, 2003.

[12] R.L. Burden, J.D. Faires. *Numerical Analysis.* Cengage Learning, Boston, MA, 2005.

[13] K. Harmon. Fungus makes zombie ants do all the work. 2009. http://www.scientificamerican.com/article.cfm?id=fungus-makes-zombie-ants

[14] O. Diekmann. Run for your life. A note on the asymptotic speed of propagation of an epidemic. *J. Differ. Equations.*, 33(1):58–73, 1979.

[15] L. MacDonald, Diary of a Landscape Architect In *Braaaiiinnnsss: From Academics to Zombies*, R. Smith?, ed., University of Ottawa Press, Ottawa, ON, 2011, pp. 311–324.

[16] D.W. Jordan, P. Smith. *Mathematical Techniques.* Oxford University Press, Oxford, UK, 2002.

[17] H.W. McKenzie, M.A. Lewis, E.H. Merrill. First passage time analysis of animal movement and insights into the functional response. *Bull. Math. Biol.*, 71(1):107–129, 2009.

[18] E. Süli and D.F. Mayers. *An Introduction to Numerical Analysis.* Cambridge University Press, Cambridge, UK, 2003.

[19] R.A. Fisher. The wave of advance of advantageous genes. *Ann. Eugenics*, 7:353–369, 1937.

[20] H. Heaton. A method of solving quadratic equations. *Am. Math. Mon.*, 3(10):236–237, 1896.

DEMOGRAPHICS OF ZOMBIES IN THE UNITED STATES

Daniel Zelterman

Abstract

We examine the relative demographic distribution of zombies in the United States as defined by the number of Google references that include a state name. Per-person zombie references are adjusted relative to the 2009 Census estimates. Statistical regression methods are used to explain the heterogeneity in zombie distribution. Our findings identify states with fewer Miss America winners and fewer shopping malls as having the greatest per-person incidence of zombies. This leads us to conclude that zombies are less common in states with a high preponderance of attractive people and that zombies either do not enjoy shopping, don't carry credit cards, avoid the bright lights of malls—or perhaps a combination of these factors.

Keywords: Hannah Montana effect; zombie retail therapy; jittered Miss America; US Census

7.1 Introduction

ZOMBIES are an infrequently studied demographic group that is not recognized by the US Census. The recently completed decennial census (as of the time of this writing) does not include a zombie category. We will not attempt to address controversies such as whether zombie status is inherited (How many of your parents were zombies? Is the genetic trait dominant, recessive or linked to other genetic traits?), a lifestyle decision (of the rich and famous, perhaps), contagious like chicken pox or a conscious decision (like joining the National Rifle Association or the Westboro Baptist Church). A web-based search for 'zombies' on the Census website[1] does

[1] http://www.census.gov/

not return a single entry. Despite this oversight, zombies are a visible demographic in the US.

This work seeks to identify and characterize the distribution of zombies across each of the fifty US states plus the District of Columbia (DC), but not including any incorporated or unincorporated territories. As seen in the previous chapter, spatial considerations are important for understanding the undead, so what better spatial consideration than the entirety of the United States (a.k.a. 'the world')? Clearly, some states have greater numbers of zombies and this may simply reflect larger populations. As a result, we adopt a zombie count relative to the living population. The living population is a well-studied demographic metric and highly reliable Census data is readily available. Perhaps a more useful denominator would be the number of deceased residents who died in each of the states, but this measure is similarly unavailable from the US Census.

The principal beneficiaries of this research will be those who are prone to zombie-avoidance behaviours. Most readers of this work would prefer not to have their brains eaten and should avoid living in high-density zombie environments such as those identified here. On the other hand, judging by the quantity of B movies available, there are clearly many individuals whose poor judgment and erratic behaviour appear to attract the attention of zombies. Such individuals will benefit from this work as well.

The aim of this work is to explain the wide variety of zombie references across the US states. A number of useful numerical characterizations of each of the fifty states were obtained from Census data with the aim of explaining the diversity of zombie distribution across the country. The data sources are described in Section 7.2. The specific demographic variables are listed in Section 7.3. Univariate analysis is summarized in Section 7.4 and multivariate statistical methods are described in Section 7.5.

7.2 Data Sources

The primary source of zombie distribution across each of the fifty US states was a Google search, performed in May 2010. For each of the fifty states, we entered the word 'zombies' followed by the state's name into Google and recorded the number of hits that were reported. These figures, rounded to the nearest 10^5, were entered into a spreadsheet database along with the additional demographic information, as described below. This spreadsheet is available to the living and zombies alike by email request to the author.[2]

An immediate difficulty became apparent when many of the Google searches returned counts of websites that were not relevant. Celebrities such as recording

[2]The author desires a yacht so please include a valid credit card number with your request. Zombies without internet access, please turn over living wealth signed with a bloody fingerprint (where fingers are available) or blood or drool DNA sample for verification of (former) identity.

Table 7.1: Univariate characteristics of variables used in this analysis. States include DC, but exclude Montana due to the Hannah Montana effect.

Variable	Mean	Std Dev	Min	Max
Google zombie references per 10^5	916	106	189	6900
State population (in millions)	6.14	6.81	0.55	37.0
Zombie references per person	21.4	13.6	5.4	65.7
Age-adjusted death rate	826	94	624	1022
Percentage of votes for Bush in 2004	52%	10%	9%	72%
Per-capita shopping malls	0.162	0.037	0.076	0.238
Percentage of cremation rates	30.7	16.3	3.2	63.3
Percentage of binge drinkers	15%	3%	8%	22%
Gun death rate	11.4	5.2	2.8	31.2
Per-capita federal tax in US$1000	7.4	4.3	3.0	30.9
Number of Miss America winners	1.66	1.85	0	6

artist Hannah Montana and football great Joe Montana were linked to zombie activities or were frequently identified as zombies themselves. Their names were often included in the internet searches, even though they had little or no connection to the state of Montana. We refer to this as the *Hannah Montana effect* and consequently found it necessary to omit the state of Montana from the analysis that follows because of the excessive association with the two. The following section describes the explanatory variables used in these analyses.

7.3 Demographic Variables Examined

All of the data examined here is readily available from public sources on the internet.[3] The following is a list of important demographic variables that we believe *a priori* to be useful in explaining the wide variance in zombie distribution across the fifty US states including DC, but not incorporated or unincorporated territories. Summary statistics of each of these demographic measures are summarized in Table 7.1, excluding Montana, as explained above.

Google zombie references per 100,000. As explained in the introduction, this is our primary outcome variable and measure of zombie distribution. We recorded the number of Google hits that included the word 'zombies' followed by

[3]To build up a down payment on the previously mentioned yacht, researchers are encouraged to purchase the data from the author rather than try to reconstruct the database firsthand.

state name. It is not clear if we can precisely attribute zombies to a single state of residence. As examples: some zombies may migrate with the seasons, perhaps maintaining separate summer and winter homes (*snow-zombies*). Similarly, younger zombies may prefer to return to the location of their former living relatives as it is the trend for some college zombies to move back home with their parents after graduation (*live-at-home adult zombies*). Older zombies may remain after their younger zombie relatives move out (*empty-nester zombies*) or else move and buy smaller homes (*downsizing zombies*), perhaps in warmer climates (*sun-belt zombies*).

State population. More populous states are likely to have greater numbers of zombies so we adjusted these rates on a per-person basis as explained in the introduction. Populations are based on the 2009 Census data. This variable is our primary outcome and is used as a dependent variable in all linear regression models.

Per-person zombie references. Values ranged as high as 125 per person in the state of Montana. This state was omitted from the analyses because of the Hannah Montana effect, described in Section 7.2. This left a maximum rate of 65.7 per-person zombie references for the state of Alaska, a value more comparable to the remaining states and DC. The mean per-person zombie reference rate is 21.4 with a standard deviation of 13.6 across states. The range was lowest (5.4) in Pennsylvania and highest in Alaska.

Age-adjusted death rate per 100,000. A higher death rate per unit population means more potential per-person zombies. Duh. Even so, a higher relative death rate may also indicate the presence of zombies who are more angry or vindictive than usual due to increased and competitive zombie populations. In a future article, we will explore the economics of unemployment rates on zombies.

Percentage of the state population voting for George Bush in 2004. It was not clear whether the presence of zombies is related to conservative political thinking or a reaction to it. This measure is related to general attitudes against taxes and the power of a centralized federal government. Witness the current rise in the US Tea Party, many of whom are zombies.[4] Zombies do not have the right to vote in the United States[5] although many long-dead individuals appear to have this privilege and regularly make use of it, especially in Chicago.

Per-capita shopping malls. When times get tough, zombies go shopping. Or so we thought initially. This is more likely a measure of income and suburban sprawl. Malls are built where there is ample real estate and also where there is a sufficiently large population close enough to drive there in large numbers with their high-balance credit cards. Shopping malls often play an important role in zombie culture. In the highly recommended 2004 classic, *Dawn of the Dead,* the survivors

[4]A recent Google search of 'zombie Tea Party' returned almost 800,000 hits and 'zombie Sarah Palin' returned almost twice as many.

[5]The US Supreme Court is oddly silent on this issue. Individual states may disenfranchise zombies by passing laws that require drivers' licences, the use of a computer terminal for voting or the ability to sign one's name.

of a particularly vicious zombie attack manage by barricading themselves in a large Midwestern shopping mall.

Cremation rate as a percentage of deaths. If the cadaver is reduced to ashes, then it is less likely to reappear at a later date as a zombie. Intuitively, a high cremation rate would probably result in a lower per-person zombie rate as well. On the other hand, perhaps zombies, dissatisfied with their present conditions, would migrate to states with high cremation rates in the hopes of being cremated. This would result in a higher zombie prevalence.

Fraction of population self-described as binge drinkers. Perhaps zombies are more likely to attack or are attracted to individuals with impaired judgment. (How many B movies have you seen where the pretty teenaged girl, making out with her boyfriend in his car, was about to "go all the way," when they were interrupted by a zombie and she ran screaming into the swamp?) Perhaps it is only through weak judgment, further reduced by alcohol, that the true nature of zombies can be appreciated and understood. An alternative theory is that zombies can more easily be seen through the haze of an alcoholic stupor.

Firearm deaths per 100,000. Guns don't kill people, people kill people, as the proponents of gun ownership in the United States are quick to point out. More gun deaths may not correlate with greater gun ownership. In the United States, gun ownership is positively correlated with attitudes towards conservative politics (see votes for George Bush, above). Gun ownership is a hot-button issue in US politics.[6] It is unclear whether the decision to have a gun in the home is a result of defence against a perceived high zombie prevalence or whether the presence of guns is effective in reducing the zombie population. It is also unclear how many firearm deaths occurred as collateral damage in a confrontation with a zombie. Also see the rate of binge drinking, above. Recent legislation in several US states has allowed patrons to carry loaded firearms into bars. This is enough reason to give up drinking entirely.

Federal taxes paid per capita. This is an indication of income levels in each state. The relationship between income and taxes is not linear because of progressive taxation, another liberal idea sold to the gullible populace. States with residents paying higher per-capita taxes also tend to have more shopping malls (see above) and are otherwise likely to be more interesting and safe places to live and work. Safe translates to fewer firearm deaths (again, please see above).

Number of Miss America winners. Beauty contests are as old as humanity. Homer described one such contest in the *Iliad* and we all know how that turned out. As a measure of the attractiveness of residents of each state, we include this useful measure of overall appearance of each state's residents. Zombies are viscerally challenged, to put it politely, and it stands to reason that states with more Miss America winners are likely to have more attractive residents overall and hence fewer zombies.

[6]The author doesn't want to go there in the event that he may someday want to run for publicly elected office.

7.4 Univariate Results

Table 7.1 summarizes each of the individual demographic variables used in our analysis. Each variable, marginally, is summarized by its mean, standard deviation, minimum and maximum. As discussed in the introduction, there are fifty observations corresponding to each of the US states except Montana, but including DC. The per-person zombie reference rates ranged from a low of 5.4 in Pennsylvania to a high of 65.7 in Alaska.

We next examined the univariate relationships between the demographic measures with the dependent variable: per-person zombie references. After omitting the observation confounded with the Hannah Montana effect, the marginal distribution of zombie references appears to be slightly more right-skewed than anticipated by a normal distribution, mostly due to the single large observation corresponding to Alaska. These values are plotted along the right margin of Figure 7.1.

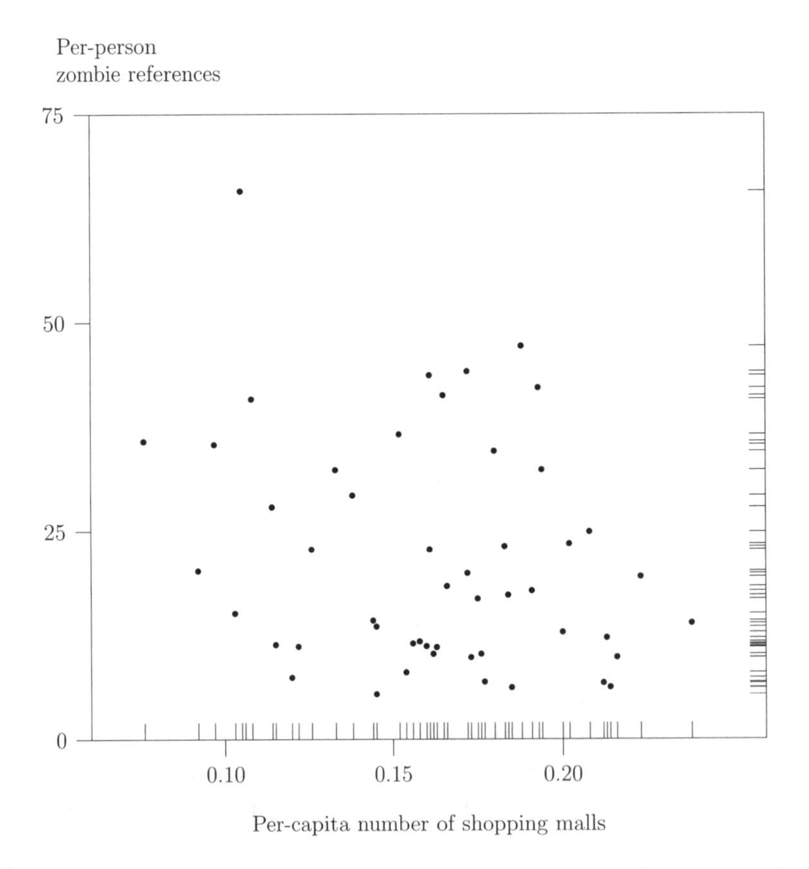

Figure 7.1: Per-person zombie references plotted against per-capita shopping malls for all fifty states except Montana. The correlation is −0.25.

The regression analyses described in this section and that following were performed using least-squares regression. This assumes that errors from the fitted linear model are normally distributed. We also examined these regressions using the logarithmic transformation of the number of per-person zombie references to correct for the right skew of the marginal data. This transformation resulted in an absence of any statistically significant findings[7] and we continued the analysis based on the original, untransformed zombie rates.

Among the demographic variables described in the previous section, the two most highly correlated with per-person zombie references are the per-person number of shopping malls and the number of Miss America winners. A level of statistical significance was achieved that causes us to reject the hypothesis that such a high correlation could have happened by chance alone. These two association measures are described separately in this section.

Figure 7.2 plots the number of Miss America winners against the per-person zombie rates. The number of Miss America winners is a discrete count. These have been *jittered* or had a small amount of random noise added to their values in this figure. The introduction of this variability spreads the values out just enough to aid in our visualization of the relationship between the two variables. We found that a jittered Miss America is helpful in this figure, but too much jittering of Miss America would only cause confusion as to the correct column membership in the data. Overjittering Miss America is to be avoided. The raw, unjittered Miss America counts were used in all statistical calculations. The jittered Miss America values are only used in this figure.

The standout value in the extreme upper left corner of this figure corresponds to the state of Alaska with 65.7 per-person zombies and no Miss America winners. Alaska also has a very large imbalance of the sexes, with many more men than women. Former vice-presidential candidate and Alaska resident Sarah Palin made many claims during the 2008 election campaign, including being to be able to field-dress a moose and being able to see Russia from her home, but did not comment on her winning the Miss America title. The three states that each had six Miss America winners are California, Ohio and Oklahoma. Each of these states had relatively low rates of per-person zombie references.

Figure 7.2 supports our hypothesis that the states with the most Miss America winners also tend to have the fewest zombies. The direct cause and effect relationship cannot be inferred, of course. The number of Miss America winners and per-person zombie reference rate may be closely tied to an unmeasured third variable such as the overall level of attractiveness of the state's population as a whole. States with an overall high ratio of attractive residents would support few zombies and result in more Miss America winners, for example.

In addition to the number of Miss America winners, we found that the per-capita number of shopping malls had a statistically significant relationship with the per-person zombie reference rate. The correlation was -0.25. The scatter

[7]Without a single p-value less than 0.05, it is very unlikely that this manuscript would ever have been published.

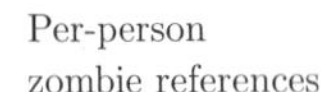

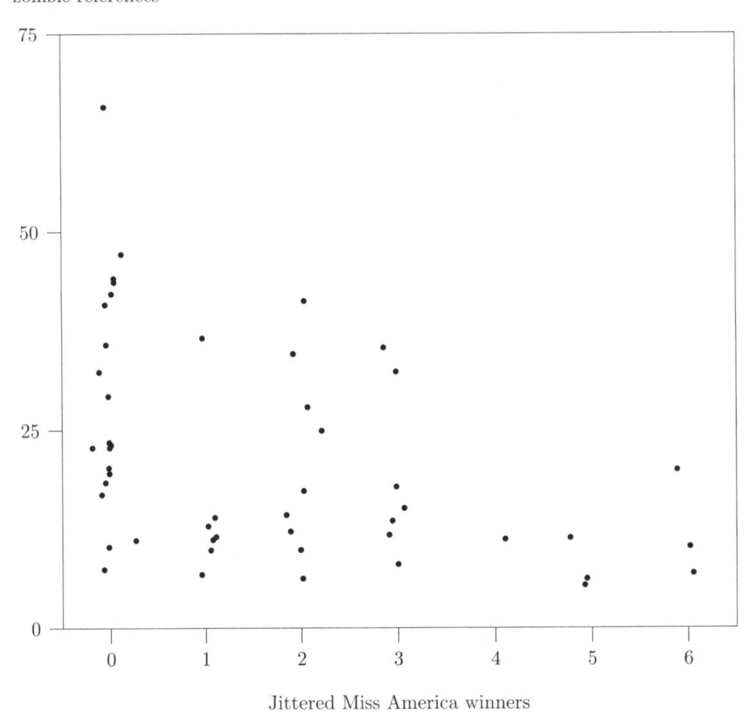

Figure 7.2: Jittered number of Miss America winners plotted against per-person zombie references. The correlation is −0.42.

plot of malls against zombies appears in Figure 7.1. Although not as striking a statistical relationship as with Miss America winners, the number of malls is also negatively correlated with the per-person zombie reference rate. Much of this relationship is due to Alaska, whose single outlier value appears in the upper left corner of Figure 7.1.

The negatively correlated relationship between malls and zombies may have several explanations in addition to the influence of Alaska in this data. It is not clear whether zombies either drive or have credit cards. Driving and high-limit credit cards are important requirements to enjoying the mall experience. Perhaps zombies just don't like hanging out and meeting up with their friends in brightly lit locations and instead prefer dimly lit cemeteries, late at night. A case can be made against *zombie retail therapy* in which the use of antidepressants is replaced with buying shoes, clothing and fashion accessories. Judging by the torn, threadbare wardrobes that zombies frequently appear to favour, this does seem to be a valid conclusion.

Shopping malls may also offer an important defence against zombie attacks. (See the reference to *Dawn of the Dead* in the previous section.) Survivors of zombie attacks barricaded in a mall are less likely to become zombies themselves. This

Table 7.2: Estimated regression coefficients for multiple linear regression of per-person zombie references on all explanatory variables.

Variable	**Parameter estimate**	**Standard error**	**t Value**	**Pr > |t|**
Intercept	70.48	35.56	1.98	0.054
Age-adjusted death rate	−0.0385	0.0354	−1.09	0.282
Percentage of votes for Bush in 2004	−6.48	23.45	-0.28	0.784
Per-capita shopping malls	−89.7	43.7	−2.05	0.047
Percentage of cremation rates	27.0	15.3	1.76	0.086
Percentage of binge drinkers	−89.4	54.5	−1.64	0.109
Gun death rate	0.891	0.541	1.65	0.107
Per-capita Federal tax in US$1000	0.00367	0.570	0.01	0.995
Number of Miss America winners	−2.790	0.885	−3.15	0.003

critical link serves to break the exponential feedback mechanism in which zombies beget ever more zombies. The dispersed population of Alaska and low per-person availability of malls suggests that this may be the case.

7.5 Multivariate Analyses

The aim of this analysis was to see if a linear combination of the explanatory variables provided additional information beyond the univariate analysis described in the previous section. The multivariate linear regression

$$\text{Mean zombie rate} = \alpha + \beta_1 x_1 + \beta_2 x_2 + \cdots + \beta_p x_p + \text{error}$$

was fit using standard statistical software.

The explanatory variables $(x_1, \ldots, x_p)$ are those described in Section 7.3 and summarized in Table 7.1. The errors were assumed to be independent and normally distributed. The regression coefficients $(\alpha, \beta_1, \ldots, \beta_p)$ were estimated using least squares. A residual plot did not reveal any apparent outliers or deviations from the assumption of constant variability. A summary of the estimated regression coefficients, their standard errors and tests of statistical significance is given in Table 7.2.

The multivariate analysis did not reveal much more than the univariate analysis described in the previous section. Specifically, the relation between zombie references and number of Miss America winners as well as per-capita shopping malls was described previously. Statistical significance levels were .003 and .047 respectively. There was no synergism or interaction between these two variables and the outcome. That is, the effect of malls and Miss America winners on zombie rates are the sum of their individual parts.

Table 7.2 does reveal some relationships with other demographic variables that are worth noting. The cremation rate has a positive correlation with the per-person zombie reference rate ($p = 0.086$). This rate did not achieve the sacred $p = 0.05$ level of statistical significance cherished by our scientific journals, but it is worthy of discussion. This finding was surprising because our *a priori* assumption was that a higher rate of cremation would reduce potential zombies to dust. It is not clear whether cremated beings can later be reconstituted whole, as zombies, or whether cremation just makes future zombies all the more vengeful. Perhaps the high rate of cremation is a result of a high zombie population and is a reaction to it by the fearful, living population.

The percent of binge drinkers and the gun death rate both nearly achieved statistical significance at the $p = 0.1$ level. The estimated regression on binge drinkers is negative. Intuitively, after seeing a zombie, many of us would start drinking heavily, but the data discounts this simple explanation. One explanation for the negative correlation is that people who have seen zombies use this as a good reason to give up drinking. Perhaps after seeing a zombie, heavy drinkers replace alcohol therapy with the solace of strong prescription drugs. Other possible explanations are that zombies avoid alcohol or that drinkers don't see and appreciate zombies when they appear through beer googles. Come to think of it, you never see zombies in singles bars.

A higher gun death rate is positively correlated with increased zombie references per person. One possible explanation is that, upon seeing a zombie, the gun owners reached for their piece while under great distress only to accidentally shoot themselves or someone else present at the time. Another explanation is that, upon learning of high zombie activity, the good citizens rush out to purchase guns that are then carelessly left lying around the home. Both of these explanations point to elevated zombie activity associated with a high gun death rate. Conversely, zombies may feel the need to confront gun owners to demonstrate that guns purchased for self-protection have no effect on zombies as we frequently see in B movies. This reasoning also shows that increased gun ownership is associated with zombies appearing in greater numbers.

7.6 Conclusion

We explored the demographic diversity of zombies across the US states and DC. Zombie populations were estimated by the number of Google hits associated with each state and normalized to per-person rates using 2009 Census data. A number of important demographic metrics were used to explain the variability across states. Among these, the number of Miss America winners and per-capita shopping malls were significantly negatively associated with zombie rates. A number of possible explanations for these findings were given. To a lesser degree, the zombie reference rate was associated with rates of binge drinking, gun death and cremation.

Appendix

A Glossary

A priori. Before having seen the data.

Correlation. A single summary measure of the strength of a linear relationship between two measurements. The correlation takes values between -1 and $+1$ with values at these extremes indicating the strongest relationship. Positive and negative values indicate the direction of this relationship: if one variable increases, does the other tend to go up or down? A correlation near zero indicates a lack of relationship between the two measurements.

Dependent variable. The variable that we wish to explain through regression methods. In the present chapter, the zombie rate per person in each US state is the dependent variable.

Explanatory variable. A variable that is used to explain the different values of the dependent variable. In the present chapter, we wish to see if states with more Miss America winners and more shopping malls are associated with more zombies references per person. In this case, Miss America and malls are the explanatory variables.

Jittering. A deliberate addition of random noise to display data more clearly. An example appears in Figure 7.2.

Least squares. Using the sum of squares of the distance between a linear fit and the data points to minimize the error in a linear approximation.

Linear regression. A method for describing the relationship between the explanatory variable and the mean of the dependent variable at each value of the dependent variable.

Standard deviation. A measure of variability between individuals. In Table 7.2, we see that some measurements are more variable than others.

Statistical significance, p-value. The (low) probability that this finding could have happened by chance alone. In linear regression, a very small p-value is evidence of a strong linear relationship between the dependent and explanatory variables. In Table 7.2, the p-values given in the last column show how strong the relationship is between each of the explanatory variables and the dependent variable, the zombie reference rate per person.

Annotated Bibliography

http://www.google.com/
Google, the well known search-engine, provided the number of zombie references associated with each state.

http://www.census.gov/
The US Census is the definitive source of demographic data.

http://www.statemaster.com/
Most of the explanatory variables in this report were obtained from StateMaster, a website with many useful characteristics of each of the states such as the number of toothless residents in each state, the average price of regular unleaded gasoline and the official state animal.

IS IT SAFE TO GO OUT YET? STATISTICAL INFERENCE IN A ZOMBIE OUTBREAK MODEL

Ben Calderhead, Mark Girolami and Desmond J. Higham

Abstract

Mathematical descriptions of physical systems usually involve parameters that cannot be measured directly. In the case of zombie outbreak models, these parameters include, for example, the propensity for human–zombie encounters to result in zombification or the rate at which seemingly dead zombies become active. The issue of using experimental data to inform our understanding of such unknown parameters has become a hot topic at the intersection of applied mathematics and applied statistics. In this chapter, we introduce the basic concepts and give some state-of-the-art computations. Our overall aim is to popularize the use of this new methodology through an eye-catching application.

8.1 Mathematical Modelling with Ordinary Differential Equations

ORDINARY differential equations (ODEs)—the kind that we meet in introductory calculus classes—have proved to be extremely useful tools for describing, in quantitative terms, how physical systems change over time. Some systems are so well understood that they follow widely accepted 'laws of motion,' such as planetary orbits, ballistic missiles, elastic springs, chemical reactions and radioactive decay. In other cases, where a range of complicated objects interact in ways that are impossible to pin down, ODEs can still be extremely effective for characterizing the main features of the system. Modelling with ODEs—that is, constructing ODEs that have some explanatory and predictive power—involves identifying the essential features of a physical system and converting these into a descriptive mathematical framework. Success stories in modelling abound throughout the physical and engineering sciences—and, of course, the world of science fiction [1].

In the case of zombie outbreaks, just as in any other modelling context, a good model allows us to

- **explain** the key mechanisms at work and how they interact;
- **predict** the future state of the system; and, more excitingly,
- **theoretically and computationally investigate** 'what-if' scenarios that might be out of the reach of experimental scientists because of technological or budgetary limitations, or ethical considerations.

To be specific, an accurate zombie outbreak model could give *quantitative* feedback to frightened individuals with concerns such as:

- Soldiers patrolling the area have reported daily observed zombie numbers for the past 5 days as 123, 127, 104, 92 and 74. Is it safe to go out yet?
- If I meet a zombie, what is my chance of fighting it off?

It could also inform brave but under-resourced action heroes who need to make tough decisions such as:

- How many soldiers should we mobilize and how many of these will survive?
- What scale of quarantine would be worthwhile?
- How effective does this cure need to be?

However, just as in any realistic modelling context, the zombie outbreak ODE models of [1] involve parameters whose values are not immediately obvious. If we picture the model as a 'black box,' then there will be dials on the box that must be tuned to the particular circumstances of the outbreak, such as the malevolence of the zombies and the pluckiness of the humans. While a few general issues can be addressed without detailed knowledge of these parameters, to make the mathematical model really useful, we must *calibrate* any unknown parameters using observed data.

Recently, the challenge of parameter estimation for ODE models has attracted the attention of the statistical inference community and some powerful tools have been developed that go far beyond the traditional 'least squares'-style point estimates [2, 3, 4]. As seen in the previous chapter, statistics allows us to answer questions of zombie prevalence across geographical locations, find correlations with previously unnoticed effects and learn the true importance of shopping malls in a zombie apocalypse. The zombie scenario therefore gives us the opportunity to raise the profile of this emerging field, and point out the advantages that arise when ideas from applied and computational mathematics and Bayesian[1] reasoning are brought together.

Using models motivated by [1], we will consider how much prior knowledge and how much observational data are needed in order to make useful inferences about a zombie outbreak. We will also introduce the *model selection* issue [3]. Suppose we hear incompatible rumours that

(a) zombies only attack alone,

(b) zombies only attack in pairs.

The two scenarios correspond to different ODE models, so to investigate which rumour is more likely to be true, we could ask which of the two models best explains the observed data. This requires us *simultaneously to calibrate and compare two or more ODE models*, a topic that has only recently been tackled systematically [3, 5].

8.2 Simple Model

In this section, we illustrate some key ideas by focusing on a very simple zombie outbreak model. Imagine a population consisting only of humans and zombies. We will let $S(t)$ and $Z(t)$ represent the number of humans and zombies at time t. In fact, $S(t)$ and $Z(t)$ will take real values, so it is more appropriate to think of these values as the concentration levels of the two species. We suppose that there is just one event that can cause a change in the levels: a zombie attacks a human and, if successful, converts that human into a zombie. We note immediately that in this simple world the population level of humans is forced to decrease to zero and the population level of zombies will correspondingly increase until all the humans are used up. This modelling assumption requires us to introduce a single parameter, a rate constant, β, that characterizes the ability of the zombies to find and infect humans. The larger the value of β, the more virulent the zombies. Simple *mass-action* modelling then leads us to the ODE

$$S'(t) = -\beta S(t)Z(t), \tag{8.1}$$
$$Z'(t) = \beta S(t)Z(t). \tag{8.2}$$

Here S' and Z' denote the time derivatives, that is,

$$S'(t) \equiv \frac{dS(t)}{dt} \quad \text{and} \quad Z'(t) \equiv \frac{dZ(t)}{dt}.$$

So the ODE system (8.1)-(8.2) specifies the rate of change of the two population levels. Adding the two equations, we see that $(S(t) + Z(t))' = 0$, so the rate of

[1]The Reverend Thomas Bayes was a British mathematician and Presbyterian minister, born at the start of the eighteenth century. He developed a theory for calculating inverse probabilities. The more general formulation in use today bears his name.

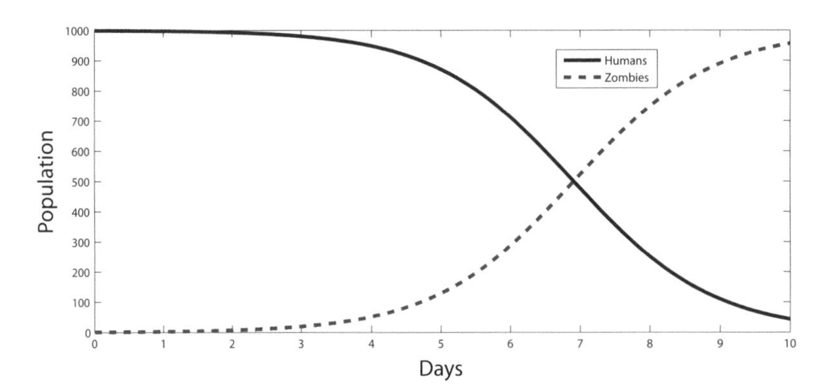

Figure 8.1: Simple zombie model with parameter $\beta = 0.001$.

change of the total population, humans plus zombies, is zero. This is intuitively obvious: each time we lose a human we gain a corresponding zombie. So $S(t)+Z(t)$ remains constant for all time. Letting the constant K denote this overall population size, we have $S(t) + Z(t) = K$. We then replace $Z(t)$ by $K - S(t)$ in (8.1) to get a single ODE

$$S'(t) = -\beta S(t)\,(K - S(t))\,. \tag{8.3}$$

This ODE fits into the class of *logistic equations.* Although nonlinear, it is sufficiently simple to admit a pencil-and-paper solution. This can be written

$$S(t) = \frac{S(0)K}{S(0) + (K - S(0))\,e^{\beta K t}}. \tag{8.4}$$

Here $t = 0$ represents the initial time when we start to monitor the populations, so $S(0)$ is the initial level of the human population.

Let us first assume that the size of the joint population (K) of humans and zombies is known, and that the initial number of humans, $S(0)$, is also known. This leaves us with one unknown parameter, β. This parameter can be interpreted as the rate at which humans are converted into zombies and is measured per zombie per day. Therefore if initially there are 1000 humans, a rate of $\beta = 0.001$ initially corresponds to $S(0) \times \beta = 1000 \times 0.001 = 1$ human being converted by each zombie every day. Note that the number of humans being converted depends on both the number of humans and the number of zombies at any particular time; by solving our differential equations, we can see how these two populations evolve relative to one another.

Figure 8.1 shows how a population of 1000 humans diminishes to less than 10% after just ten days, based on zombies attacking and converting humans at a rate of one human per zombie per day, where the initial zombie population consists of a single zombie. We can see the effect of doubling the rate constant to $\beta = 0.002$ in Figure 8.2. In this case, the population dwindles to less than 10% after just five days; by the seventh day, the humans have effectively died out.

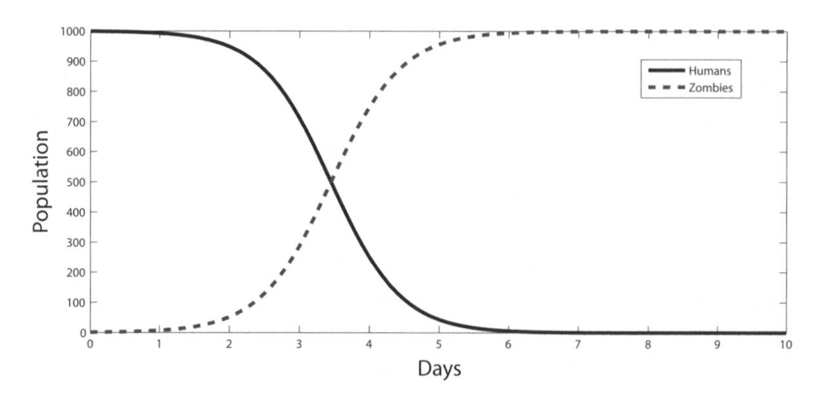

Figure 8.2: Simple zombie model with parameter $\beta = 0.002$.

These predictions have been made under the assumption that we know the exact rate at which zombies attack and convert humans. Given an initial number of humans and zombies, along with a rate constant, our model tells us how the populations evolve. In a more realistic scenario, we would not know the rate constant, but would instead have (usually inaccurate) observations of how the human population changes over a period of time. The goal in this case is then reversed: given some observations regarding the number of humans at certain time points, we want to estimate the rate constant β such that our model best describes the situation as we see it. Indeed, estimating these rate constants is often very challenging and is known as the *inverse problem*. Once we have inferred the rate parameter β, we can quantify the likelihood of future scenarios, based on the knowledge that our model adequately describes the past.

It can be argued that there is no single 'true' value of β because the model, which is based on simplifying assumptions, does not capture every detail of the physical system. More fundamentally, there will also be measurement errors and uncertainties in the data; in the case of a zombie attack, humans in hiding might go unrecorded and zombies might lurk unnoticed in dark corners. Therefore there may well be many values of β that are approximately as good as each other at describing the data. It makes sense then to determine the probability distribution over the *most likely* values for β as opposed to a single 'best' value. Statistical inference, and in particular *Bayes's theorem*, gives us a mathematical framework in which to carry out these calculations in terms of probability distributions. We refer to [6] for an accessible introduction to the Bayesian framework and [7] for a more comprehensive treatment.

There are three levels of inference that we might ultimately wish to carry out. The first determines the parameters with which the model plausibly describes the data. This is the probability of the free parameters $\boldsymbol{\theta} = [\theta_1, ..., \theta_n]$ given some data $\mathbf{Y}$ and a particular model M, which can be written as $P(\boldsymbol{\theta}|\mathbf{Y}, M)$. In our example, there is just one parameter, $\boldsymbol{\theta} = \beta$, and $\mathbf{Y}$ is a vector of observations at a number of time points. The second level of inference sheds light on the uncertainty associated with our choice of model; this is the probability of a particular model M given the data $\mathbf{Y}$, written as $P(M|\mathbf{Y})$.

Finally, the third level of inference describes the probability of a *prediction* given the data; this prediction may be based on multiple plausible models that are weighted according to their relative probabilities. In order to estimate these probability distributions, we can make use of Bayes's theorem, which gives us a method of combining prior knowledge and newly obtained data in a mathematically consistent manner. Key to the Bayesian approach is the idea that observations are inherently uncertain, so that a single data point is assumed to be but one sample from some underlying probability distribution. This distribution is represented as the likelihood of the data given a model and its current set of parameters, $P(\mathbf{Y}|\boldsymbol{\theta}, M)$. The prior distribution, $P(\boldsymbol{\theta}|M)$, which is often simply referred to as 'the prior,' characterises our initial knowledge or belief regarding plausible values of the parameters.

Let us therefore first define our prior and likelihood distributions for the model describing simple human/zombie interactions. A prior can be constructed by considering a reasonable time scale for the process; we could argue that all the action will not be over in a single day. This gives an upper limit of 1 for the rate constant β; with that value, $S(0) \times \beta = S(0)$, so all $S(0)$ humans could be converted by one zombie on the very first day. Likewise, a reasonable minimum rate is $\beta = 0$, in which case zombies never convert humans. With no further information, it is reasonable to take the view that, *a priori*, any rate constant between 0 and 1 is equally likely. So we choose our prior on β to have a uniform distribution over this range.

The likelihood is a measure of the goodness of fit between the data and the output of the model. The choice of which probability distribution to employ should depend on the problem context. For some modelling scenarios—where, for example, the observed data are the number of counts occurring within a particular time interval—the choice of a Poisson distribution is appropriate. Alternatively, if the observed data are obtained from estimates that might be affected by a large number of small unknown random factors, then, due to the Central Limit Theorem [6, 7], the associated error can be well-approximated by a Gaussian ('normal') distribution. In the case of modelling our zombie attack, we shall assume that the estimated population levels are subject to small unknown errors, with the final estimates combining local intelligence, large numbers of individual sightings and so on. We also assume that the errors at different observation times are independent. We therefore define the likelihood function to be the following quantitative measure of the agreement between the model output and the observed data over all time points:

$$L = P(\mathbf{Y}|\boldsymbol{\theta}, M) = \prod_t N_{Y(t)}(S(t), \sigma^2). \tag{8.5}$$

Here

- $P(\mathbf{Y}|\boldsymbol{\theta}, M)$ represents the probability of an observation given a model M with parameters $\boldsymbol{\theta}$

- $Y(t)$ is the observation at time t
- $S(t)$ is the output of the model at time t, given parameters $\boldsymbol{\theta}$
- $N_x(\mu, \sigma^2)$ is the density function for a Gaussian with mean μ and variance σ^2; that is,

$$N_x(\mu, \sigma^2) \equiv \frac{\exp(-(x-\mu)^2/(2\sigma^2))}{\sqrt{2\pi\sigma^2}}.$$

The variance σ^2, which we can think of as the inherent level of uncertainty in the data, can either be estimated and fixed in advance, or inferred along with the other parameters. We note that the product (8.5) corresponds to an overall normal distribution. In the case of our simple zombie model, $S(t)$ is easily calculated from (8.4). For more complex models, however, it will be necessary to compute a numerical solution for the ODE model.

In summary, for a particular combination of model parameters and initial conditions, given an observation of the number of surviving humans at a known point in time, we compute the likelihood by taking a Gaussian density centred on the model prediction and then finding the value of the density function at the observed value. We repeat this for each data point and multiply the answers together.

Bayes's theorem allows us to update our initial belief about the parameter values, as defined by our prior, by taking the data into account. Our updated knowledge is then quantified by the posterior distribution $P(\boldsymbol{\theta}|\mathbf{Y}, M)$ (first level of inference) by combining our prior distribution with the likelihood function in the following mathematically consistent manner,

$$\begin{aligned} P(\boldsymbol{\theta}|\mathbf{Y}, M) &= \frac{P(\mathbf{Y}|\boldsymbol{\theta}, M)P(\boldsymbol{\theta}|M)}{P(\mathbf{Y}|M)} \\ &\propto P(\mathbf{Y}|\boldsymbol{\theta}, M)P(\boldsymbol{\theta}|M). \end{aligned}$$

Noting that the marginal likelihood $P(\mathbf{Y}|M)$ is constant for a particular model M, we see that it can be calculated as the integral of the likelihood times the prior over all parameter values,

$$\begin{aligned} P(\mathbf{Y}|M) &= \int\int \ldots \int P(\mathbf{Y}, \boldsymbol{\theta}|M) d\theta_1 d\theta_2 \ldots d\theta_n \\ &= \int P(\mathbf{Y}|\boldsymbol{\theta}, M)P(\boldsymbol{\theta}|M) d\boldsymbol{\theta}. \end{aligned}$$

Performing the second and third levels of statistical inference over ODE models is challenging precisely because we have to estimate this integral, which is generally analytically intractable and high dimensional. Only recently has it been demonstrated that this integral can be efficiently and accurately estimated using a technique called thermodynamic integration [5]; we shall employ this later in the paper to discriminate between competing model hypotheses describing zombie populations and to quantify predictions of future behaviour.

Finally, we need a method of sampling from the posterior distribution, which in our case is analytically intractable, since we can't calculate the marginal likelihood for our models based on nonlinear differential equations. Fortunately, we can instead employ the Metropolis-Hastings algorithm to draw a series of random samples from an approximation of the posterior. This Markov chain Monte Carlo method was first developed by nuclear physicists at Los Alamos Laboratory, New Mexico in the 1950s, who were investigating methods of simulating the random behaviour of neutrons in the fissile material of atomic bombs [8]. Later it was realized that the method could be used more generally to simulate from virtually any probability distribution and a generalization was provided by Hastings [9]. Modern computing power is now making this technique feasible across a tremendous variety of applications. In particular, in Bayesian statistics, it allows us to draw accurate samples from the posterior distribution even if the marginal likelihood $P(\mathbf{Y}|M)$ is not known. The basic approach is relatively straightforward to implement: given a current value of our parameter, β_c, we can randomly generate a new state, β_n, from a *proposal distribution* $Q(\beta_n|\beta_c)$, which depends only on the current state. For efficiency, the proposal distribution should be as similar as possible to the target distribution we wish to sample from; this ensures that we do not waste too much time proposing unlikely values. In practice, since we generally have little information about the posterior distribution in advance, it suffices to employ a Gaussian distribution with mean β_c and some variance which we can choose to achieve an acceptance rate of between 20% and 40%. This new state is accepted with probability

$$\min\left\{\frac{P(\beta_n)Q(\beta_c|\beta_n)}{P(\beta_c)Q(\beta_n|\beta_c)}, 1\right\},$$

where $P(\beta)$ is the probability distribution we wish to sample from, which in our case is the posterior $P(\beta|\mathbf{Y}, M)$. Loosely, this technique allows us to search through the set of possible parameter values, spending most of our time in the 'hot spots' where parameter values are most promising.

We are now ready to infer a posterior distribution over the parameter values given some data. We will do this using 'artificial' data so that we can judge the performance of the algorithm under controlled circumstances. More precisely, we evaluate the solution (8.4) of the differential equation for a chosen value of β at a number of time points and add some Gaussian-distributed noise with known variance to the solution to generate some experimental data. We generate four data sets this way, as shown in Figure 8.3. We then treat these data as though they came from the model with an unknown value of β and see what quality of inference is possible. Figure 8.4 shows how the posterior distribution over β becomes more sharply peaked around the true value as the number of data points increases from three to fifty. The noise induces a noticeable bias when using just three data points, although the posterior is reasonably large at the true value of 0.001. Figures 8.5 and

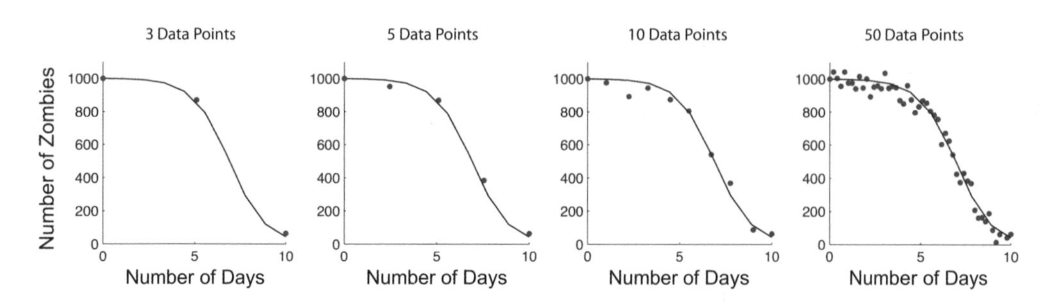

Figure 8.3: 3, 5, 10 and 50 data points generated from the simple zombie model over ten days with parameter $\beta = 0.001$ and Gaussian-distributed noise with a standard deviation (SD) of 50.

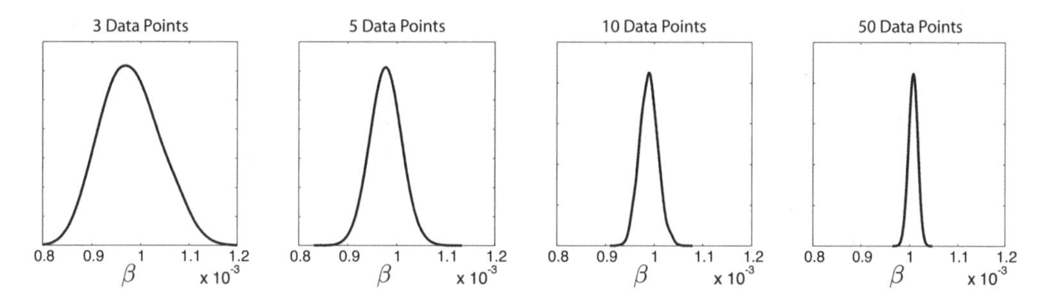

Figure 8.4: Posterior output from the simple zombie model with parameter $\beta = 0.001$ and Gaussian-distributed noise with a standard deviation of 50, as shown in Figure 8.3. As the number of data points increases, the posterior becomes more sharply peaked around the true value of β.

8.6 show how, as one might expect, as we add less noise to the data, the posterior distribution becomes less diffuse, indicating a greater confidence in the range of values for which the model could plausibly describe the data. We can also examine the effect of our prior on the posterior. Changing the prior from uniform over $[0, 1]$ to uniform over $[0, 0.01]$ has very little effect on the posterior of β, as shown in the left and middle pictures of Figure 8.7. If, however, we were to badly misspecify the prior by, for example, setting it to be a sharply peaked Gaussian distribution over the wrong value, we would observe a biased and rather skewed posterior distribution as in the picture on the right in Figure 8.7. This type of misspecified prior can often be diagnosed by comparing the prior and posterior.

Having shown how Bayesian inference can be applied in a very simple ODE setting, in the next section we will move on to a more realistic model where (a) there is more than one unknown parameter and (b) the ODE solution cannot be written down explicitly.

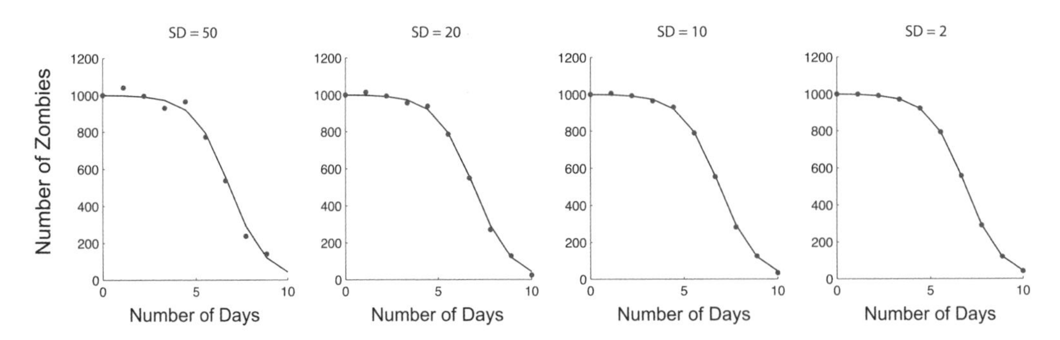

Figure 8.5: Ten data points generated from the simple zombie model over ten days with parameter $\beta = 0.001$ and Gaussian-distributed noise with a standard deviation of 50, 20, 10 and 2.

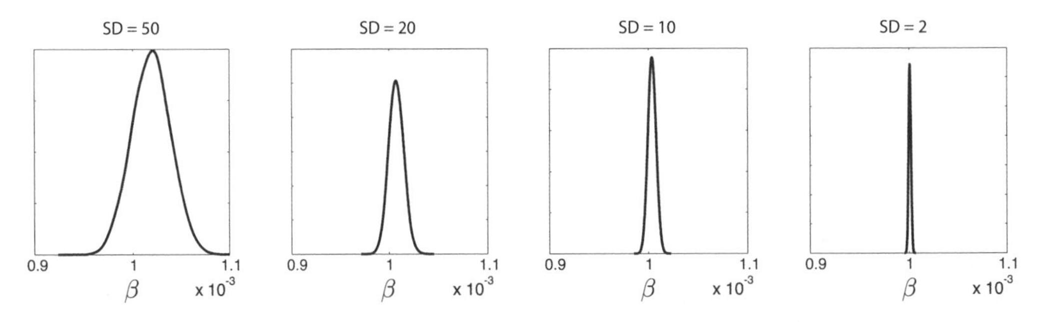

Figure 8.6: Posterior output from the simple zombie model with parameter $\beta = 0.001$ and ten data points. As the standard deviation of the added noise decreases, the posterior becomes more sharply peaked around the true value of β.

8.3 More Realistic Model

The model derived in section 2 of [1] splits the overall population into three classes. At each time t we have

- susceptibles (humans), $S(t)$,
- zombies, $Z(t)$,
- removed ('dead' zombies), $R(t)$; these may return as zombies.

As in our simple model (8.1)-(8.2), humans—now called *susceptibles* to be consistent with the traditional population-dynamics literature—are liable to attack by zombies; such an attack may convert the human into a zombie. However, we now allow for the possibility that a human can prevail in a human–zombie encounter. In this case the human emerges unscathed and the zombie joins the removed class. The term *removed* is used rather than *dead* because a human can never completely kill a zombie; at some later time, the zombie in question could return to regular zombie

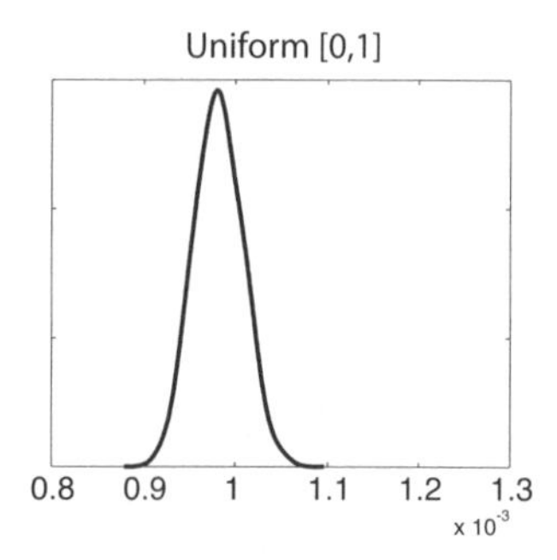

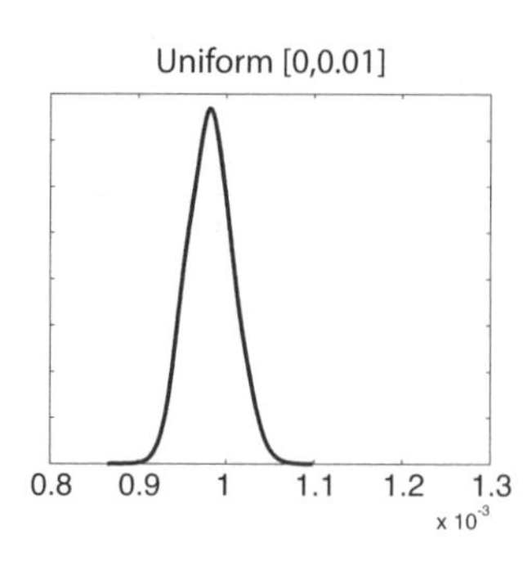

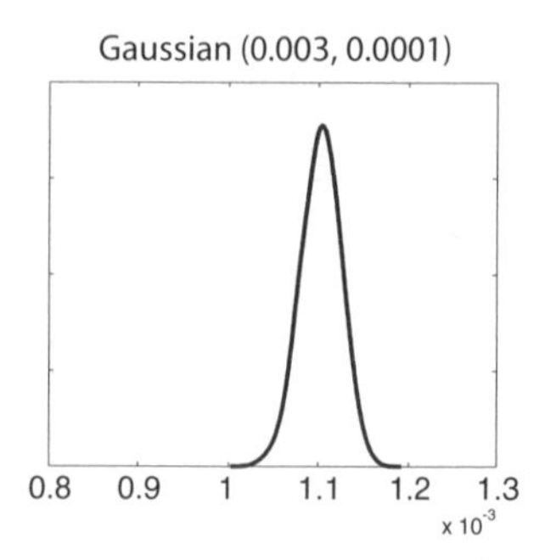

Figure 8.7: Posterior output from the simple zombie model with parameter $\beta = 0.001$, three data points and Gaussian noise with standard deviation 20. Changing the uniform prior from the range $[0, 1]$ to $[0, 0.01]$ has little effect on the posterior. However, a misspecified Gaussian prior sharply peaked around 0.003 with standard deviation 0.0001 results in a biased posterior distribution based on this small amount of data.

status. These modelling assumptions lead us to the following ODE system from [1]:[2]

$$S' = -\beta SZ \tag{8.6}$$
$$Z' = \beta SZ + \zeta R - \alpha SZ \tag{8.7}$$
$$R' = \alpha SZ - \zeta R. \tag{8.8}$$

The (constant, positive) parameters in the model are:

- β, a rate constant for a susceptible/zombie encounter causing a susceptible to turn into a zombie. This parameter arose in the simple model of Section 8.2. A larger value of β applies when zombification is more likely.

- α, a rate constant for a susceptible/zombie encounter causing a zombie to be removed. A larger value of α applies when removal is more likely.

- ζ, a rate constant for removed zombies returning to the zombie state. A larger value of ζ applies when removed individuals are more likely to revert to zombie status.

Here we consider a zombie population attacking a mid-sized town, such as Falkirk in Scotland. As in Section 8.2, we generate artificial data from the ODE model and investigate the quality of the inference. We infer the initial conditions for the zombie and human species, as well as all parameter values. We assume that there are no removed individuals initially, around 40,000 humans living in the town and around 10,000 zombies attacking from Paisley. Given daily observations over a period of three, five, seven and nine days, we consider the *predictive model*

[2]For simplicity, we consider only the short time scale regime ($\Pi = \delta = 0$ in [1]), so no new humans are born and no humans die from any other causes.

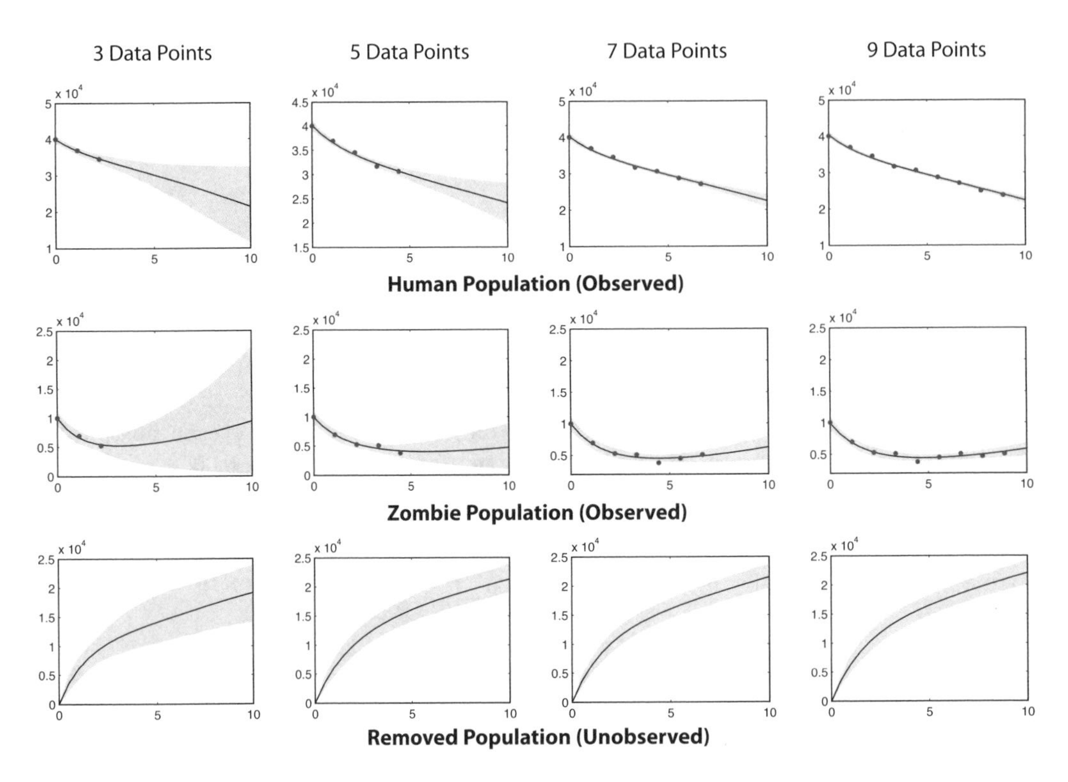

Figure 8.8: Posterior output from the complex zombie model (8.6)-(8.8) with parameters $\beta = 0.00001$, $\alpha = 0.00002$, $\zeta = 0.1$ and Gaussian-distributed noise with a standard deviation of 500. As the number of data points increases, the uncertainty in the posterior model output decreases.

output; that is, two standard errors or a 95% confidence for the output of the ODE at each time point over the ten days after the first zombie attack, shown in Figure 8.8. The solid line shows the model output, given the posterior mean of the inferred parameters and initial conditions. Figure 8.9 also shows the inferred initial conditions, which in this example are relatively insensitive to the number of observed data points. As the number of observed data points increases, however, we see that the uncertainty in the predictive model output over subsequent days does indeed decrease. This is also seen clearly when we consider the predictive posterior model output for Day 15, given observations over three, five, seven and nine days, as shown in Figure 8.10. With data for just three days, we learn very little about the population sizes on Day 15 after the zombie attack. Given an additional observation on each of the following two days, however, we can predict with much greater certainty that the number of surviving humans is likely to be between 10,000 and 25,000. As we collect more data over the subsequent days, our predictions become more confident, as seen from the more sharply peaked posterior distributions over each population. Indeed, after seven days, the predicted number of humans

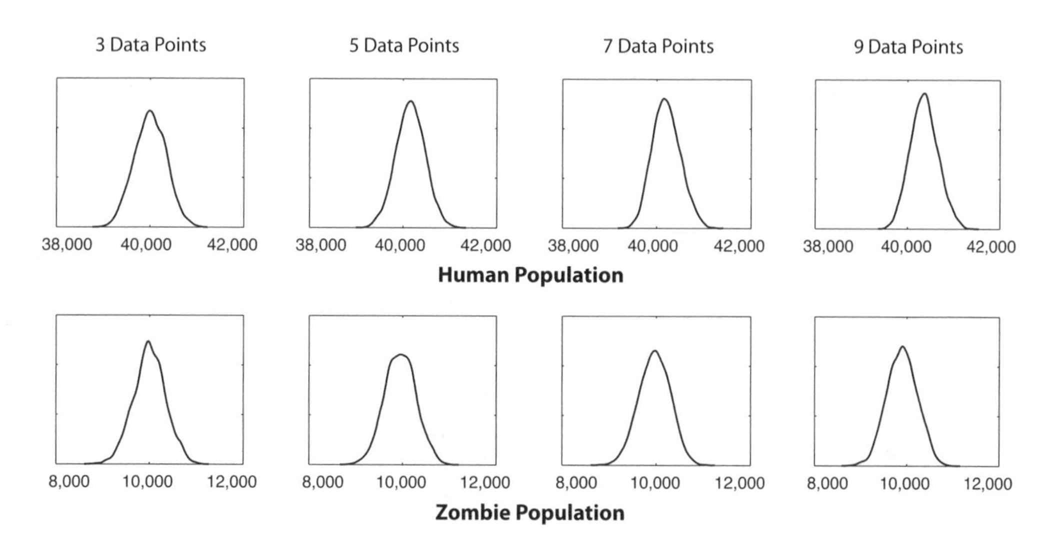

Figure 8.9: Posterior distributions of the inferred initial conditions from data generated by the complex zombie model with parameters $\beta = 0.00001$, $\alpha = 0.00002$, $\zeta = 0.1$ and Gaussian-distributed noise with a standard deviation of 500. The initial conditions are relatively insensitive to the number of data points observed.

is roughly between 10,000 and 18,000; after nine days, it is between 13,000 and 17,000. These predicted ranges are tending towards the 'true' number of humans, determined by the system of ODEs to be 14,790. Likewise, the predicted numbers of zombies and removed individuals tend towards their 'true' values of 10,426 and 24,784, respectively.

8.4 Model Selection

We now consider a second level of inference, in which there is uncertainty not only in the parameters, but also in the specified model. Suppose that during the zombie attack on Falkirk, there are rumours that the zombies will only attack in pairs. To simplify things, we suppose that, if the rumours are true, a zombie will always try to flee if it meets a human alone; in the same vein, a human who meets a pair of zombies will always try to flee. So:

- when a single zombie encounters a susceptible, either both remain unscathed or the zombie becomes removed;
- when a pair of zombies encounters a susceptible, either all remain unscathed or the susceptible succumbs to zombification.

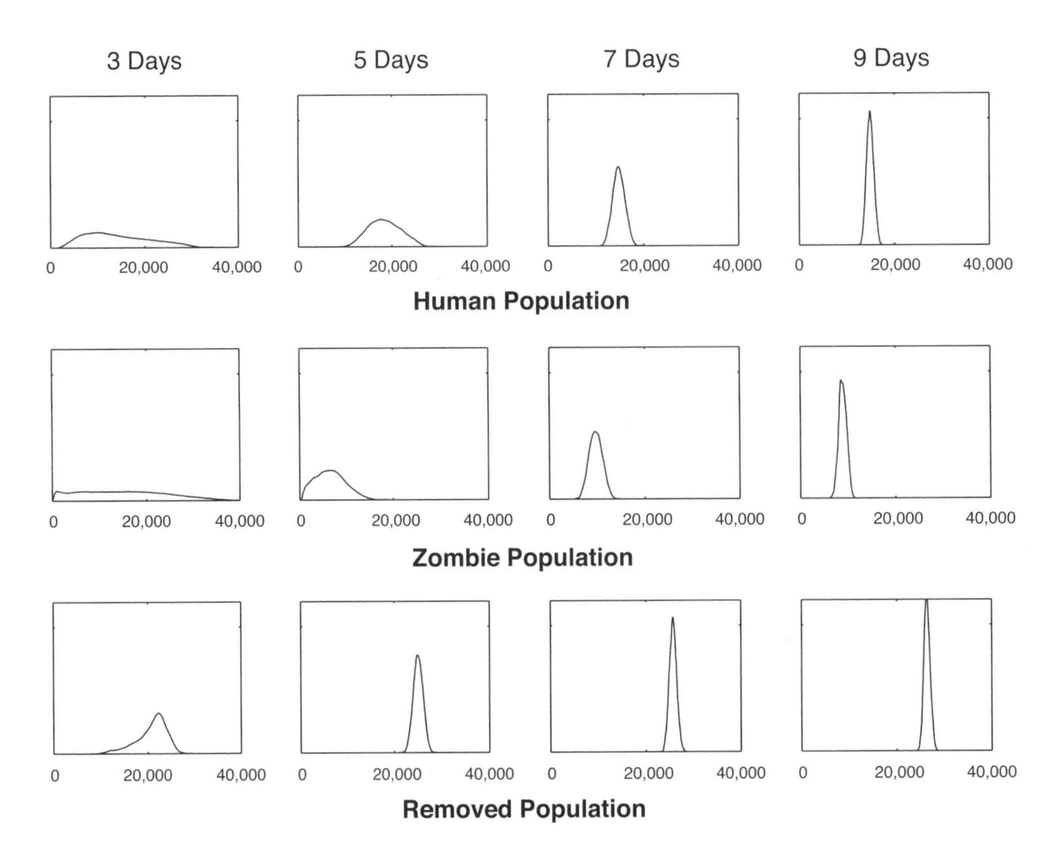

Figure 8.10: Predicted levels of humans, zombies and removed on Day 15 having observed levels of humans and zombies for three, five, seven and nine days.

This sounds like a plausible model for behaviour and, applying similar arguments to those used for (8.6)-(8.8), we arrive at the alternative set of ODEs:

$$S' = -\beta S Z^2 \tag{8.9}$$

$$Z' = \beta S Z^2 + \zeta R - \alpha S Z \tag{8.10}$$

$$R' = \alpha S Z - \zeta R. \tag{8.11}$$

We refer to (8.6)-(8.8) as Model 1 and (8.9)-(8.11) as Model 2.

Given some data, we can now infer the parameters for each model. Observations over nine days were generated by simulating from Model 1, in which zombies attack individually, and adding Gaussian-distributed noise with standard deviation 500. We wish to determine whether an inference algorithm will allow us to conclude that Model 1 describes these data better than Model 2, in the absence of information about the rate constants. Figures 8.11, 8.12 and 8.13 show the posterior model outputs for the two models, setting the standard deviation of the noise to be 500, 1000 and 2000, respectively. Visually trying to assess which is the better model is difficult, since the posterior output covers most of the data points for both models.

Table 8.1: Interpretation of Bayes factors.

$\mathbf{B_{12}}$	**Evidence against alternative**
1 to 3	Not worth more than a bare mention
3 to 10	Substantial
10 to 100	Strong
> 100	Decisive

We therefore resort to calculating Bayes factors such that B_{12} represents the weight of statistical evidence in favour of Model 1 over Model 2. This is computed as the ratio of the marginal likelihoods for the two competing models:

$$B_{12} = \frac{P(\mathbf{Y}|M_1)}{P(\mathbf{Y}|M_2)}. \tag{8.12}$$

We recall that calculating the marginal likelihood involves estimating the integral of the likelihood times the prior over all values of the parameters, which is an extremely challenging task. We employ the technique of thermodynamic integration, which has recently been shown to provide accurate, low-variance estimates of this quantity [10], whereas other seemingly simpler methods, such as the Posterior Harmonic Mean estimator, could fail to produce usable results [5].

Table 8.1 is a useful guide for interpreting the evidence provided by the estimated Bayes factors [11]. Table 8.2 shows the results of the marginal log-likelihoods estimated ten times for each model. In each case, the log Bayes factors correctly identify Model 1 as the one used to produce the data. As the standard deviation of the added noise increases from 500 to 1000 to 2000, the weight of evidence as indicated by the log Bayes factors decreases, although the evidence remains substantially in favour of the correct model.

Table 8.2: Summary of marginal likelihoods for each model.

Model	**Noise SD**	**Marginal Log-Likelihood (± Standard Error)**	**Log Bayes Factor $\mathbf{Log(B_{12})}$**
Model 1	500	−152.7 (± 0.1)	31.5 (± 1.8)
Model 2	500	−184.2 (± 1.7)	
Model 1	1000	−158.5 (± 0.1)	16.9 (± 1.1)
Model 2	1000	−175.4 (± 1.0)	
Model 1	2000	−167.1 (± 0.1)	9.9 (± 3.5)
Model 2	2000	−177.0 (± 3.4)	

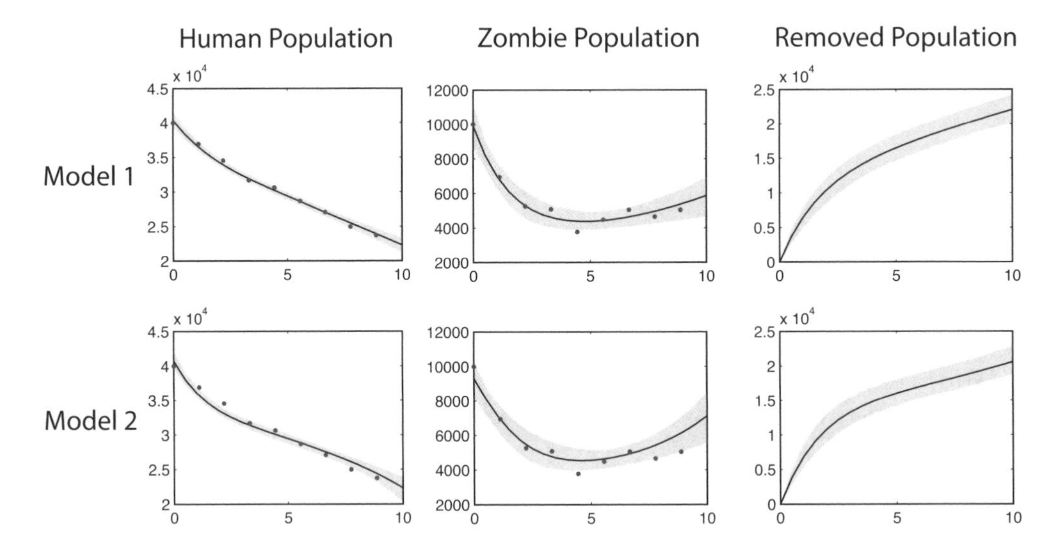

Figure 8.11: Posterior output for the two competing model hypotheses with standard deviation of noise set to 500.

8.5 Is It Safe to Go Out Yet?

Having introduced the Bayesian statistical machinery for performing inference over systems of differential equations to describe zombie attacks, we now return to a scenario posed in the introduction.

- Soldiers patrolling the area have reported daily observed zombie numbers for the past five days as 123, 127, 104, 92 and 74. Is it safe to go out yet?

Suppose that we are holed out in the basement of a local convenience store with access to enough Irn Bru and Tunnock's Caramel Wafers to survive for around fifty days. Knowing those zombie levels over the past five days, should we hold tight or chance an escape?

We now know how to address this question through the use of Bayesian modelling. We assume Model 1 to be a fair representation of the interaction between the zombie population, human population and the removed population. (Of course, if we had multiple plausible models, we could once again do a full model comparison by calculating Bayes factors.) We perform parameter inference over the model given our data of daily observations of zombies and plot the predictive model output, shown on the left of Figure 8.14.

In this case, the 95% confidence output from the model includes a wide variety of outcomes, and the uncertainty in the estimate naturally increases with time. We cannot rule out the scenarios where (a) there is relatively little impact on the human population or (b) zombies take over completely within the next month! The mean number of zombies continues to decrease over the next five days, until Day 10 when

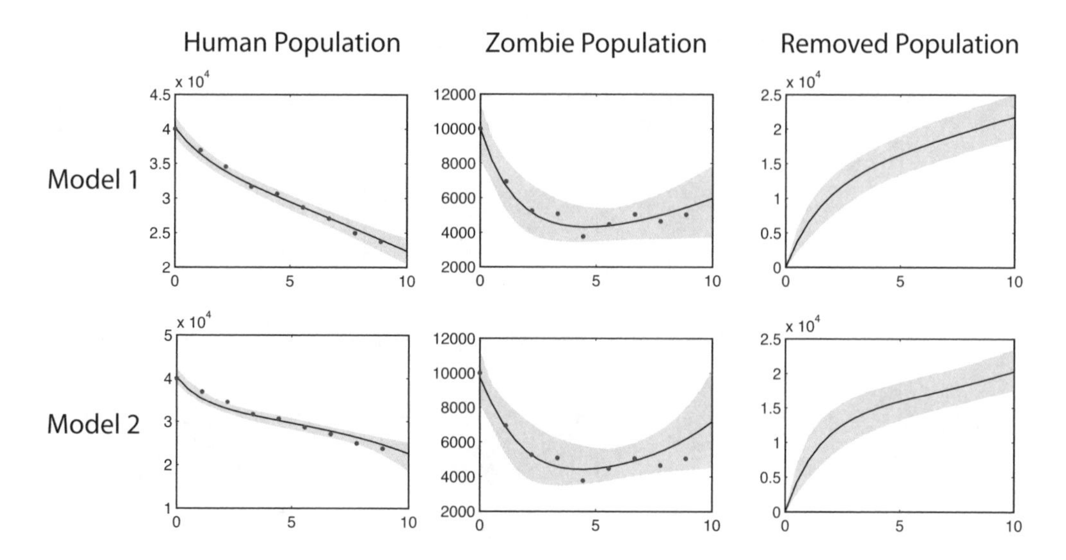

Figure 8.12: Posterior output for the two competing model hypotheses with the standard deviation of noise set to 1000.

it begins to increase. We can therefore argue in favour of making an early exit, after which there is much more uncertainty that the number of zombies in the village will still be relatively low.

If we have additional data on the human population in the village over the past five days, then we once again perform parameter inference over our model including the data for both humans and zombies. The additional data and predictive model output are shown on the right of Figure 8.14. Given these extra data, we can say with much more certainty that the number of zombies in the village will remain low for a longer period of time, making it perhaps less urgent for us to make our getaway immediately. In this case, it is feasible to sit tight a little longer and enjoy our surroundings before gathering up supplies and attempting an escape.

8.6 Discussion

Mathematical modelling of natural phenomena has a long and illustrious history. Differential equations have the potential to describe and predict the behaviour of many physical and engineered systems. However, any mathematical model represents an abstracted summary that cannot be relied on to capture all characteristics of interest. A well-known quote that is paraphrased from [12] and attributed to George Box says that "All models are wrong, but some are useful." Modelling involves compromises and it generates an inherent level of uncertainty. Moreover, the task of identifying unknown or unmeasured model parameters introduces further uncertainty. In cases where model predictions may be used to guide policy—for example, in economics, weather forecasting and epidemiology—a systematic and

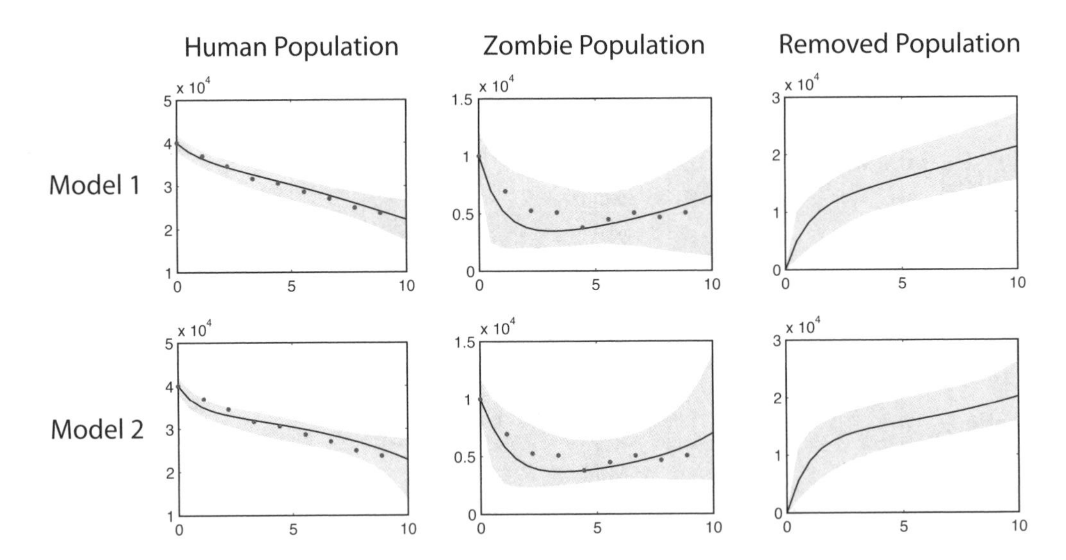

Figure 8.13: Posterior output for the two competing model hypotheses with the standard deviation of noise set to 2000.

consistent treatment of all levels of model uncertainty is vital. We have demonstrated through simple examples in an eye-catching context that the Bayesian statistical framework provides an appropriate means with which to capture information and to reason under uncertainty. We have shown how Bayesian inferential methodology may be extremely useful, not only when calibrating statistical models based on systems of differential equations, but also in comparing different models and judging which model best explains the observed data.

Since we prepared this chapter, in 2010, there has been a dramatic increase in the level of research activity at the interface between computational/applied mathematics and applied statistics. For example, the American Statistical Association and the Society for Industrial and Applied Mathematics have combined forces to found a *Journal on Uncertainty Quantification*, and the Institute of Mathematics and its Applications in the United Kingdom now publishes the journal *Information and Inference*, through Oxford University Press.

An outstanding challenge in this area is the calibration of models with a large number of unknown parameters, that requires efficient sampling in very high dimensions. Added complications arise when parameters are highly correlated. The underlying 'forward' problem of solving initial-value ODEs numerically is a well-studied topic which has spawned very sophisticated software tools. However, in the inference context it is particularly important to simulate the ODE quickly, since massive numbers of solves may be required, and there is potential for exploiting the special nature of the problem: (a) highly accurate solutions are not required and (b) the ODE may be solved repeatedly for very similar parameter values. Another key issue is the development and analysis of new Markov chain Monte

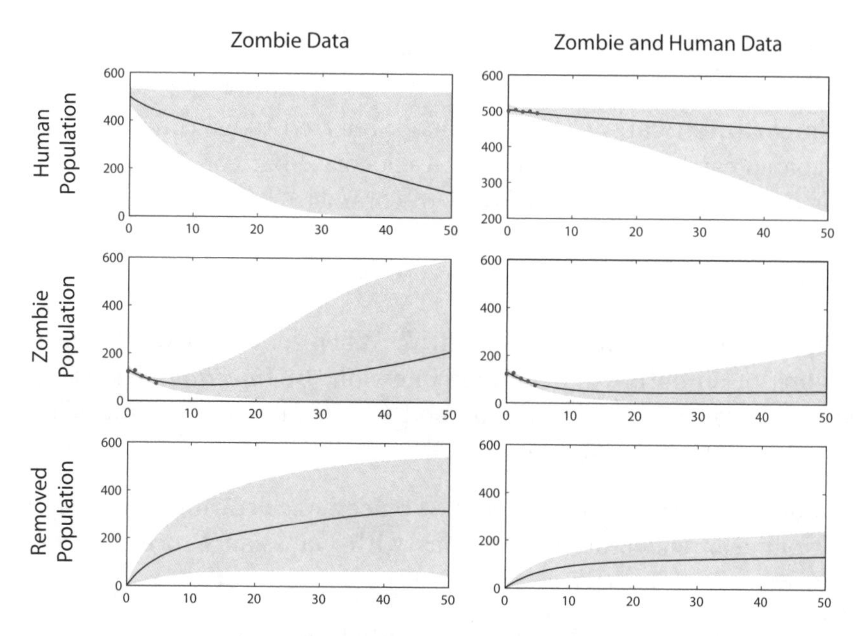

Figure 8.14: The plots on the left-hand side show the predictive model output from the three-parameter model given observations of zombies attacking a small village. The plots on the right-hand side show the output given additional observations of human population numbers. (Note the five data points in three of the graphs.)

Carlo methodologies to provide unbiased and low variance estimates of the quantities required for model comparison. Overall, this blossoming field brings together ideas from applied mathematics, statistics and computer science, and it offers many opportunities for interdisciplinary research.

Appendix

A Glossary

Bayes's theorem. A mathematical rule for quantifying how the chance that a theory is true is affected by a new piece of evidence.

Irn Bru. A carbonated soft drink produced in Scotland.

Likelihood. A measurement of the chance of an event taking place.

Markov chain. A Markov chain is a stochastic process with the property that the probability of its next state depends only on its previous state.

ODE (ordinary differential equation) model. One or more equations describing the rate of change of a physical system.

Random variable. A quantity whose value is subject to chance.

Statistical inference. The act of drawing conclusions from data that are subject to fluctuations.

Tunnock's caramel wafer. (More formally referred to as Tunnock's milk chocolate coated caramel wafer biscuit.) A chocolate bar consisting of five layers of wafer, interspersed with four layers of caramel.

References

[1] P. Munz, I. Hudea, J. Imad, R.J. Smith?. When zombies attack!: Mathematical modelling of an outbreak of zombie infection. In *Infectious Disease Modelling Research Progress* J.M. Tchuenche and C. Chiyaka, eds., Nova Science, Happuage, NY 2009, pp. 133–156.

[2] B. Calderhead, M. Girolami, N. Lawrence. Accelerating Bayesian Inference over Nonlinear Differential Equations with Gaussian Processes. *Adv. Neural Inf. Process. Syst.* 21, 217–224, 2009.

[3] V. Vyshemirsky, M. Girolami. Bayesian Ranking of Biochemical System Models. *Bioinformatics* 24:833–839, 2008.

[4] V. Vyshemirsky, M. Girolami. BioBayes: A software package for Bayesian inference in Systems Biology. *Bioinformatics* 24:1933–1934, 2008.

[5] B. Calderhead, M. Girolami. Estimating Bayes factors via thermodynamic integration and population MCMC. *Comput. Statist. Data Anal.* 53:4028–4045, 2009.

[6] D. Sivia, J. Skilling. *Data Analysis: A Bayesian Tutorial.* Oxford University Press, Oxford, UK, 2006.

[7] E.T. Jaynes. *Probability Theory: The Logic of Science.* Cambridge University Press, Cambridge, UK, 2003.

[8] D.B. Hitchcock. A History of the Metropolis–Hastings algorithm. *Amer. Statistician* 57:245–257, 2003.

[9] W.K. Hastings. Monte Carlo sampling methods using Markov chains and their applications. *Biometrika* 57:97–109, 1970.

[10] N. Friel, A. Pettitt. Marginal likelihood estimation via power posteriors. *J. R. Stat. Soc. Ser. B Stat. Methodol.* 70:589–607, 2008.

[11] R. Kass, A. Raftery. Bayes Factors. *Amer. Stat. Assoc.* 90:175–196, 1995.

[12] G.E.P Box, N.R. Draper. *Empirical Model-Building and Response Surfaces.* Wiley, New York, NY, 1987.

THE SOCIAL ZOMBIE: MODELLING UNDEAD OUTBREAKS ON SOCIAL NETWORKS

Laurent Hébert-Dufresne, Vincent Marceau, Pierre-André Noël, Antoine Allard and Louis J. Dubé

> **Scully**: *Zombies are just projections of our own repressed cannibalistic and sexual fears and desires. They are who we fear that we are at heart. Just mindless automatons who can only kill and eat.*
> **Mulder**: *Well, I got a new theory. I say that when zombies try to eat people, that's just the first stage. You see, they've just come back from being dead, so they're gonna do all the things they missed from when they were alive. So, first, they're gonna eat. Then they're gonna drink. Then they're gonna dance and make love.*
>
> — The X-Files, Season 7, Episode 19

Abstract

According to Mulder's theory, the zombies will eventually fall on each other and make love. However, be it for love or evil, the cold, hard reality remains that the actions of the undead, just like those of the living, are structured by simple constraints of a social or spatiotemporal nature. In this chapter, we consider the underlying social network of the living and the horde behaviour of the undead. This model is then further improved by considering the adaptive nature of social interactions: people usually tend to avoid contact with zombies. Doing so captures the co-evolution of the human social network and of the zombie outbreak, which encourages humans to naturally barricade themselves in groups of survivors to better fight the undead menace. And then? Better stack goods, arm yourself and be patient, for the undead hordes are there to stay—hopefully dancing and making love.

9.1 Introduction

We constantly encounter networks in our everyday lives. We might be on our way to work or school (roads and public transport networks), updating a Facebook status (internet, online social networks, World Wide Web), meeting friends or getting a high-profile job (acquaintance networks), calling abroad or getting directions using a GPS (satellite networks) or simply turning on the radio (electrical and information networks). By improving our understanding of the structure of such networks, we are increasingly able to derive as much benefit as possible from the advantages that networks have to offer, while efficiently protecting ourselves from the disadvantages.

Being social animals, people interact with each other for a variety of reasons: friendships, family bonds, sexual partnerships, business relations and so on. At the population level, these interactions sum up to form a giant web: the *social network*. How individuals are connected to one another in the social network depends on the nature of the considered interactions. Within this structure, people may exchange or transmit information, opinions or infectious diseases, to name a few, while the underlying social network shapes the propagation dynamics.

Studying social networks has been quite useful in the recent past to understand and, to a certain extent, predict the propagation of infectious diseases in human populations. While research has mostly been focused on containing and fighting 'traditional' emerging infectious diseases such as HIV and influenza, another threat has been grossly underestimated and is now imminent: the zombie apocalypse.

Although zombies may all look broadly similar and have the same desires, previous chapters have shown that they do not necessarily behave in the same way. The location of zombies matters and their proximity to you is important. While not social animals, zombies nevertheless have their own social behaviour.

As the hands of the clock push us inexorably closer to the End Time, experts believe that preparation is now imperative to ensure any future for humanity. Using concepts from network theory and contact network epidemiology, we present in this chapter a novel approach that models zombie outbreaks and human counterattack actions. This model will help authorities to conceive and test different resistance strategies beforehand, and be as prepared as one possibly can for "when there will be no more room in Hell and the dead will walk the Earth".[1]

9.2 Modelling a Zombie Invasion: A Practical Guide

To model a zombie invasion, we first need to introduce some basic notions of network theory. We will then define rules governing how people and zombies should interact.

[1] G.A. Romero. *Dawn of the Dead*, directed by G.A. Romero. 1978.

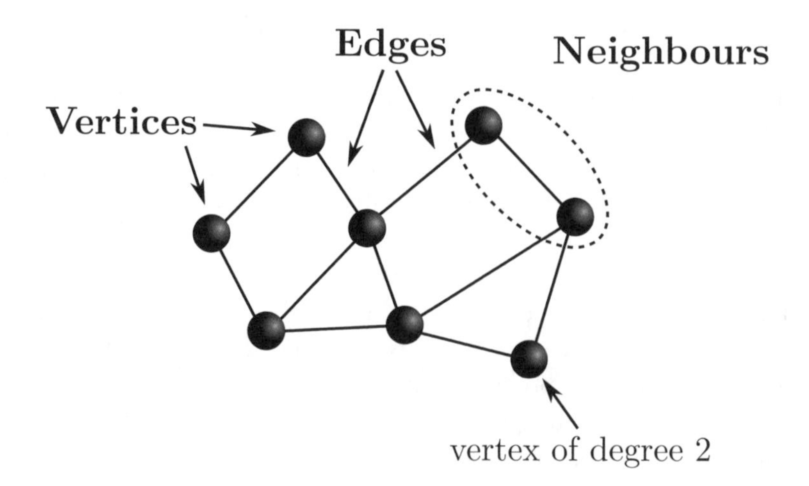

Figure 9.1: A graph of eight vertices and eleven edges. Neighbours are two vertices that share an edge; the degree of a vertex is the number of neighbours it has.

9.2.1 Contact Networks 101

A contact network is represented by a *graph*: a set of dots linked by lines. Dots and lines are generally referred to as *vertices* and *edges*, respectively, and we will use this nomenclature throughout this chapter. Figure 9.1 shows a simple example of a graph.

When modelling social networks, individuals are represented by vertices and two individuals are linked by an edge if they interact with each other. In the case of a zombie invasion, any kind of direct interaction can lead to infection. Therefore social interactions such as friendship, family bonds or acquaintances from school or work are taken into account in the model.

Two vertices sharing an edge are said to be *neighbours* and the number of neighbours of a given vertex is called its *degree*. A property of real social networks that plays a major role in disease propagation is their *degree distribution*, noted as $\{p_k\} = [p_0, p_1, p_2, \ldots]$. Each p_k gives the fraction of the vertices that have k neighbours, which is also the probability that a randomly selected vertex is of degree k. For example, the degree distribution of the graph shown in Figure 9.1 is $\{p_k\} = [0, 0, 1/2, 1/4, 1/4, 0, 0, \ldots]$.

9.2.2 Building a Contact Network

The structure of a network contains *a lot* of information. To acquire this information through a survey, we would have to ask each vertex who its neighbours are. While this might be simple for a very small network, it is usually not possible for larger populations. Indeed, it would be much simpler to survey a small fraction of the population and then figure out what information this gives us about the whole network.

Now suppose that we have the results from a survey that asked 100 people how many neighbours they have (i.e., what is their degree). If the sole other information you have about the network is that there are a total of 10,000 people in it, how do you fill in the blanks? How do you obtain the degrees of the other 9,900 other vertices and how do you join all the vertices together?

The first question is easy to answer by requiring consistency: the degree of all the 10,000 vertices should be chosen such that if we asked any randomly chosen 100 of these vertices to answer the same survey, the result should be similar to the one previously obtained. In practice, this is usually done by obtaining the degree distribution $\{p_k\}$ of those 100 people who answered the survey and then assume that the remaining 9,900 follow the same probability distribution. We can thus pick them one by one, assigning degree k to each new vertex with probability p_k.

The second question, how to join these vertices together, may be less intuitive. We could design a complicated process based on friendship, chances of encounters in the street, who is going to which supermarket . . . but if we do not have this information, should we make it up?

In all cases, *the best choice* is to use the little we do know, and nothing that we do not. In our case, this comes out as forcing vertices to have the degrees that we earlier chose while randomly assigning edges between them. One possible algorithm for doing this goes as follows: for each of the N vertices in the network, place in a bag k pieces of paper each bearing a tag uniquely identifying that vertex of degree k. Shuffle, draw two pieces of paper, assign an edge between the two corresponding vertices, destroy the drawn pieces of paper and repeat until the bag is empty. When we need to build a network from this information, we will use a procedure very close to the one just described. Of course, everything will be automatized in a computer program. The networks considered in the remainder of this chapter will be constructed by this algorithm and therefore simply defined by their size N and their degree distribution $\{p_k\}$.

9.2.3 The Rules of the Game

Until now, all the vertices in the network were intrinsically the same. This is clearly undesirable for our model since we do not expect a zombie to behave the same way as a healthy human. However, it is probably acceptable to say that all zombies behave as 'a typical zombie,' and that all humans behave as 'a typical human.'

We will differentiate individuals into three states: those who have not been infected yet (denoted S for *survivors*), those who have been infected (denoted Z for *zombies*) and dead zombies (denoted R for *removed*[2]). A zombie can only bite neighbours that are in state S; conversely, a survivor can only become a zombie if one or more of her neighbours is a zombie.

---

[2]Or rotting in peace.

Let us now add numbers to this description. During an infinitesimal time period $[t, t+dt)$, a survivor with one Z-neighbour has a probability $\alpha \cdot dt$ to be bitten by the latter and to become a zombie. For a survivor with n Z-neighbours, the probability would be $n \cdot \alpha \cdot dt$.

Survivors also have their say. During the same infinitesimal time period $[t, t+dt)$, a zombie with m S-neighbours has a probability $m \cdot \beta \cdot dt$ to be definitively[3] killed by one of them, henceforth being permanently confined to the R state.

We now have everything in hand to model a simple zombie invasion. More complexity can be considered and Section 9.5 will explore some realistic additions to our basic model. For now, we will limit ourselves to these simple rules.

9.2.4 Monte Carlo Simulations

Now that the rules of the game are set, we can write a computer program to diligently use them. This program has three main tasks: build a network, set the initial conditions and apply the zombie propagation rules.

We have already seen in Section 9.2.2 how to build a network of known size N from its degree distribution $\{p_k\}$. Once the network is built, the initial state of each of its vertices must be chosen. We will specify this initial condition through the proportion ϵ of the population that starts out as zombies. In other words, all vertices are initially survivors except for ϵN randomly selected ones that are zombies. There are initially no removed individuals.

In order to apply the rules themselves, we discretize time into small intervals of length $\Delta t > 0$. Although this quantity is finite (not infinitesimal), we use it instead of dt in the probabilities of Section 9.2.3. If Δt is sufficiently small, the resulting dynamics should be a good approximation of the continuous time dynamics.[4]

During each of these time intervals, we count how many Z-neighbours each of the survivors has. There is then a probability $n \cdot \alpha \cdot \Delta t$ for each survivor with n Z-neighbours to become a zombie at the next time interval. Similarly, for each zombie, we count the number m of its S-neighbours. The zombie will then be killed (sent to the R state) at the next time interval[5] with probability $m \cdot \beta \cdot \Delta t$. This is repeated for as many time intervals as required; that is, until there are no remaining edges linking S and Z vertices together.

We now have a complete procedure (program) providing the state of each vertex at any time t. Since the network construction, assignment of initial conditions and propagation rules are all probabilistic in nature, two different realizations of the process (executions of the program) will typically lead to different results. A process of this kind, relying on randomness, is called a *Monte Carlo simulation*.

[3]Without any possible comeback to 'life.'

[4]Note that there are better ways to apply these kind of rules, but this will be sufficient at present.

[5]All changes are applied simultaneously before the next time step.

Usually, one wants to perform many different Monte Carlo simulations using the same parameters in order to obtain reliable statistics about the model's predictions, such as the mean number of zombies at a given time. How many simulations are required? Depending on the problem and the precision required, the answer may range from a few hundred to billions or more.

While numerical simulations are a rather easy way to obtain results, they also have their disadvantages. For instance, it may happen that the simulations take too long on average or that the total number of simulations to be performed becomes prohibitively high. But perhaps the greatest flaw of such a 'brute force' approach is the lack of insight that we gain from it. We do get results, but they do not offer a good grasp on the underlying mechanics. To address this issue, we will use a completely different approach in the next section that provides better insights into the dynamics of the invasion.

9.3 Mathematical Zombies: A Theory of the Undead

In this section, we basically want to do the same thing as in Section 9.2.4, but avoid its principal drawbacks. The alternative approach is based on rewriting the problem in terms of a set of ordinary differential equations (ODEs). Some approximations are needed, but the results are in perfect (well, almost perfect) agreement with those obtained by the more direct approach of Section 9.2.4. As a bonus, however, what is actually happening in the system comes to light (no pun intended).

9.3.1 The Variables of the System

We shall now write an ensemble of equations that will follow the propagation of a zombie outbreak on a social network. One might first be inclined to simply follow the behaviour of the fractions of the population at time t who have survived, $S(t)$, of those who are now zombies, $Z(t)$, and of those who are now left rotting, $R(t)$. However, this approach pretty much nullifies the incentive for using a network structure in the first place: i.e., to consider the importance of heterogeneous human behaviour on the evolution of the population.

Logically, we at least need to follow the behaviour of survivors $S_k(t)$ and zombies $Z_k(t)$ for each degree k. Since we define our networks through their degree distribution, throwing degrees away would be a waste. However, this is still far from perfect. Consider how different the behaviour of an individual hidden with ten friends would be from the behaviour of an individual surrounded by ten zombies! We therefore need to go a step further and differentiate survivors and zombies according to both the state and the number of their neighbours. To that effect, let $S_{m,n}(t)$ and $Z_{m,n}(t)$ be the proportion of individuals at time t who are survivors or zombies and are currently in contact with m survivors and n zombies.

Note that the total fraction of survivors and zombies at time t can easily be obtained from these quantities by summing over both m and n:

$$S(t) = \sum_{m,n} S_{m,n}(t)$$
$$Z(t) = \sum_{m,n} Z_{m,n}(t).$$

Moreover, the fraction of removed can also be obtained from the fact that $S(t)$, $Z(t)$ and $R(t)$ are fractions of populations; as such, they should all sum to 1:

$$R(t) = 1 - S(t) - Z(t). \tag{9.1}$$

Hence the knowledge of $S_{m,n}(t)$ and $Z_{m,n}(t)$ for all times t solves our model.

9.3.2 Moment-Closure Approximation

The astute reader might have noticed the slight problem that we do *not* know everything about the network. The following example illustrates the essence of this problem. You are currently one of the $S_{1,0}$: a survivor with one S-neighbour and no Z-neighbours. You are therefore safe—or are you? What if your neighbour becomes a zombie? If he has some Z-neighbours, this may very well occur! On the other hand, if your neighbour is a $S_{1,0}$ like you, you form an isolated pair protected from the zombie invasion. Paranoia and suspicion are key factors when it comes to surviving a zombie apocalypse, and you are thus entitled to know who, apart from you, is connected to your friends.

In Section 9.3.1, we could very well have chosen a more complicated set of variables to explicitly take these situations into account. However, no matter how much you know about the network, you could always know more: who are the neighbours of the neighbours of your neighbours? And what about their neighbours?

Increasing complexity has a cost: it can ruin our ability to obtain anything useful. Arguably, a wise choice is to track only the neighbours of each vertex and then infer the neighbourhood of these neighbours from the total available information.

Considering our previous example, let us try to guess the state of your unique survivor friend. Once again, when you do not know much, the best choice is to use all the information you have and nothing else. Should we say that he is in the state $S_{m,n}$ with m and n chosen at random? No, because we know more than that. For example, if we know that $S_{5,2}(t) = 0$, your neighbour is certainly not[6] in the state $S_{5,2}$ since nobody is in that state right now. The probability that your neighbour is in the state $S_{m,n}$ at time t is thus proportional to the population in the state $S_{m,n}(t)$.

Moreover, your neighbour cannot have $m = 0$ (i.e., no S-neighbour) since you are there. In fact, the probability that your neighbour is in the state $S_{m,n}$ at time

[6] Well, '*almost surely not*' would be more accurate, but this is another story.

t is proportional to m: the more S-neighbours he has, the more likely you are one of them.[7]

This chain of reasoning leads to an expression for the probability that an individual is in the state $S_{m,n}$ at time t given that he is a survivor and has at least one S-neighbour. This probability must be proportional to $m \cdot S_{m,n}(t)$; requiring normalization, this probability is then

$$\frac{mS_{m,n}(t)}{\sum\limits_{m',n'} m'S_{m',n'}(t)}.$$

It is now quite easy to obtain quantities such as the mean number of Z-neighbours that each S-neighbour of a survivor has:

$$\langle z(t)\rangle_{s,s} = \frac{\sum\limits_{m,n} n \cdot mS_{m,n}(t)}{\sum\limits_{m',n'} m'S_{m',n'}(t)}.$$

This is called a *moment-closure approximation*: basically, the idea is to guess the higher moments of a distribution such that they are consistent with the already-known lower-order moments.

Other moments may be obtained the same way. Noting that $\langle i(t)\rangle_{j,k}$ is the mean number of neighbours in state i of a vertex in state j, itself being a neighbour of a vertex in state k, we have

$$\langle s(t)\rangle_{z,s} = \frac{\sum\limits_{m,n} m \cdot mZ_{m,n}(t)}{\sum\limits_{m',n'} m'Z_{m',n'}(t)}$$

$$\langle s(t)\rangle_{z,z} = \frac{\sum\limits_{m,n} m \cdot nZ_{m,n}(t)}{\sum\limits_{m',n'} n'Z_{m',n'}(t)}$$

$$\langle z(t)\rangle_{s,z} = \frac{\sum\limits_{m,n} n \cdot nS_{m,n}(t)}{\sum\limits_{m',n'} n'S_{m',n'}(t)}.$$

The problem raised at the beginning of this section is now solved. We still do not know everything about the network, but with these quantities at hand, we have enough to write differential equations governing the evolution of $S_{m,n}(t)$ and $Z_{m,n}(t)$.

[7]This is why, on average, your friends always have more friends than you do.

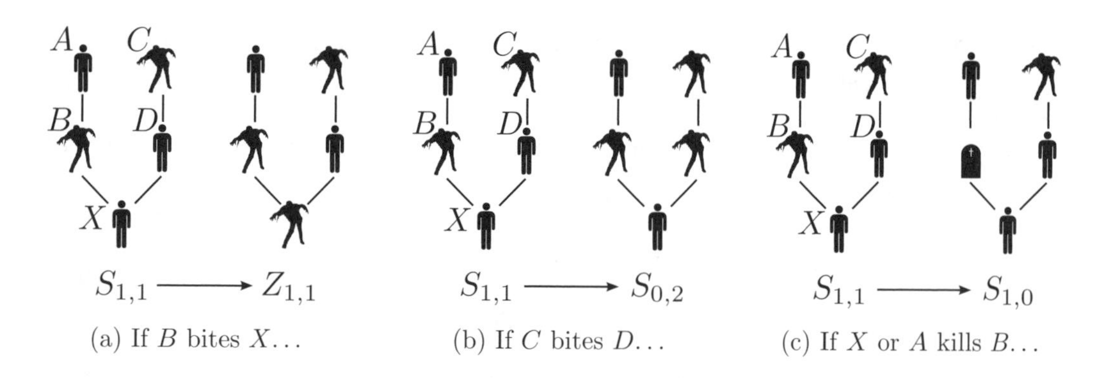

Figure 9.2: Processes affecting the population of survivors. Transitions are shown for the individual known only as X.

9.3.3 Writing the Dynamical System

We now consider every possible processes through which an individual might go from one state to another, thereby changing the population of its initial and new states. We then write ODEs providing the rate of change of $S_{m,n}$ and $Z_{m,n}$ for the time-dependent susceptible and zombie vertices.[8]

Here are the transitions that can affect a vertex X, together with their occurrence probability during a time interval $[t, t + dt)$.

1. *X, a survivor, is bitten by a zombie.* As seen in Section 9.2.3, this event causes X to become a zombie. Hence the transition $S_{m,n} \rightarrow Z_{m,n}$ occurs with a probability $n\alpha S_{m,n}dt$. An example is shown in Figure 9.2A.

2. *A surviving neighbour of X is bitten.* Since the state in which X resides depends on the state of each of its neighbours, the state of X has to change when one of its S-neighbours becomes a zombie. We therefore have the transition $S_{m,n} \rightarrow S_{m-1,n+1}$ with probability $m\alpha\langle z\rangle_{s,s}S_{m,n}dt$ if X is a survivor and $Z_{m,n} \rightarrow Z_{m-1,n+1}$ with probability $m\alpha\langle z\rangle_{s,z}Z_{m,n}dt$ if X is a zombie. Notice the use of $\langle z\rangle_{s,s}$ and $\langle z\rangle_{s,z}$ in order to determine the average number of Z-neighbours that the S-neighbours of X have. An example is shown in Figure 9.2B.

3. *B, a zombie, is killed by a survivor.* This causes B to become removed through the transition $Z_{m,n} \rightarrow R$ with probability $m\beta Z_{m,n}dt$. Since we do not explicitly track removed individuals [but instead obtain them through equation (9.1)], we simply decrease the population of $Z_{m,n}$ (Figure 9.2C).

4. *A zombie neighbour of X is killed.* This is a side effect of event 3. We have the transition $S_{m,n} \rightarrow S_{m,n-1}$ with probability $n\beta\langle s\rangle_{z,s}S_{m,n}dt$ if X is a survivor

[8]From here on, we will lighten the notation by removing the time dependence "(t)" on quantities that clearly vary in time.

and $Z_{m,n} \to Z_{m,n-1}$ with probability $n\beta\langle s\rangle_{z,z} Z_{m,n} dt$ if X is a zombie. An example is shown in Figure 9.2C.

The net effect of these transitions on $S_{m,n}$ and $Z_{m,n}$ gives rise to a system of ODEs of the form

$$\begin{aligned}\frac{d}{dt}S_{m,n} &= \langle z\rangle_{s,s} \cdot \alpha\left[(m+1)S_{m+1,n-1} - mS_{m,n}\right] \\ &\quad + \langle s\rangle_{z,s} \cdot \beta\left[(n+1)S_{m,n+1} - nS_{m,n}\right] - n\alpha S_{m,n}\end{aligned} \tag{9.2}$$

$$\begin{aligned}\frac{d}{dt}Z_{m,n} &= \langle z\rangle_{s,z} \cdot \alpha\left[(m+1)Z_{m+1,n-1} - mZ_{m,n}\right] \\ &\quad + \langle s\rangle_{z,z} \cdot \beta\left[(n+1)Z_{m,n+1} - nZ_{m,n}\right] + \alpha n S_{m,n} - m\beta Z_{m,n}.\end{aligned} \tag{9.3}$$

The degree distribution $\{p_k\}$ and the initial fraction of zombies ϵ are introduced into the system through the initial conditions

$$S_{m,n}(0) = (1-\epsilon)\, p_{m+n} \binom{m+n}{n} \epsilon^n (1-\epsilon)^m$$

$$Z_{m,n}(0) = \epsilon\, p_{m+n} \binom{m+n}{n} \epsilon^n (1-\epsilon)^m.$$

The binomial coefficients come from the fact that a vertex of degree k has probability $\binom{k}{n}\epsilon^n(1-\epsilon)^{k-n}$ to have n Z-neighbours.

Using these initial conditions, we can solve (i.e., by means of numerical integration[9]) the system of ODEs given by equations (9.2) and (9.3) to trace the future of a zombie outbreak on our social network.

9.4 Results: Does Humanity Have the Slightest Chance?

We now compare the approaches we obtained in Sections 9.2.4 and 9.3.3 before extracting meaning from them. But, most importantly, this is where we see if humanity has a chance against a zombie invasion.

9.4.1 Choosing the Victims

Before proceeding, we must decide the size of the population to model. A small city of $N = 10{,}000$ people should be large enough, and say that 1% of them are already zombies ($\epsilon = 0.01$), none of which have yet been killed.

What is the degree distribution $\{p_k\}$ in that city? If we say that everybody has an equal chance to be connected with everybody and, in addition, we know λ, the

[9]For those not familiar with numerical integration techniques, see the introductory notes presented in the appendix of this chapter.

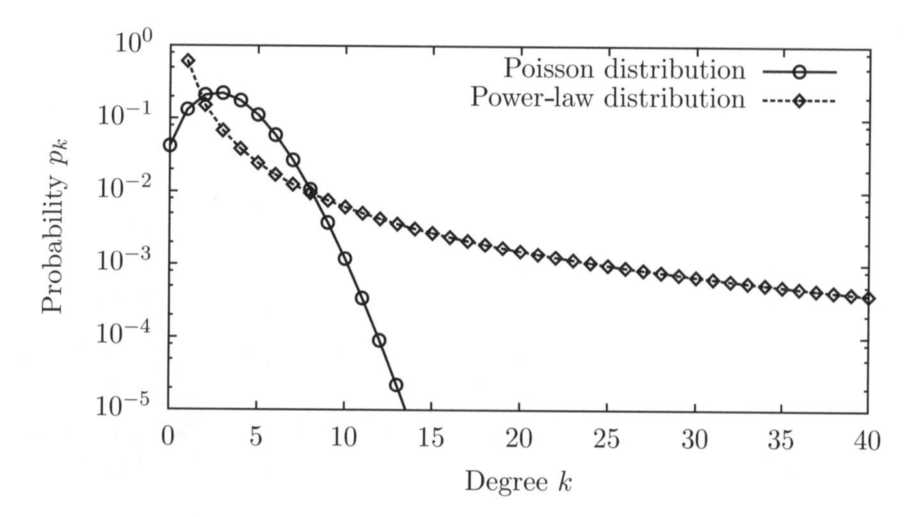

Figure 9.3: Two degree distributions are considered on this semi-log plot. Although both share the same average degree $\lambda = 3.17$, the power-law distribution ($\tau = 2$, $k_{\max} = 100$) falls very much slower than the Poisson distribution. The number of friends each individual has in the Poisson network does not vary much from one person to another. In the power-law distributed network, some people *really* have a lot of friends.

mean number of acquaintances people have, then $\{p_k\}$ is provided by the Poisson distribution for large populations (Figure 9.3):

$$p_k = \frac{\lambda^k e^{-\lambda}}{k!}.$$

However, real human populations behave very differently: some people have many more neighbours than the average. There are not many of them, but they disproportionately influence any spreading processes occurring on the network. The *power-law* distribution, with parameter $\tau > 0$ and truncated at $k_{\max}$, namely

$$p_k = \begin{cases} 0 & \text{if } k = 0 \\ \dfrac{k^{-\tau}}{\sum_{k'=1}^{k_{\max}} (k')^{-\tau}} & \text{if } 1 \leq k \leq k_{\max} \\ 0 & \text{if } k > k_{\max}, \end{cases}$$

exemplifies this behaviour (Figure 9.3).

We will use these two distributions and collect the corresponding results. However, to make a fair comparison (to "compare apples with apples"), we use $\lambda = 3.17$, $\tau = 2$ and $k_{\max} = 100$ such that both distributions have the same average degree λ.

9.4.2 Body Count

Our two approaches, Monte Carlo simulations and integration of our system of ODEs, are applied to the two populations differs in degree distribution. Figure 9.4

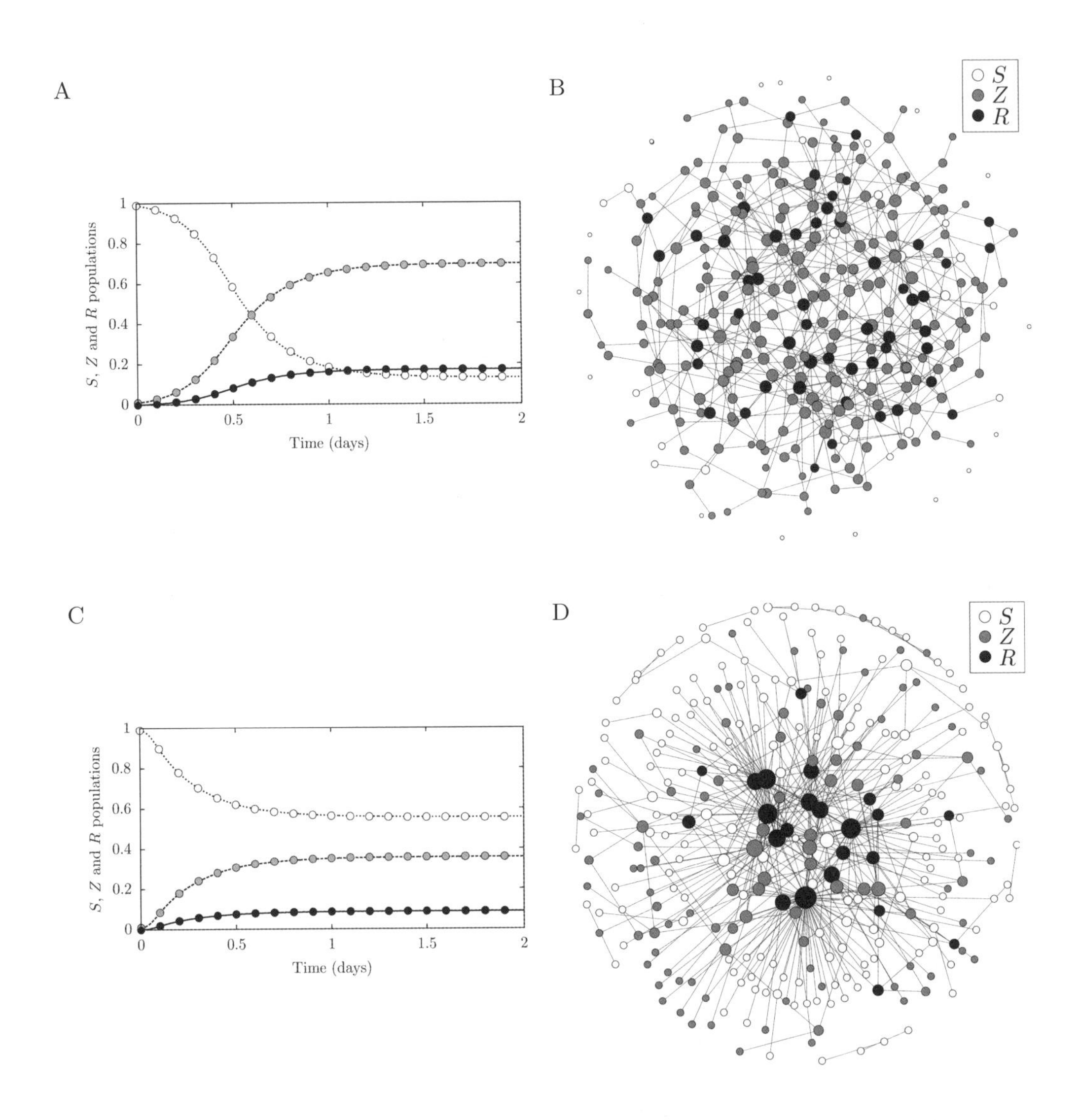

Figure 9.4: Time evolution of the zombie invasion for networks with (A) Poisson and (C) power-law degree distributions (same average degree $\lambda = 3.17$). Curves (dotted S, dashed Z and solid R) show results for the ordinary differential equation formalism while symbols (clear S, grey Z and black R) represent those obtained through averaging over 100 Monte Carlo simulations. Figures (B) and (D) show the final state of a single Monte Carlo simulation performed under the same conditions as (A) and (C) respectively, except that a smaller network size (300 vertices) was used in order to improve visibility. The visual representations are produced by the GUESS software using an algorithm that distinguishes vertices by their state (colour) and their degree (size).

shows[10] typical results for the parameters $\alpha = 5$ and $\beta = 1$. These parameters represent a (sadly likely) case where zombies are much better at biting survivors than survivors are at killing zombies.

The first observation is the *striking* agreement between Monte Carlo simulations (symbols) and the integrated system of ODEs (curves) for both populations (9.4A and 9.4C). What are the implications of such a match? When a sufficiently small time increment Δt is used for Monte Carlo simulations, the results should be arbitrarily close to an exact realization of the rules we chose to model (both the network structure and the propagation rules). Thus if the integrated system of ODEs agrees with the simulations, then they in turn accurately reflect the chosen rules.

Achieving this level of accuracy, in spite of simplifications and approximations made while building the system of ODEs, means that the ignored information does not have much impact on the propagation dynamics of the whole system after all. We have therefore gained the important insight that everything of importance has been included in the model.

A second observation further refines this insight: there are significant differences between the results for the two degree distributions. In a non-network model, these choices would both give the same results since they lead to identical average degree for vertices. This fact alone justifies the need for a network-based approach to the problem.

A third important observation is that both scenarios explored in Figure 9.4 eventually reach an equilibrium where survivors and zombies coexist. This may be surprising to some readers who may be too closed-minded to consider cohabitation with the undead, but it is an unavoidable consequence of the set of rules we chose and has previously been observed in some particular cases.[11,12]

No equilibrium can exist in this model as long as there are survivors with zombie neighbours: the survivor would eventually either kill the zombie or be bitten by the zombie. Hence survivors can coexist with zombies in the equilibrium state only if the neighbours of their neighbours of their neighbours . . . well, all of them, are survivors *or* if the only paths leading to zombies have to go through removed vertices. Therefore subnetworks of survivors and subnetworks of zombies can coexist in the same network as long as they are separated by a no man's land of (probably) rotting and (really) dead corpses. In order to get rid of zombies, human populations should therefore develop very efficient tools to eliminate them at the onset of the invasion, or otherwise should bring themselves to consider and to accept such untypical cohabitation.

An observation that illustrates the uniqueness of a zombie invasion is that things are actually *improving* (there are more survivors at the end) with a power-law

[10]The network snapshots were produced using Eytan Adar, *GUESS: The Graph Exploration System*, http://graphexploration.cond.org.

[11]E. Wright and S. Pegg. *Shaun of the Dead*, directed by E. Wright. Working Title Films, 2004.

[12]G.A. Romero. *Land of the Dead*, directed by G.A. Romero. 2005.

degree distribution compared to the Poisson degree distribution. The presence of high-degree vertices, often called 'hubs,' usually makes things worse for 'normal' diseases. Consider for example the impact of sex workers on sexually transmitted infections or of hospitals and schools on pulmonary infections. Once infected, these hubs quickly dispatch the disease to a lot of people, some of these also being hubs, and this results in many individuals getting infected.

Things are different for zombies, and the difference lies in the very fact that survivors may obliterate zombies. Once a hub is zombified, it can surely bite its many neighbours, but these neighbours also get a chance to eradicate the zombie hub. Most of the big vertices, hubs, are thus eliminated by their neighbours (see Figure 9.4D). These removals effectively reduce the average degree of the remaining vertices, since their degree was originally large compared to the mean. In a more balanced situation such as the Poisson distribution (Figure 9.4B), hubs are mostly absent. When a zombie dies, it is just an average zombie and may very well be replaced by another. Had we chosen $\beta = 0$ (or a negligibly small value), the usual effect of hubs would have been observed.

9.5 The Social Zombie: Adapting for Realism

There are some fundamental issues that arise concerning the dynamics we chose to model. For example, survivors will fight with any number of zombie neighbours, until the undead perish or until the survivors themselves join the nightmarish hordes. In real life, an individual will probably choose to take flight when the situation becomes too dire and survivors will then try to regroup while avoiding zombies. On the other hand, the undead are also usually seen in hordes. Zombies are always inclined to give a hand to their fellow undead friends and help them hunt their remaining targets. This cooperative behaviour between zombies will be used to emulate the typical formation of zombie hordes.

While the networks we used up to now were *static*—the same edges were always linking the same vertices—the introduction of new rules will allow the structure of the network itself to change. By using *adaptive networks*, we are moving one step further away from the homogeneous approximation. This level of details comes at additional cost, but as it will be shown, it is well worth the spending.

9.5.1 The New Rules of the Game

The rules introduced in Section 9.2.3 have to be adapted. Survivors are now allowed to run away when facing a zombie and to form groups into barricades for a better defence against the threat. On the other hand, zombies may form hordes, improving their collective biting abilities.

During an infinitesimal time interval $[t, t + dt)$, a survivor facing a zombie has a probability γdt of fleeing from it and stopping at the next individual she encounters (see Figure 9.5). The survivor then disconnects from the zombie and reconnects to

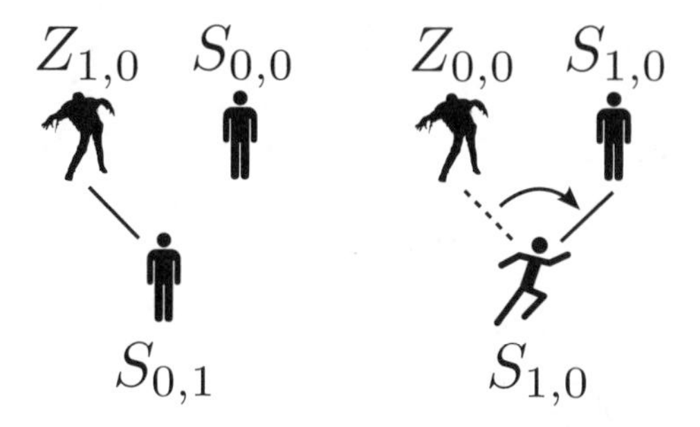

Figure 9.5: A case of flight. A survivor runs away from a zombie (ceasing to be its neighbour) and reconnects to a random new neighbour, here another survivor.

either a survivor ($S_{m,n} \to S_{m+1,n-1}$) or a zombie ($S_{m,n} \to S_{m,n}$). As in Section 9.3.3, this causes 'side effect transitions' both to the zombie he flees from and to that zombie's new neighbour. The new neighbour will be in a state whose probability is directly proportional to the population of that state.

Both rules—for survivors killing zombies or zombies biting survivors—are adapted such a way that an individual is better at attacking the enemy when she is helped by neighbours of her own type (S or Z). During the time interval $[t, t+dt)$, a zombie with n fellow flesh-eaters can bite a survivor with probability $\alpha(1 + h \cdot n)dt$, just as a survivor defending herself alongside m friends will kill a zombie with probability $\beta(1 + h \cdot m)dt$. We will refer to h as the *H-factor*,[13] a parameter determining how useful friends are in combat. Side-effect transitions are affected likewise.

The derivation of the system of ODEs is left as an exercise. Here we provide only the final result:

$$
\begin{aligned}
\frac{d}{dt}S_{m,n} = & \left(h\langle z\rangle_{z,s} + 1\right)\langle z\rangle_{s,s}\alpha\left[(m+1)S_{m+1,n-1} - mS_{m,n}\right] \\
& + \left(h\langle s\rangle_{s,z} + 1\right)\left(\langle s\rangle_{s,z} - 1\right)\beta\left[(n+1)S_{m,n+1} - nS_{m,n}\right] \\
& + (1 + hm)\,\beta\left[(n+1)S_{m,n+1} - nS_{m,n}\right] - \left(h\langle z\rangle_{z,s} + 1\right)n\alpha S_{m,n} \\
& + \gamma\frac{S}{S+Z}\left[(n+1)S_{m-1,n+1} - nS_{m,n}\right] + \gamma\langle z\rangle_s\frac{S_{m-1,n} - S_{m,n}}{S+Z}
\end{aligned}
\tag{9.4}
$$

$$
\begin{aligned}
\frac{d}{dt}Z_{m,n} = & \left(h\langle z\rangle_{z,s} + 1\right)\left(\langle z\rangle_{s,z} - 1\right)\alpha\left[(m+1)Z_{m+1,n-1} - mZ_{m,n}\right] \\
& + \left[(n-1)h + 1\right](m+1)\alpha Z_{m+1,n-1} - (hn + 1)\,m\alpha Z_{m,n} \\
& + \left(h\langle s\rangle_{s,z} + 1\right)\langle s\rangle_{z,z}\beta\left[(n+1)Z_{m,n+1} - nZ_{m,n}\right] \\
& - \left(h\langle s\rangle_{s,z} + 1\right)m\beta Z_{m,n} + \left(h\langle z\rangle_{z,s} + 1\right)n\alpha S_{m,n} \\
& + \gamma\left[(m+1)Z_{m+1,n} - mZ_{m,n}\right] + \gamma\langle z\rangle_s\frac{Z_{m-1,n} - Z_{m,n}}{S+Z}.
\end{aligned}
\tag{9.5}
$$

[13]'H' may stand for *hordes, help, hunting, hellbent* or even *Hades* itself. Pick your poison.

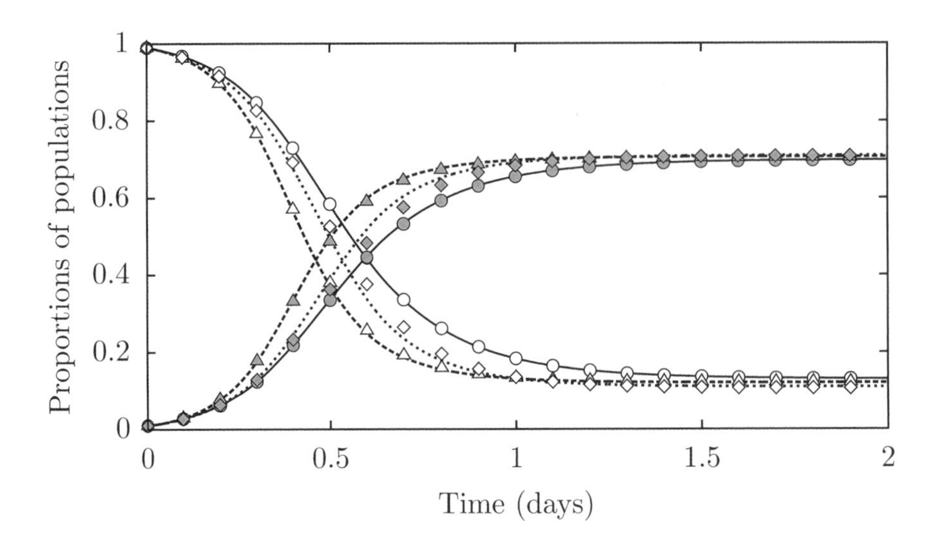

Figure 9.6: Monte Carlo (symbols) and integration of the system of ODEs [equations (9.4) and (9.5)] (curves) results for the new set of rules. The 'old' parameters are the same as those of Figure 9.4A, and the Poisson degree distribution was used. For new parameters, three different cases are studied: $h = 0$ and $\gamma = 0$ (solid curves and ○); $h = \lambda^{-1}$ and $\gamma = 0$ (dashed curves and △); and $h = \lambda^{-1}$ and $\gamma = 2$ (dotted curves and ◇). Survivors are shown in white and zombies in grey.

For convenience, we defined the average number of Z-neighbours a survivor has:

$$\langle z \rangle_s = \sum_{m,n} n S_{m,n},$$

which appears in the equations since the probability that any individual will encounter a survivor running away from a zombie depends on that quantity.

Note that the new system encompasses the previous model since equations (9.4) and (9.5) reduce to equations (9.2) and (9.3) when one sets the flight rate equal to zero ($\gamma = 0$) and chooses a null H-factor ($h = 0$).

9.5.2 Hope Is the Last Thing to Die, Isn't It?

Figure 9.6 is obtained by performing Monte Carlo simulations and integrating the system of ODEs corresponding to the new rules introduced in Section 9.5.1. From a quick comparison of the three different cases studied, it is clear that the H-factor (i.e., the cooperation effect) plays out to the advantage of the zombies. We could expect it to be the other way around if survivors were more efficient at killing zombies than the latter are at biting (if only . . .). On the other hand, allowing survivors to flee (setting $\gamma > 0$) from the zombies gives them a better life expectancy, but only postpones the inevitable. In fact, all three scenarios eventually lead to an equilibrium where around 15% of the population manages to survive by barricading themselves against the undead hordes. Again, our results suggest similar conclusions: human populations should develop efficient weapons and fighting tactics in order to eradicate these nightmarish invaders before they take over the world.

Notice that the fit between the Monte Carlo simulations and the analytical predictions is somewhat less satisfying in the scenario where flight is possible. Great insights can be gained from this simple observation: something we either approximated or neglected when obtaining the system of ODEs becomes important when flight is allowed.

The culprit is most likely correlation. In our system of ODEs, when a survivor flees from a zombie and then encounters vertex X, we add a new S-neighbour to vertex X. Later on, we consider this survivor as the *average survivor*. This is where the shoe pinches: he is not the average survivor, but the average survivor who just ran away from a zombie, and this additional information tells us something about his state. He would probably not have been able to get this far had he been in contact with 100 zombies, nor would he be fleeing had he not been in contact with at least one zombie in the first place. The fact that our equations do not consider these correlations is probably the principal cause of the observed differences with the Monte Carlo simulations.

In order to improve the results, one could embark on the tedious endeavour of including the state of the neighbours of the neighbours of each vertex in the description of the system. Doing so would require something like

$$S_{\substack{m_{0,0}, m_{0,1}, m_{0,2}, \ldots, m_{1,0}, m_{1,1}, \ldots \\ n_{0,0}, n_{0,1}, n_{0,2}, \ldots, n_{1,0}, n_{1,1}, \ldots}} \quad \text{and} \quad Z_{\substack{m_{0,0}, m_{0,1}, m_{0,2}, \ldots, m_{1,0}, m_{1,1}, \ldots \\ n_{0,0}, n_{0,1}, n_{0,2}, \ldots, n_{1,0}, n_{1,1}, \ldots}},$$

where each $m_{i,j}$ represent the number of survivor neighbours who themselves have i survivor and j zombie neighbours ($n_{i,j}$ playing the same role for zombie neighbours).

Since a differential equation would be required for each of these quantities, it is easy to imagine how quickly the complexity of this sort of models can explode. However, considering that the error observed in the results was not that critical in the first place, the simpler approach that we have used was appropriate to the problem we chose to tackle.

9.6 A Conclusion on Networks . . .

This was only the tip of the iceberg. As we have witnessed in this chapter, network modelling (namely, the inclusion of a topology in a model of interaction between a large number of elements) is useful when considering the effects of local dynamics on the global evolution of the system. What does this mean for zombie modelling? The model developed here allowed us to consider multiple local effects:

Heterogeneity of behaviours . . . and fates. Some individuals can be (or get) disconnected from the rest of the world while others might end up in the middle of the zombie apocalypse.

Local environment. Individuals are affected not by the global state of the world, but only by their immediate environment (this can allow a single zombie to start a very virulent but localized outbreak).

Social behaviours. Networks do not have to be static: humans can flee from the zombies and regroup, while the undead will hunt in hordes.

However, a multitude of other similar effects could have been (and should be) incorporated in a more complex model:

Types of individuals. Not all humans will react the same in a zombie invasion (for example, the behaviour of a soldier is likely to differ from that of a child) and network modelling is perfectly suitable for considering such heterogeneity.

Social structure. We have given very basic behaviour to both the living and the dead, but the formation of more complex social structure could be taken into account.

Note that, while network modelling is a natural approach to the description of individual and heterogeneous behaviours, the inclusion of structure in the problem usually complicates its treatment. Furthermore, complexity usually increases rapidly in network theory. This means that our little exercise in network modelling would have been an entirely different story had we considered an even more complex system.

9.7 . . . and Zombies

What have we learned from this exercise that could help us survive the imminent zombie apocalypse?

First, we have learned that, in a scenario where the set of rules that we have chosen applies, any invasion must be stopped at its very beginning; otherwise, a significant fraction of the population will be either dead or zombified, thus marking the end of the supremacy of humanity on Earth. We therefore stress once again the importance of developing efficient and powerful anti-zombie weapons and defences.[14]

Secondly, the inclusion of a cooperation effect in our model has been shown to *accelerate* the zombie invasion (if $\alpha > \beta$), while our results show that flight only slows down the progression of the invasion without significantly modifying its outcome. Therefore if we consider fighting the undead in single combat, we should use state-of-the-art fighting techniques in order to use the cooperation effect to our advantage (i.e., $\beta > \alpha$).[15]

Finally, within the context in which our model has been developed, it has been shown that any invasion can be stopped, but cannot be reversed. Once a succesful

[14]In our model, such equipment would result in reducing α and increasing β.

[15]Again, in our model, such a strategy could be studied by using different values of the H-factor for survivors and zombies.

barricade is established, we can expect to find a certain number of roaming, isolated zombies. To wipe out these remaining threats, we believe it is necessary to consider more than fighting at the individual level. Optimistically, efforts should be deployed in developing antidotes and vaccines. On the other hand, the use of weapons of mass destruction should be considered once a 'stable' condition is reached. In light of the results presented in this chapter, such strategies may well be our only chance to take these poor souls back to Hell from whence they spawned . . . until next time.

> *Yea, though I walk through the valley of the shadow of death, I will fear no evil: for thou art with me, you mathematical-physicist; thy model and thy simulation they comfort me.*[16]

Appendices

A Numerical Integration

In this chapter, we have created a model based on a large set of nonlinear ordinary differential equations (ODEs) to describe the dynamics of a zombie outbreak on a social network. Solving this system of ODEs by analytical means would be impossible due to its size and complexity. Thus we have to rely on *numerical integration* to approximate the analytical solution for the time evolution of our system.

To illustrate the basics of numerical integration, let us consider a simple one-dimensional ODE system:

$$\frac{dx(t)}{dt} = f\big(x(t)\big) \tag{9.6}$$

with the initial condition $x(t_0)$, and for which we do not know any analytical solution. Consider now the following intuitive reasoning, using an analogy to unidimensional motion. Suppose $x(t)$ is the position of a particle at time t. At this particular time, its speed (in general, the rate of change of the variable) is given by $f\big(x(t)\big)$. Starting from the known initial condition, we are now interested in finding where this particle will be at time $t_0 + \delta t$ based on the knowledge of the position $x(t_0)$ and speed $f\big(x(t_0)\big)$ at time t_0. For δt sufficiently small, the speed of the particle will remain approximately constant during this short time step. This results in a linear motion, given by:

$$x\left(t_0 + \delta t\right) = x\left(t_0\right) + \delta t f\big(x(t_0)\big). \tag{9.7}$$

This procedure can then be used iteratively to generate an approximate solution at later discrete times:

$$x\left(t_0 + 2\delta t\right) = x\left(t_0 + \delta t\right) + \delta t f\big(x(t_0 + \delta t)\big).$$

[16] Any resemblance with Psalm 23:4 of the Original King James Bible is purely coincidental.

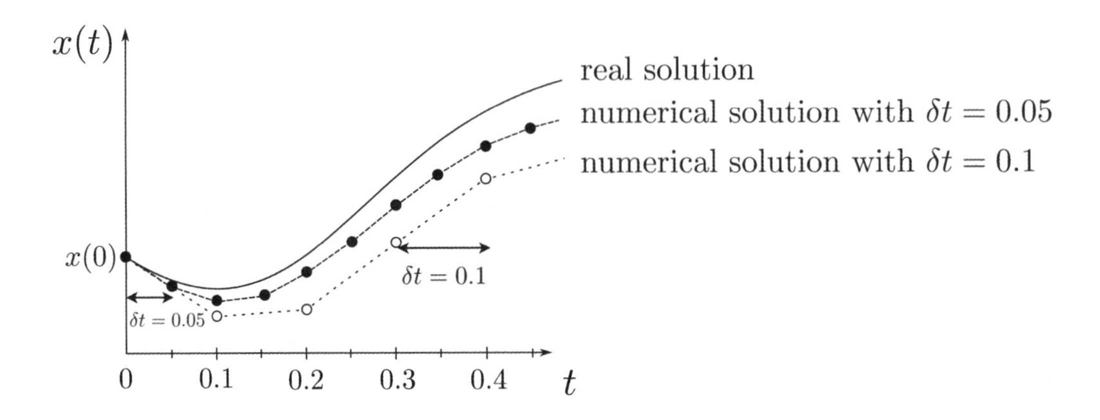

Figure 9.7: Schematization of Euler's method.

This numerical integration scheme is known under the name of *Euler's method* and is schematically illustrated in Figure 9.7. With this method, the approximate solution has a *local error*, i.e., the error made at each time step δt, of order $\mathcal{O}\left(\delta t^2\right)$. Practically, it means that the numerical approximation gets closer to the real solution as we diminish δt. Formally, this is because equation (9.7) corresponds to the first two terms of a Taylor series expansion of $x(t_0 + \delta t)$ around t_0:

$$x(t_0 + \delta t) = x(t_0) + \left.\frac{dx(t)}{dt}\right|_{t_0} \delta t + \left.\frac{d^2x(t)}{dt^2}\right|_{t_0} \delta t^2 + \left.\frac{d^3x(t)}{dt^3}\right|_{t_0} \delta t^3 + \mathcal{O}(\delta t^4).$$

Consequently, if we integrate the equation (9.6) for a total time $T = N\delta t$, the resulting *global error* will be of order $\mathcal{O}(\delta t)$.

Euler's method can also be easily generalized to system of ODEs of more than one dimension, as it was the case in our problem:

$$\vec{x}\,(t_0 + \delta t) = \vec{x}\,(t_0) + \delta t \vec{f}\big(\vec{x}(t_0)\big)$$

$$\text{where}\quad \vec{x}(t) = \begin{pmatrix} x_1(t) \\ x_2(t) \\ \vdots \\ x_M(t) \end{pmatrix} \quad \text{and} \quad \vec{f}\big(\vec{x}(t)\big) = \begin{pmatrix} f_1\big(x_1(t), x_2(t), \ldots, x_M(t)\big) \\ f_2\big(x_1(t), x_2(t), \ldots, x_M(t)\big) \\ \vdots \\ f_M\big(x_1(t), x_2(t), \ldots, x_M(t)\big) \end{pmatrix}.$$

Other, more accurate, methods are also available. One of the most frequently used families of algorithms for ODE integration is based on the Runge–Kutta method, that uses several points in the interval $[t, t + \delta t]$ to approximate the effective slope of $x(t)$. In this chapter, we have used an even more complicated technique[17] based on a Runge–Kutta algorithm which has an estimated global

[17] Sometimes called the *Dormand–Prince method* or *Runge–Kutta 4–5*.

error of order $\mathcal{O}(\delta t^4)$ and uses an adaptive time step δt in order to keep the error small even when the slope of $x(t)$ is very steep. The interested reader can find a wealth of information in the book *Numerical Recipes* by Press *et al.*

B Glossary

Adaptive network. A network whose topological structure evolves adaptively with the dynamics taking place on it, thus creating a feedback loop. In other words, the dynamics *on* the network (zombie outbreak) influences the dynamics *of* the network (social contacts), which in turn affects the way the dynamics will further evolve *on* the network.

Degree. The number of neighbours of a given vertex.

Degree distribution. The set of probabilities $\{p_k\}$ with which a vertex chosen at random in the graph has a degree k.

Edge. A line joining two vertices in a graph. In a contact-network model, it represents a social contact between two individuals (such as a bond of family or friendship).

Graph. Mathematical abstract representation of a network, consisting of dots (vertices) and lines (edges).

Moment-closure approximation. A mathematical technique used to approximate higher-order moments of a system with the knowledge of lower-order moments. It is often used to express systems of equations in a closed form.

Monte Carlo simulation. A computer algorithm (simulation) relying on the repeated use of random numbers.

Neighbours. Two vertices linked by an edge.

Network. An interconnected system of 'things.' Those things can be people, electrical components, words, computers, etc. The way they are connected defines the topological structure of the network.

Ordinary differential equation (ODE). A relation involving functions of *only one* independent variable and one or more of the derivatives of those functions with respect to that variable.

Static network. A system where the connections between elements do not change with time and its topological structure remains fixed.

Vertex. A dot in a graph. In a contact-network model, it often represents an individual (alive, dead or in between) in a population.

Further Reading

General literature on networks and network theory

- M.E.J. Newman, A.-L. Barabási, D.J. Watts. *The Structure and Dynamics of Networks*. Princeton University Press, Princeton, NJ, 2006.
- A. Barrat, M. Barthélemy, A. Vespignani. *Dynamical Processes on Complex Networks*. Cambridge University Press, Cambridge, UK, 2008.
- M.E.J. Newman. *Networks: An Introduction*. Oxford University Press, Oxford, UK, 2010.

Contact network epidemiology

- M.E.J. Newman. Spread of epidemic disease on networks. *Phys. Rev. E* 66(1):016128, 2002.
- M.J. Keeling, K.T.D. Eames. Networks and epidemic models. *J. R. Soc. Interface*, 2, 295–307, 2005.
- L.A. Meyers. Contact network epidemiology: Bond percolation applied to infectious disease prediction and control. *Bull. Am. Math. Soc.*, 44, 63–86, 2007.
- P.-A. Noël, et al. Time evolution of epidemic disease on finite and infinite networks. *Phys. Rev. E*, 79:026101, 2009.
- A. Allard, P.-A. Noël, L.J. Dubé, B. Pourbohloul. Heterogeneous bond percolation on multitype networks with an application to epidemic dynamics. *Phys. Rev. E*, 79:036113, 2009.
- L. Hébert-Dufresne, et al. Propagation dynamics on networks featuring complex topologies. *Phys. Rev. E*, 82:036115, 2010.

Adaptive networks

- T. Gross, C.J. Dommar D'Lima, B. Blasius. Epidemic dynamics on an adaptive network. *Phys. Rev. Lett.*, 96:208701, 2006.
- T. Gross, B. Blasius. Adaptive coevolutionary networks: A review. *J. R. Soc. Interface*, 5:259–271, 2008.
- V. Marceau, et al. Adaptive networks: coevolution of disease and topology. *Phys. Rev. E*, 82:036116, 2010.

Numerical methods

- W.H. Press, S.A. Teukolsky, W.T. Vetterling, B.P. Flannery. *Numerical Recipes: The Art of Scientific Computing*, Cambridge University Press, Cambridge, UK, 2007.

ZOMBIE INFECTION WARNING SYSTEM BASED ON FUZZY DECISION-MAKING

Micael S. Couceiro, Carlos M. Figueiredo, J. Miguel A. Luz and Michael J. Delorme

Abstract

This chapter presents a biological system where fuzzy decision-making is used to produce information about imminent attacks. To model a form of biological growth, a mathematical model of a zombie infection studied in Munz et al. [1] is used. The model consists of a zombie attack using biological assumptions based on popular zombie movies. The output of the fuzzy system will provide both a linguistic and numeric quantification of the warning in order to perform an analysis of the outbreak and development of the zombie infection. Furthermore, the warning system based on fuzzy decision-making will be used to control the zombie outbreak, thus switching among quarantine, treatment and even impulsive eradication of zombies.

10.1 Introduction

FIGHTING infectious disease has been part of the human experience since the beginning of recorded history. Even with the significant advances in treating and preventing disease, humans are still susceptible to many infectious diseases. Surveillance systems are integral to modern epidemiology and important for pointing out specific infectious-disease outbreaks. These systems can be described as public health warning systems and typically use routinely collected information. Warning systems are used for hazardous natural events (e.g., hurricanes, volcano eruptions or tsunamis). In contrast, little attention has been given to the development of such versatile systems for infectious-disease epidemics [2]. The goal of a disease-warning system would be to provide information about the disease outbreak, thus widening the range of feasible response options.

Fuzzy decision-making is well placed as a warning system for infectious disease outbreaks. Traditionally, infectious disease experts attempt to include uncertainty when thinking about disease outbreaks. The strength of using fuzzy logic is that uncertainty can be included in the decision process. Furthermore, some studies have shown the applicability and efficiency of fuzzy logic in spatial and non-spatial processes [3, 4, 5, 6, 7]. In the fuzzy approach, vagueness and imprecision associated with qualitative data can be represented in a logical way, using linguistic variables and overlapping membership functions in the uncertain range. The goal of fuzzy decision methodology is to make an optimal decision within a set of constraints.

In this chapter, we implement and analyze a warning system based on fuzzy decision-making in order to prevent or control a zombie epidemic. As previous chapters have shown, zombies can quickly overwhelm a network or overtake your spatial location. Faced with a zombie outbreak, decisions have to be made quickly, often without access to full information.

The key difference between the mathematical models of a zombie outbreak [1] and other models of infectious disease is that the dead can come back to life. Other applications include allegiance to political parties, diseases with a dormant infection (e.g., Epstein–Barr virus) or a return of the plague, as seen earlier.

However, this is not the first time that decision-making systems have made contact with zombies. The game *Resident Evil 3: Nemesis* implemented a Live Selection Mode in which, at certain points in the game, it prompted the player to choose between one of two possible actions [8]. The player needed to choose wisely and choose fast, since there was only a limited amount of time to make the decision (Figure 10.1). Most live selection modes occur when zombies or the Nemesis (i.e., a master zombie) finds the player. Choosing a certain option could incapacitate the enemy. However, if no decision has been selected before the time ends, the player either incurs damage or is forced to fight the enemy. For instance, somewhere in the game, a large number of zombies ambush the player, forcing her to decide between running to the emergency exit or increasing the electricity output of the power transformer (see Figure 10.1). Running to the emergency exit causes the player to bust open a rusty door facing the master zombie Nemesis himself, whereas increasing the power output introduces the zombies to the wonders of voltage: their blood boils and their heads pop off, thus freeing the path.

In this chapter, we want to go a little bit further towards the development of an autonomous warning system based on fuzzy decision-making, allowing us to automatically determine the current danger level and act accordingly, as human analysis of this information may be compromised.

A review of previous work [1] and basic principles of fuzzy logic comprise the base of the current research. The development of various zombie outbreak warning systems are presented with multiple approaches used to account for varying

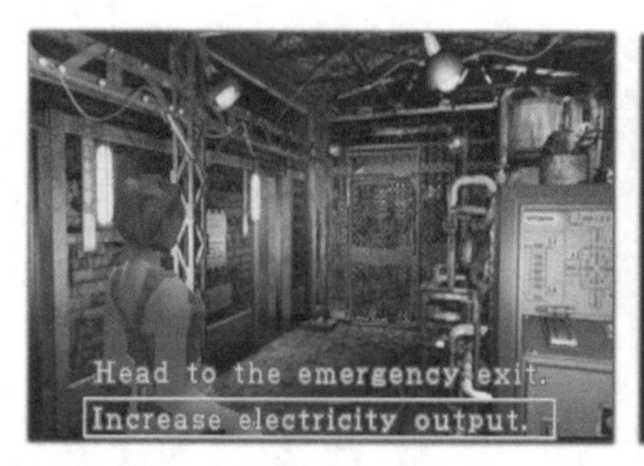

Figure 10.1: Screenshots from the *Resident Evil 3: Nemesis* game showing the first contact between decision-making systems and zombies. Images are copyright the Capcom Co. Ltd.

zombie and information scenarios. Simulations of zombie outbreaks and the different warning systems are presented and discussed.

10.2 Zombie Mathematical Models

The mathematical models of the zombie outbreak that were developed in [1] were used as a starting point for this study. They follow the same basic compartmental structure first developed by Kermack and McKendrick [9].

10.2.1 The Base Model

The first and simplest model, which we shall call the 'Base Model' (Figure 10.2), considers three different classes: Susceptible (S) people (i.e., humans), Zombies (Z) and Removed individuals (R). People move from the S to Z class through encounters with zombies. Susceptible people can also move to the removed class through natural death. Zombies can move to the removed class if they are 'defeated' when they encounter a human. In the model, it is assumed that those in the R class can be resurrected and become zombies spontaneously. This describes the zombies that appear in the film *Night of the Living Dead* [10]. Susceptible individuals are replenished at a constant rate.

10.2.2 Incorporating Latent Infection

The second model, 'Model with Latent Infection' (Figure 10.3), adds a latent class of infected individuals. This captures a common feature of zombie infection, the existence of a period of time (approximately 24 hours) after susceptible humans have been bitten and before they succumb to their wounds and become zombies [11]. The main difference is that members of the S class, once infected, are first moved to the Infected (I) class and remain there for a period of time. I-class individuals can still

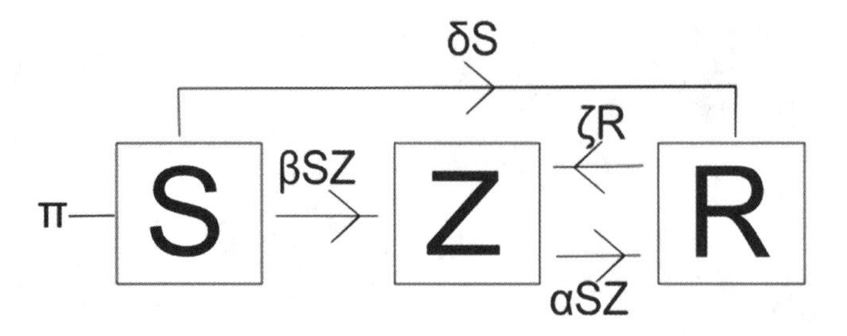

Figure 10.2: Zombie Outbreak Base Model from Munz et al. [1]. Reprinted with permission.

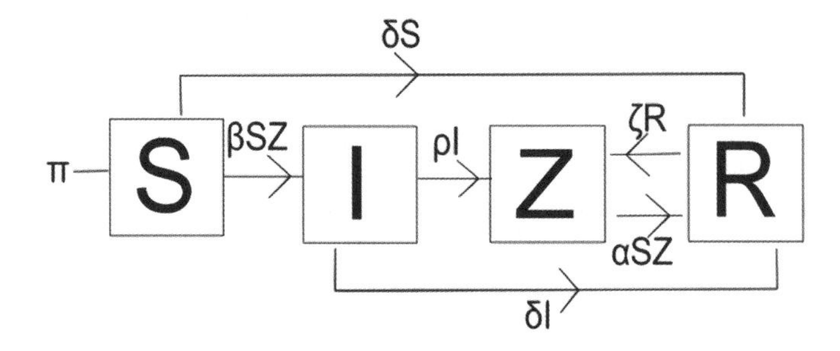

Figure 10.3: Zombie Outbreak 'Model with Latent Infection' from Munz et al. [1] Reprinted with permission.

die a 'natural' death before they become members of the Z class; otherwise, they become zombies.

10.2.3 Zombie Outbreak with Quarantine of Zombies and Infected Individuals

In order to contain the outbreak, Munz et al. decided to model the effects of partial quarantine of zombies. In the 'Model with Quarantine' (Figure 10.4), it is assumed that quarantined individuals are removed from the population and cannot infect new individuals while they remain in this condition. The Quarantine (Q) class only contains members of the Infected or Zombie populations. If quarantined individuals try to escape they are killed and enter the R class.

10.2.4 Medical Treatment for Zombies

The 'Model with Treatment' (Figure 10.5) adds a treatment that would allow the zombie individual to return to human form. However, the cure does not provide immunity. In this model, it does not matter if the zombie was created through spontaneous resurrection from the dead, or through an encounter with a zombie. Any zombie may be cured and enter the susceptible class. Since there is a treatment,

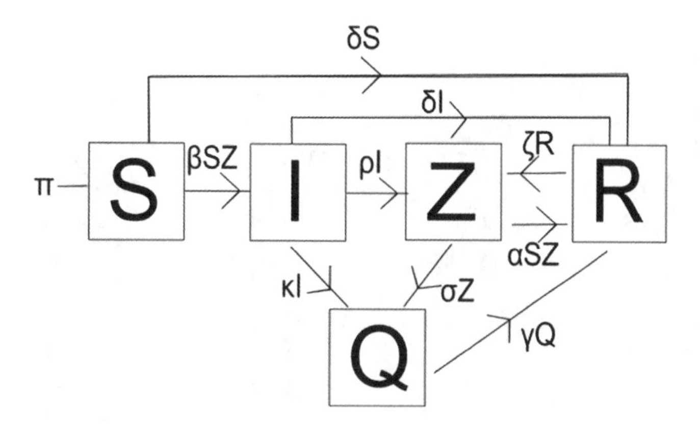

Figure 10.4: Zombie Outbreak 'Model with Quarantine' from Munz et al. [1]. Reprinted with permission.

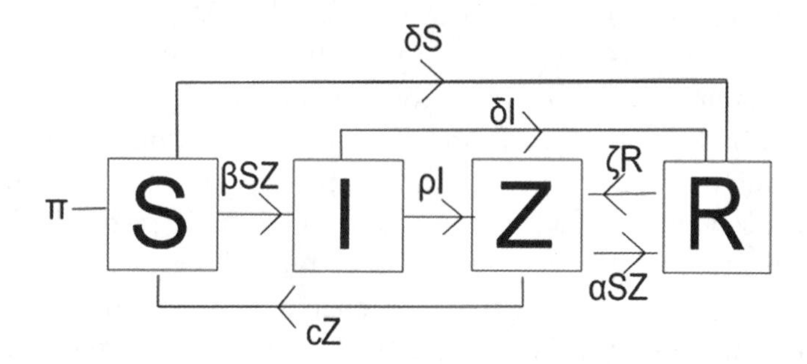

Figure 10.5: Zombie Outbreak 'Model with Treatment' from Munz et al. [1]. Reprinted with permission.

the Q class is no longer necessary. The cure will allow individuals from the Z class to return to their original human form (S class), regardless of how they became zombies in the first place.

10.3 Introduction to Fuzzy Systems

In recent years, the number and variety of applications of fuzzy logic has increased significantly [12, 13, 14]. Fuzzy logic is a practical alternative for numerous challenging control applications, since it provides a convenient method for constructing nonlinear controllers using heuristic information. Fuzzy-logic controls can be used for a variety of systems beyond a simple closed-loop system. Fuzzy logic concerns decision-making while taking into account the relative importance of precision (Figure 10.6). In order to properly apply fuzzy logic, it is important to know about its creation.

Figure 10.6: Zombie attack example illustrating a) precision and b) significance.

Fuzzy logic was first proposed by Lotfi A. Zadeh of the University of California at Berkeley in 1965 [15]. A fuzzy system is a static nonlinear mapping between its inputs and outputs (i.e., it is not a dynamic system). The term 'fuzzy' refers to the method's ability to deal with imprecise or vague information. Although the input information to the system may be imprecise, the results of fuzzy analysis are not. The field of fuzzy logic has a solid foundation of research that allows for meaningful application of its principles. Fuzzy logic describes a control system in linguistic terms in order to define the relationship between the input information and the output action instead. This linguistic definition is different from other fields of analysis that would use complex mathematical equations.

Fuzzy sets and fuzzy logic are used to quantify the meaning of linguistic variables, values and rules that are specified accordingly within the scope of the study. This specification is the translation from the rules of language to the rules of mathematics. Fuzzy sets need membership functions, mathematical equations that can take certain shapes. Reasonable functions are often linear functions, such as triangular or trapezoidal functions, because of their simplicity and efficiency when considering computational issues. Depending on the application and user, many different membership functions could be used.

In Figure 10.7, the membership function $\mu_A(x)$ describes the membership of the elements x of the base set X in the fuzzy set A, while $\mu_B(x)$ describes the membership of the elements x of the base set X in the fuzzy set B. A point x_0 may simultaneously belong to more than one fuzzy set. For example, in Figure 10.7, $\mu_A(x_0) = 0.75$ and $\mu_B(x_0) = 0.25$. Note that any value in a membership function is always in the unit interval $[0, 1]$. Membership functions are subjectively specified in an ad hoc (heuristic) manner from experience or intuition. The input or output of a fuzzy system is simply the range of values the inputs and outputs can take.

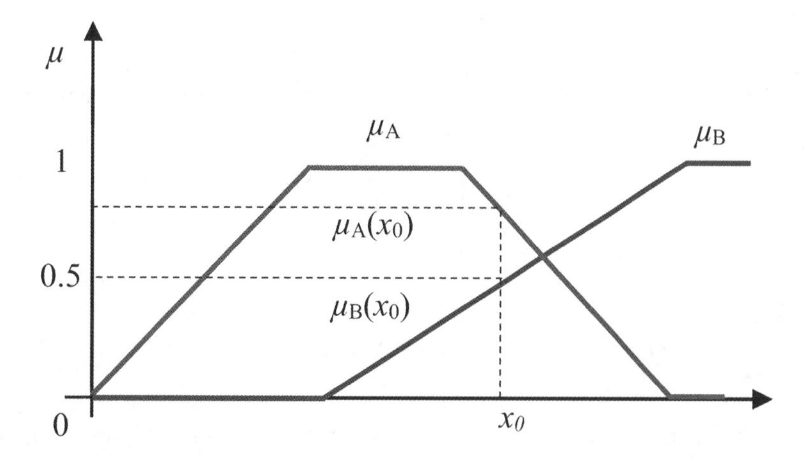

Figure 10.7: Membership grades of x_0 in the sets A and B.

In order to process the input to get the output reasoning, there are six steps involved in the creation of a rule-based fuzzy system (Figure 10.8):

1. Identify the inputs and their ranges, and name them;
2. Identify the outputs and their ranges, and name them;
3. Create the degree of the fuzzy membership function (a measure of 'vagueness') for each input and output;
4. Construct the rule base that the system will operate under;
5. Decide how the action will be executed by assigning strengths to the rules;
6. Combine the rules and defuzzify the output.

The inputs and outputs are 'crisp': they are real numbers, not fuzzy sets. Fuzzy sets are used to quantify the information in the rule base and the inference mechanism operates on fuzzy sets. Hence it is necessary to specify how the fuzzy system will convert its numeric inputs into fuzzy sets so that they can be used by the fuzzy system; this process is called fuzzification.

This process involves the use of certain defined rules. To specify rules for the rule base, a linguistic description is used. Hence linguistic expressions are needed for the inputs and outputs and the characteristics of the inputs and outputs. For instance, temperatures are not always given in °C or °F, but in linguistic terms like *cold*, *warm* or *hot*. Since linguistic values are not precise representations of the underlying quantities they are describing, linguistic rules are not precise either. They are simply abstract ideas about how to achieve good control that could mean different things to different people. However, they are at a level of abstraction that humans are often comfortable with in terms of specifying how to control a certain process.

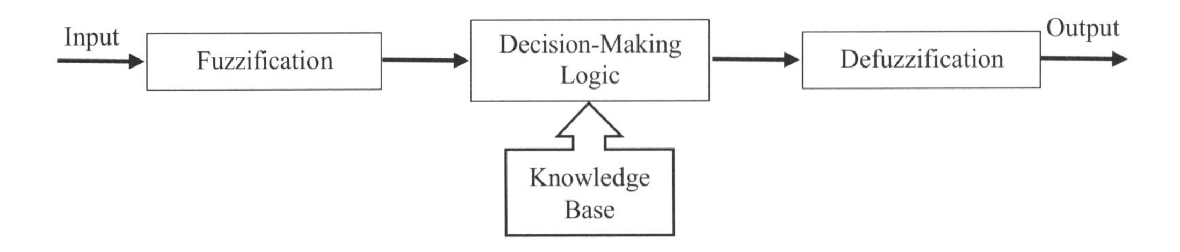

Figure 10.8: Generic representation of a fuzzy system.

The problem in applying this is that the appropriate fuzzy operator may not be known. For this reason, fuzzy logic usually uses IF-THEN rules. Rules are usually expressed in the form:

IF variable is property, THEN action

The decision-making logic (Figure 10.8) determines how the fuzzy-logic operations are performed and, together with the knowledge base, determines the outputs of each fuzzy IF-THEN rule. As in Figure 10.7, the propositional variables are replaced by fuzzy propositions, and the implication can be replaced by fuzzy union, fuzzy intersection, fuzzy complement or a combination.

There are many fuzzy implication methods that can be chosen, as in Table 10.1. Mamdani minimum inference method is one of the most often used to represent the AND connective. With this method, in order to perform fuzzification, the minimum value is selected among the available variables. For instance, for Figure 10.7, at x_0, the value 0.25 would be chosen (since $\mu_A(x_0) = 0.75$ and $\mu_B(x_0) = 0.25$).

After fuzzy reasoning, there is a linguistic output variable that needs to be translated into a crisp value. This process is called defuzzification. This is equivalent to translating the output from the fuzzy domain back into the crisp domain.

Some defuzzification methods (such as the Centre-of-Gravity method) tend to produce an integral output by considering all the elements of the resulting fuzzy set with the corresponding weights. Other methods (such as the Centre-of-Maximum method) take into account just the elements corresponding to the maximum points of the resulting membership functions. The following defuzzification methods are of practical importance and the most commonly used [16].

- The Centre-of-Area (CoA) method is often referred to as the Centre-of-Gravity (CoG) method, because it computes the centroid of the composite area representing the output.

- In the Centre-of-Maximum (CoM) method, only the peaks of the membership functions are used. The defuzzified crisp compromise value is determined by finding the place where the weights are balanced. Thus the areas of the

Table 10.1: Various operators for fuzzy logic implication.

Name	**Truth Value**
Mamdani	$\min\{\mu_A(x), \mu_B(x)\}$
Kleene–Dienes	$\max\{1 - \mu_A(x), \mu_B(x)\}$
Lukasiewicz	$\min\{1 - \mu_A(x) + \mu_B(x), 1\}$
Reichenbach	$1 - \mu_A(x) + \mu_A(x) \cdot \mu_B(x)$
Wilmott	$\max\{1 - \mu_A(x), \min\{\mu_A(x), \mu_B(x)\}\}$

membership functions play no role and only the maxima are used. The crisp output is computed as a weighted mean of the term membership maxima, weighted by the inference results.

- The Mean-of-Maximum (MoM) is used only in some cases where the CoM approach does not work. This occurs whenever the maxima of the membership functions are not unique and the question is which one of the equal choices one should take.

10.3.1 Zombie Infection Warning System

Based on the mathematical models of the previous section, we design a simple warning system to produce alarms if certain conditions occur in the diseased population; in other words, a simple warning system to quantify the zombie infection. The use of decision-making systems to produce alarm information about the spread of infectious diseases has been an interesting subject aimed at preventing possible outbreaks [17, 18].

The warning system uses as inputs $x_1(t)$, the number of infected individuals, and $x_2(t)$, the number of uninfected individuals. The output of the warning system is an indication of what type of warning condition occurred, along with the certainty that this warning condition has in fact occurred.

While uninfected individuals always correspond to the S class, the infected individuals are in several different classes according to the mathematical model in consideration. Assuming that we are considering the Base Model (Figure 10.2), the infected individuals correspond only to the Z class. If we are assuming the Model with Latent Infection (Figure 10.3) or Model with Treatment (Figure 10.5) then the infected individuals correspond to the $Z + I$ classes. Finally, if we are dealing with the Model with Quarantine (Figure 10.4), the infected individuals correspond to the $Z + I + Q$ classes.

To specify the warning type, we started by using conventional (non-fuzzy) logic and decision regions. Different alarms are described in Table 10.2.

Table 10.2: The zombie infection warning system.

Warning Level	Description
Alert	The number of infected individuals is greater than a specified amount α_1, the 'unsafe' level
Caution	The number of infected individuals is unsafe and the number of infected individuals is greater than the number of uninfected individuals by a specified amount α_2, but does not outnumber uninfected individuals by a specified amount α_3
Critical	The number of infected individuals is unsafe and the number of infected individuals outnumbers uninfected individuals by a specified amount α_3

The 'Alert' warning occurs as long as $x_1(t) > \alpha_1$, where $\alpha_1 > 0$ specifies a decision region where the warning should be given. This means that even if higher warning levels are reached, there is still a minimum Alert level warning. The Caution warning occurs if $x_1(t) > \alpha_1$ and $x_2(t)+\alpha_2 \leq x_1(t) < x_2(t)+\alpha_3$, where $\alpha_3 > \alpha_2 > 0$. Finally, the Critical warning only occurs if $x_1(t) > \alpha_1$ and $x_1(t) \geq x_2(t)+\alpha_3$. Figure 10.9 represents the decision regions related with the three different alarms.

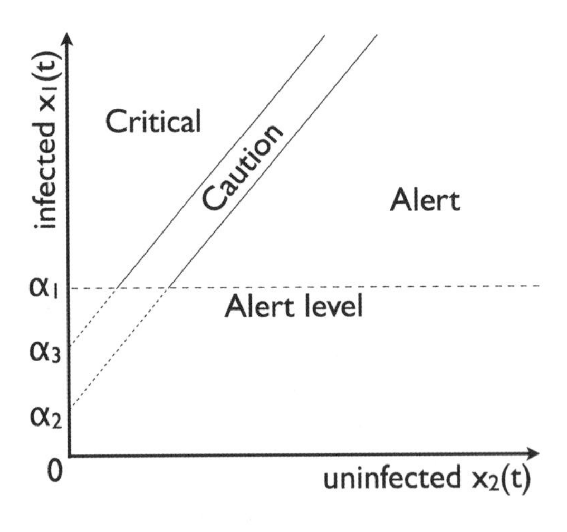

Figure 10.9: Decision regions for the zombie infection.

A system can be implemented just by using the above inequalities, taking $x_1(t)$ and $x_2(t)$ as input variables and outputting an indication of which kind of warning has occurred. Then, as the zombie mathematical model evolves, the values of $x_1(t)$ and $x_2(t)$ change and different warning conditions will hold (when none hold, there

Table 10.3: Fuzzy rules.

Rule 1	IF $x_1(t) > \alpha_1$ THEN warning is Alert
Rule 2	IF $x_1(t) > \alpha_1$ AND $x_1(t) \geq x_2(t) + \alpha_2$ AND $x_1(t) < x_2(t) + \alpha_3$ THEN warning is Caution
Rule 3	IF $x_1(t) > \alpha_1$ AND $x_1(t) \geq x_2(t) + \alpha_2$ AND $x_1(t) \geq x_2(t) + \alpha_3$ THEN warning is Critical

is no warning). This methodology could be applied to systems with a high number of warnings, if it was deemed necessary. In the next section, we will implement a fuzzy decision-making system by using fuzzy logic to improve this decision-making system.

10.3.2 Decision-Making System Using Fuzzy Logic

In the previous section, a non-fuzzy decision-making system was used with rigid criteria used for decision-making. We implement a fuzzy decision-making system by using fuzzy logic to 'soften' the decision boundaries. This system also allows for a graduated warning system that gives outbreak notifications in a more timely fashion. Here we focus on the design of a fuzzy decision-making system to provide warnings about the spread of a zombie infection.

A fuzzy decision-making system is a fuzzy controller built to make decisions about a model. It is a form of artificial—that is, non-biological—decision-making. The fuzzy decision-making systems can be used in robotics (e.g., path and mission planning or navigation), medicine (e.g., diagnostic systems or health monitoring), business (e.g., credit evaluation or stock market analysis) and many other computer decision-making systems.

The fuzzy system used to provide warnings for the spread of zombie infection consists of the three rules listed in Table 10.3.

Fuzzy logic will now be used to quantify the rules of Table 10.3. The meaning of each of the premise terms needs to be quantified and then we will be able to use the standard fuzzy-logic approach to quantify the meaning of the AND in the premises.

The premise terms (the three alarms) will be quantified with the membership functions shown in Figure 10.10, which are used to interpret the meaning of the input data and its strength (i.e., the relationship between the numbers of infected and uninfected individuals). Notice that the positioning of the membership functions in Figure 10.10B and Figure 10.10C is dependent on the value of $x_2(t)$; hence, to compute the certainty of the statement $x_1(t) \geq x_2(t) + \alpha_2$ and $x_1(t) < x_2(t) + \alpha_3$, the membership function is first placed with the given value of $x_2(t)$ and then the certainty of the statement (i.e., its membership value) is computed. The shifting of the membership functions can be avoided by simply making the two inputs to the fuzzy system $x_1(t)$ and $x_1(t) - x_2(t)$ rather than $x_1(t)$ and $x_2(t)$.

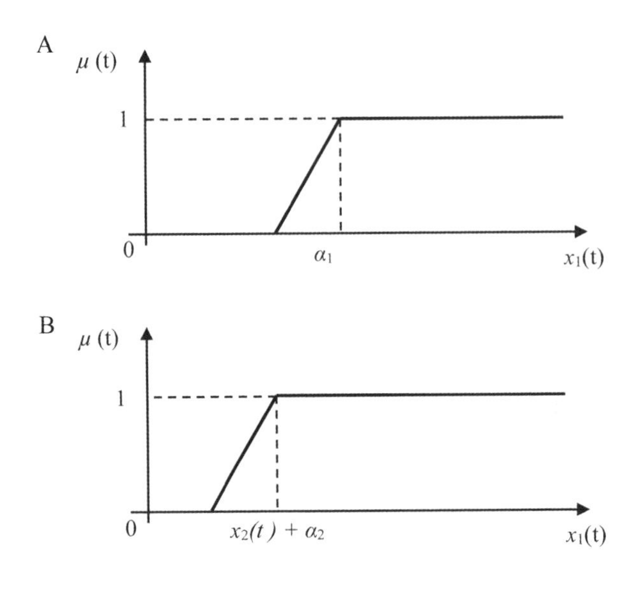

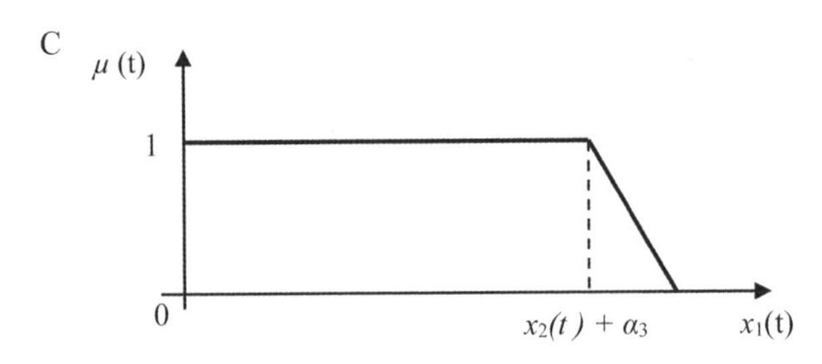

Figure 10.10: Membership functions representing the premise terms. A: $x_1(t) > \alpha_1$. B: $x_1(t) \geq x_2(t) + \alpha_2$. C: $x_1(t) < x_2(t) + \alpha_3$.

We use fuzzy logic to quantify the consequences of the three rules. To do this, suppose that we let the universe of discourse for warnings be the interval of the real line $[0, 10]$. Then we simply use the membership functions shown in Figure 10.11, where the membership function on the left represents Alert, the one in the middle represents Caution and the one on the right represents Critical. The Mamdani-Minimum will be used to quantify the premise and implication. Defuzzification will be performed using the CoG method, since it is a continuous method and one of the most frequently used in control engineering and process modelling.

10.4 Experimental Results

Considering an initial population of 500,000 individuals (who all start in the S class) we define $(\alpha_1, \alpha_2, \alpha_3) = (100{,}000, 50{,}000, 100{,}000)$. These values determine when the system produces each type of warning. The values used in the mathematical

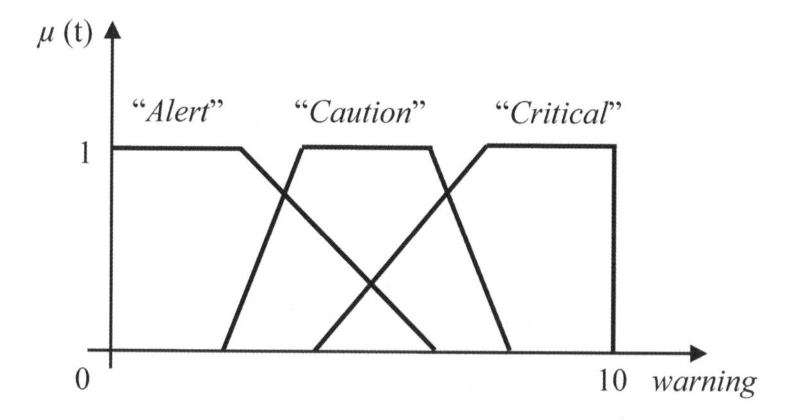

Figure 10.11: Membership functions to quantify the consequences.

models are the same as Munz et al. [1]. We will use the Model with Latent Infection since it results in a more 'realistic' possibility. The zombie model parameters are $(\alpha, \beta, \zeta, \delta) = (0.005, 0.008, 0.0001, 0.001)$. The results under the previous conditions are illustrated in Figure 10.12 where the logic warning system gives the decision boundaries.

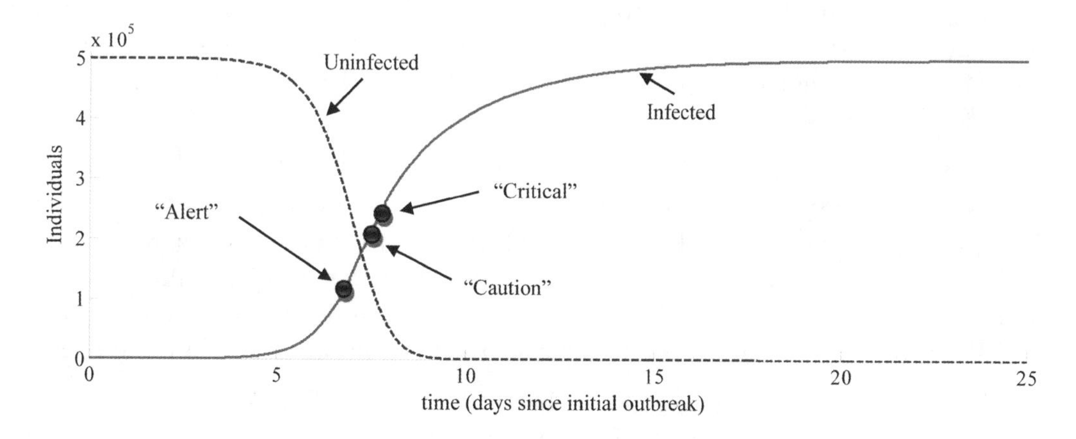

Figure 10.12: An outbreak of zombies with latent infection with the logic warning system.

Figure 10.13 shows the relevant acting zone of the warning system response (where a zombie warning system has occurred) under the supervision of the fuzzy decision-making system. The warning signal $y(t)$ (after CoG defuzzification) assumes values from 0 (which represents no warning) to 5 (which represents the worst scenario where both Alert and Critical membership functions are active). Note that the Alert membership function is always active whenever either Caution or Critical is active.

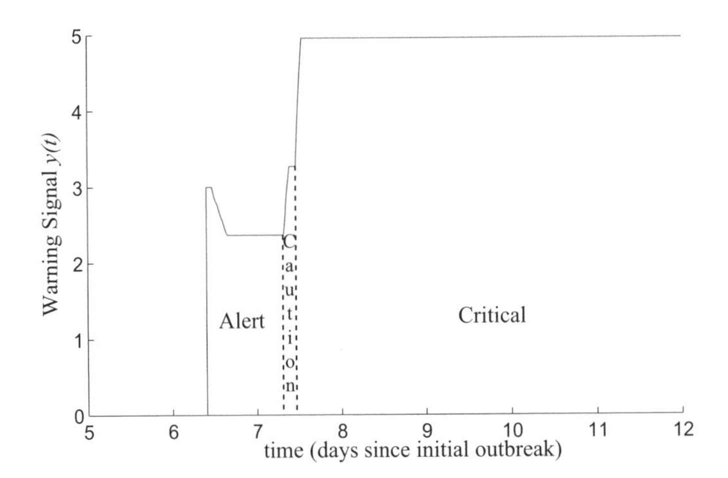

Figure 10.13: Warning signal obtained (after CoG defuzzification) in an outbreak of zombies with latent infection.

Having a numeric quantification $y(t)$ of the warning system is a significant advantage for containing the outbreak. We will use the warning system based on fuzzy decision-making to control the zombie epidemic. Depending on the warning signal $y(t)$, we will try to contain the outbreak using the mathematical models of the zombie epidemic presented earlier. We will then follow the next flow chart (Figure 10.14).

'Impulsive Eradication' [1] is the last resort for controlling the zombie population by strategically destroying it at such times as the resources permit (as suggested in Brooks [19]). We will repeatedly attack and try to destroy more zombies, resulting in an impulsive effect [20]. If this worst-case scenario happens, then we will repeatedly apply impulsive eradication every two days. Figure 10.15 shows the system response under the control of warning system following the previous flow chart (Figure 10.14).

This methodology reduces the number of infected individuals and allows stabilizing and maintaining a certain number of uninfected (in the particular case, around 75,000 individuals, which is nearly a sixth of the initial population). In Figure 10.16, it can be confirmed that all the alarms were triggered, allowing the outbreak to be contained.

Figure 10.16 shows the relevant acting zone of the warning system under the supervision of the fuzzy decision-making system. As shown, due to the epidemic control process, the warning signal passes from no warning at all to Alert in the beginning of the outbreak, quickly rising to Caution and then Critical. As the zombie outbreak is becoming contained, the warning signal changes from Critical to Caution, and only then to Alert.

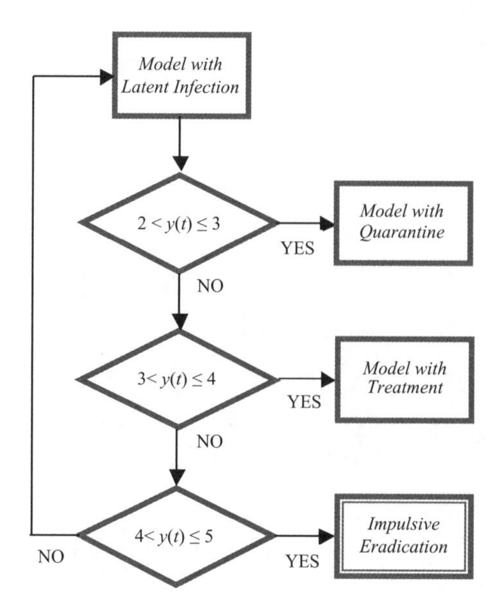

Figure 10.14: Schematic representation of the control process in order to contain the zombie outbreak. Here $y(t)$ represents the warning level.

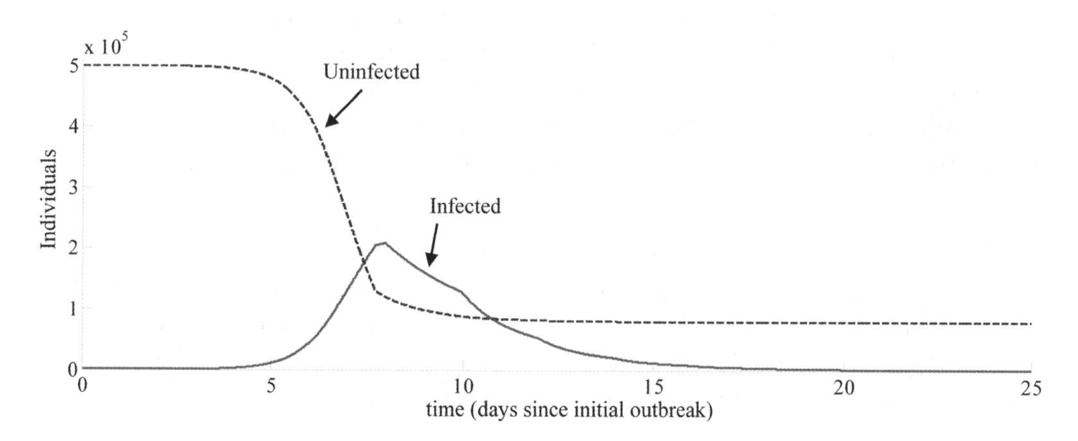

Figure 10.15: A zombie outbreak controlled by a warning system based on fuzzy decision-making.

10.5 Conclusions

The computational intelligence approach used in this work focuses on a warning system based on fuzzy decision-making. The decision-support system was designed to prevent and control a zombie epidemic. However, by running a simulation over the model of the controlled zombie outbreak, it was possible to confirm the decrease of the population from 500,000 to nearly 75,000 individuals which, although

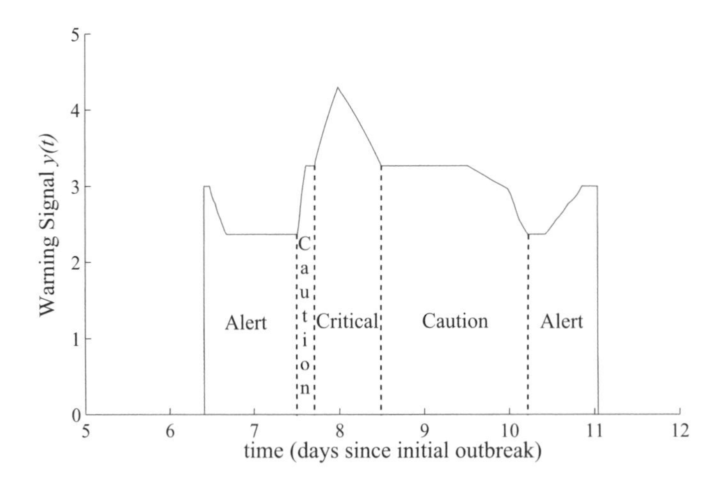

Figure 10.16: Warning signal obtained (after CoG defuzzification) in a controlled zombie outbreak.

preventing the eradication of humans in that area, obviously isn't the best-case scenario. The number of victims could be minimized by changing the membership functions as well as the specified alert levels, or by changing the approach and using the impulsive eradication sooner (in the Caution zone for instance) since neither quarantine nor treatment are viable options and would only delay the inevitable.

However, such a system would probably not be implemented because it would be hard to quantify the number of both infected and uninfected individuals in a certain area as the outbreak occurs. As the outbreak evolves, efficient control methods would be needed and those methods shouldn't rely only on the control of humans, as they are in the process of being eradicated.

In order to overcome that issue, the use of autonomous systems (robots) to quantify the number of infected individuals could be a viable option, as they would be immune to a zombie epidemic. Furthermore, robots could also be helpful assisting humans in the zombie impulsive eradication task [21]. However, we must be vigilant against robots disregarding their mandate of eradicating zombies and turning on their human masters [22]. In this scenario, instead of just fighting walking dead creatures with a hunger for human flesh and brains, the human race may have the additional challenge of defending itself from giant mechanical vacuum cleaners ready to wipe every living creature off the face of the earth. That, however, is a problem for another day.

Appendices

A Acknowledgements

This work was supported by a Ph.D. scholarship (SFRH/BD/73382/2010) granted to the first author by the Portuguese Foundation for Science and Technology (FCT). Also, this work was made possible by the support and advice of Prof. Robert J. Smith? from the University of Ottawa.

B Glossary

Defuzzification. The process of obtaining a crisp value from a fuzzy result.

Degree. A measure of 'vagueness.' The degree of a fuzzy membership, also known as truth degree, may be (mis)interpreted as a probability. Just like a probability, the degree is expressed by a real number in the interval $[0, 1]$. However, while probability is a mathematical model of 'ignorance,' fuzzy logic uses truth degrees as a mathematical model of the 'vagueness.'

Fuzzification. The process of making a crisp quantity fuzzy by converting deterministic variables into linguistic variables.

Fuzzy decision-making. A technique that allows one to obtain a decision that is optimized in the sense that some set of goals is attained, while observing a simultaneous set of fuzzy constraints.

Fuzzy inference. The process of formulating the mapping from a given input to an output using fuzzy logic. The mapping then provides a basis from which decisions can be made, or patterns discerned. Mamdani's fuzzy inference method is the most commonly seen fuzzy methodology.

Fuzzy logic. A mapping from an input space to an output space in a natural and human fashion. By resorting to fuzzy reasoning, one may build this understanding into the process. The process of fuzzy logic starts by gathering a crisp set of input data and converting it to a fuzzy set using fuzzy linguistic variables, fuzzy linguistic terms and membership functions. Afterwards, an inference is made based on a set of rules. The resulting fuzzy output is mapped to a crisp output using the membership functions, in the defuzzification step.

Linguistic variable. The input or output variable of the system whose values are words or sentences from a natural language, instead of numerical values. A linguistic variable is generally decomposed into a set of linguistic terms.

Membership function. A function that is used in the fuzzification and defuzzification steps of a fuzzy logic system, to map the non-fuzzy input values to fuzzy linguistic terms and vice versa. A membership function is used to quantify a linguistic term.

References

[1] P. Munz, I. Hudea, J. Imad, R.J. Smith?. When zombies attack!: Mathematical modelling of an outbreak of zombie infection. In *Infectious Disease Modelling Research Progress*, J.M. Tchuenche and C. Chiyaka, eds., Nova Science, Happuage, NY 2009, pp. 133–156.

[2] Committe on Climate, Ecosystems, Infectious Disease, and Human Health. In *Under the Weather: Climate, Ecosystems, and Infectious Disease*. National Academy Press, Washington, DC, 2001.

[3] P.A. Burrough. Fuzzy Mathematical Methods for Soil Survey and Land Evaluation. *J. Soil Science*, 40:477–492, 1989.

[4] A.B. McBratney, J.J. De Gruijter. A Continuum Approach to Soil Classification by Modified Fuzzy k-means with Extragrades. *J. Soil Science*, 43:159–175, 1992.

[5] E. Triantaphyllou, C.T. Lin. Development and Evaluation of Five Fuzzy Multiattribute Decision-Making Methods. *Int. J. Approx. Reasoning*, 14:281–310, 1995.

[6] A.X. Zhu. A Personal Construct-Based Knowledge Acquisition Process for Natural Resource Mapping. *Int. J. Geogr. Inf. Science*, 13(2):119–141, 1999.

[7] L.A. Andriantiatsaholiniaina. S.A.F.E: Sustainability Assessment by Fuzzy Evaluation. European Association of Environmental and Resource Economist (EAERE). Tenth Annual Conference, University of Crete, Department of Economics, Rethymnon, Greece, 30 June–2 July, 2000.

[8] Y. Kawamura. *Resident Evil 3: Nemesis*, directed by K. Aoyama. Capcom, 1999.

[9] W.O. Kermack, A.G. McKendrick. A contribution to the mathematical theory of epidemics. *Proc. R. Soc. London Ser. A*, 115:700–721, 1927.

[10] G.A. Romero, J.A. Russo. *Night of the Living Dead*, directed by G.A. Romero. The Walter Reade Organization, 1968.

[11] M. Brooks *The Zombie Survival Guide: Complete Protection from the Living Dead.* Three Rivers Press, New York, NY, 2003.

[12] P. Yang, Z. Cao, X. Dong. Fuzzy Identity Based Signature with Applications to Biometric Authentication.*Comput. Electr. Eng.*, 37:(4):532–540, 2011.

[13] L.A. Zadeh. Toward Extended Fuzzy Logic—A First Step. *Fuzzy Sets Syst.*, 160:3175–3181, 2009.

[14] M.S. Couceiro, N.M.F. Ferreira, J.A.T. Machado. Hybrid Adaptive Control of a Dragonfly Model. *Commun. Nonlinear Sci. Numer. Simul.* 17(2):893–903, 2012.

[15] L.A. Zadeh. Fuzzy Sets *Inform. Contr.*, 8:338–353, 1965.

[16] I.S. Shaw. *Fuzzy Control of Industrial Systems: Theory and Applications.* Kluwer Academic Publishers, Dordrecht, Netherlands, 1998.

[17] K.M. Passino, S. Yurkovich. *Fuzzy Control.* Addison-Wesley, Reading, UK, 1998.

[18] E.I. Papageorgiou, et al. A fuzzy cognitive map based tool for prediction of infectious diseases. *IEEE International Conference on Fuzzy Systems*, 2009.

[19] M. Brooks. *World War Z: An Oral History of the Zombie War.* Three Rivers Press, New York, NY, 2006.

[20] D.D. Bainov. P.S. Simeonov. *Systems with Impulsive Effect.* Ellis Horwood Ltd, Chichester, UK, 1989.

[21] D.H. Wilson. *How to Build a Robot Army: Tips on Defending Planet Earth Against Alien Invaders, Ninjas, and Zombies.* Bloomsbury, New York, NY, 2007.

[22] D.H. Wilson. *How to Survive a Robot Uprising: Tips on Defending Yourself Against the Coming Rebellion.* Bloomsbury, New York, NY, 2005.

IS THERE A ZOMBICIDAL MANIAC NEAR YOU? YOU'D BETTER HOPE SO!

Nick Beeton, Alex Hoare and Brody Walker

Abstract

They are pale, pasty, emaciated, soulless creations of horror. From behind sunken eyes, their slow, feeble minds struggle to process anything but the most basic of data. But despite this, they are capable of devastation and terror on an overwhelming scale. No, not supermodels; we speak of zombies. But do not worry, there will still be modelling involved. In this chapter, we use agent-based modelling to attempt to answer some questions about zombie-ism; in particular, the nature of the transmission dynamics and how to best protect communities with limited access to firearms. With access to the latest in zombie modelling technology (i.e., computer games), we were able to address these questions, while also identifying any potential, shall we say, unforeseen consequences.

11.1 Introduction

ZOMBIES. Ghouls. The undead. Whatever you call these creatures, it cannot be denied that the rise of the zombie is a problem that has troubled the imaginations of people throughout the world [1, 2, 3, 4]. Unfortunately, the great majority of data on zombie behaviour patterns exist merely as conjecture and speculation from people who have probably never even seen a zombie, let alone killed one. This has led to a growing gap between the readiness of the general population and that of certain private individuals, which needs to be rectified. Otherwise, in the case of a major zombie outbreak, large numbers of people fleeing the infected zones will simply add to the numbers of the living dead [4, 5, 6, 7].

The history of zombie outbreaks is a long and varied one. What few historical records exist cannot be properly verified [8], although tales about the hungry dead can be seen as far back as the tale of Gherib and Agib in *One Thousand and*

One Nights. Zombie outbreaks have continued, unchecked by governmental forces, commonly attributed to roving bands of locals [3, 9], with exceedingly large casualty lists. Despite this lack of evidence, over the course of the twentieth century, rumours started circulating throughout one field of mass information: Hollywood. Since 1932, with *White Zombie*, Hollywood has slowly allowed zombie movies that used real-world data to describe possible outbreak scenarios, albeit altered to avoid creating a mass panic.

The beginning of the current and most serious outbreak was first reported in Australia [10]. However, many of these first cases were among the country's politicians, and so no one noticed until it was too late. Zombie-ism spread quickly and met with very little resistance. Many have cited Australia's ban [11] of Left 4 Dead 2 [12] as evidence of zombie sympathizers preventing the population from developing much-needed zombie-hunting skills. The threat posed by this zombie epidemic is great, with reports that the undead have now reached Asia [13, 14, 15]. The situation is escalating, to the extent where entire communities are disappearing each day [1, 16, 17].

As the rest of world prepares itself for the inevitable, we find that there are still many unknown factors concerning the undead. Reports from the affected regions are imprecise and often contradictory, making preparations difficult [5]. In the name of science, we have taken it upon ourselves to gather as much information as possible about the epidemic, including considerable amounts of 'pre-emptive research' with the perusal of all known research materials. Many a long night was spent watching zombie films to garner more information for this study, although we should admit that most of the research actually took place some years before this study was ever conceived. We have also sent in a team of specially selected researchers to discover as much as we can about the nature of zombies and the biological processes involved.

Unchecked, a small zombie infestation can grow rapidly and is capable of consuming large cities in a matter of days [5]. For this reason, the most pressing concern for governments in currently uninfected regions is how to best protect their people in an efficient manner. As we have seen, zombies may not be evenly distributed in a space, may be moving about erratically and may require fast decisions, often with imperfect information. We will attempt to address these issues using an agent-based modelling (ABM) approach.

An agent-based model is a simulation composed of several entities or agents. Each entity can be given a particular behaviour and act in a specific way. This makes agent-based models ideally suited for this kind of problem, in which behaviour plays an important role. With this approach, we will explore the nature of transmission of zombie infection as compared to traditional human and wildlife diseases, and estimate the effects of introducing armed and trained zombie hunters as a mechanism for 'disease' control.

11.2 Data Collection

To collect such a large volume of data—and to do it in a cost-effective manner—we had to resort to creatures only marginally more social, hygienic and intelligent than zombies: Ph.D. students. Under the guise of an all-expenses-paid trip to an exotic location, the students were sent to gather information about the zombie horde. Interestingly, those with a large diet of horror films and violent video games have proven to be ideally suited to the task, whereas those without have met with some . . . difficulties (results soon to be published [18]). These students were able to provide data on the average running speed of a zombie, the zombification process, tensile strength of zombie skin, food preferences and several other biological characteristics unique to zombies.

Behavioural studies were carried out by social researchers, but only those conforming to the 'emo' subculture had any success in integrating with the zombie society. The researchers found that the zombies had no hierarchy and exhibited a pack-like behaviour. Unfortunately, more detailed behavioural data could not be obtained, with the zombies showing a particular disdain for answering surveys (again, results soon to be published [19]).

11.3 Agent-Based Model

We developed an agent-based model using the program Garry's Mod [20] based on the game Half-Life 2, using the Source engine [21, 22]. This allowed us to give our agents a higher level of artificial intelligence than is possible with other programs (also, we got to play a computer game and call it work).

Within the model, we created three different classes of individuals: 1) humans, 2) zombie hunters and 3) zombies (Figure 11.1). The human class represents the general populace, who are unarmed and unable to put up any resistance against a zombie attack. The hunters are equipped with a combat Acme shotgun, capable of firing six rounds before needing to reload. Shotguns have been found to be the most effective anti-zombie weapon due to their potential damage, the lack of requirement for accuracy and their design for easy use in close quarters [23, 24]. The zombies have been given characteristics based on the data collected.

Capturing and replicating behaviour is very important for this kind of model. Based on the observed behaviour of humans in the zombie infested regions, we assume that, when confronted by a zombie, humans will do one of two things: 1) run for their lives, or 2) cower in fear (Figure 11.1C). When not faced with any zombie threat, the humans will either wander randomly about the map (Figure 11.2) or stand idle, unaware of the potential chaos unfolding around them.

For the hunters, they will fire upon a zombie if within range and only reload their weapons when there are no zombies nearby. If they run out of ammo, they will attempt to run away to a safe distance before reloading. When not confronted with a zombie threat, hunters will wander randomly about the map.

Figure 11.1: A: A human surrounded by a pack of zombies moments before becoming zombified. B: The heroic zombie hunter, armed with a brand new Acme shotgun (the best shotgun around) and ready for action. C: Humans cowering in fear as zombies begin their attack, demonstrating why running is a much better survival strategy. Art assets are copyright Valve Software (Source engine) and Garry Newman (Garry's Mod).

Zombie behaviour is similar: without any humans within sensory range, the zombie will wander aimlessly, independent of other zombies around. When a human enters its range, the zombie will then proceed to chase them down. The zombies used in this model are assumed to be 'fast zombies' [2, 5] that can move as fast or faster than a sprinting human, as opposed to 'slow zombies' [4, 9, 25] that can move only at a shuffling pace. However, the zombies will always go after the nearest human. So, as in many similar scenarios, a potential survival technique is to ensure that you can run faster than the person next to you. However, note that, for simplicity, in the model presented here, all humans run at the same speed. Once a human is caught, infection is assumed to occur via a virus spread solely through fluid contact such as a bite or scratch [25].

Many of the parameters and details of this model are not available to publish, since the software used to create these scenarios is provided by Valve Software and Garry Newman. The code to replicate our experiments, however, is freely available by emailing the authors (our email addresses are given in the

Figure 11.2: The staging area, or 'map,' on which the zombies, humans and hunters interact. Art assets are copyright Valve Software (Source engine) and Garry Newman (Garry's Mod).

biographies at the end of this book). An example run of the simulation can be found at http://tinyurl.com/gmodzombiemodel

11.4 Stochastic Model

Very little research has been done to date on the transmission dynamics in zombie infection. From studies of zombie films, it is suspected that transmission is unrelated to classical studies in human and even wildlife diseases, partially due to the unwavering, breathtaking level of stupidity of the zombies—and, on that note, many humans. Perhaps more importantly, the predator–prey style of infective interaction between zombies and humans is of a unique nature, with infection converting the human to a zombie instead of merely killing the victim.

To study the potential forms of transmission for the zombie infection, we used a stochastic model for the purposes of direct comparison with our agent-based model. We used a very simple ordinary differential equation as a model to represent the very simple 'zombie-infects-human' interaction:

$$\frac{dS}{dt} = -f(S, Z)$$
$$\frac{dZ}{dt} = f(S, Z)$$

where S represents the number of susceptible humans, Z represents the number of zombies, and $f(S, Z)$ is the transmission function. To convert this system into a stochastic form for direct comparison with our 'experimental' results, we use a stochastic model with a strong mixing assumption [26]. The infection dynamics can thus be modelled as a Poisson process: a zombie conversion process occurring continuously and independently at a constant rate depending directly on the transmission function. As a result, the time between conversion events is modelled by an exponential distribution:

$$P(X \to X', \tau) = f(S, Z)e^{-f(S,Z)\tau} d\tau,$$

where X is the state (S, Z) and X' is the state $(S - 1, Z + 1)$. We simulate different traditional transmission functions using this method by repeatedly using the probability distribution above to determine the timing of each infection for each transmission function until all the humans are converted to zombies.

11.5 Transmission Dynamics

Our aim is to determine the likelihood that the agent-based model we used has an underlying transmission function related to those found in the disease-modelling literature [1]. To achieve this, we need a measure of how well our agent-based model fits a particular transmission function and, importantly, we also need to measure how well the stochastic model fits. Though the stochastic model is based on the transmission function, due to randomness it will not be a perfect fit to the data.

For the agent-based model, we made ten runs each pitting one zombie against 24, 49, 74 and 99 humans in turn, randomly placed on the map; we ran simulations until the zombies overran the human population. These simulations gave enough diverse data about $\frac{dZ}{dt}$ (calculated by using a centred first-order derivative from the discrete data), S, Z and N to make effective comparisons with each of the four functions. We performed linear regressions and collected coefficient of determination, or R^2, values.

For the purpose of direct comparison, we ran the stochastic model using the same initial conditions and sampled in exactly the same way as the agent-based model. We then calculated values $\frac{dZ}{dt}$ from the sampled data, also in the same way.

Four transmission functions are used, based on a literature review [27] and some preliminary analysis. We calculated maximum likelihood parameter estimates using nonlinear least-squares optimization with confidence intervals (CIs), then using a likelihood-ratio test. Using the obtained maximum likelihoods and the number of parameters for each model, we calculated corrected Akaike Information Criterion [28] (or AICc) values for each model. The AICc values are a measure of the relative goodness of fit of a statistical model that take into account the number of its parameters; they are used to rank candidate models in order to determine which best fit the data. The lower the AICc value, the better the fit, so the model with the lowest AICc is considered the best model. In Table 11.1, N represents $S + Z$, the total number of humans and zombies combined. The asymptotic contact function simulates a power relationship with the additional possibility of the transmission being anywhere between density-dependent ($\epsilon = 0$) and frequency-dependent ($\epsilon = 1$).

Table 11.1: Some candidate transmission functions for the spread of a zombie plague. The best AICc score is highlighted in bold.

Transmission function	**Mass-action**	**Frequency-dependent**	**Power relationship**	**Asymptotic contact**
	βSZ	$\frac{\beta SZ}{N}$	$\beta S^p Z^q$	$\frac{\beta S^p Z^q}{1-\epsilon+\epsilon N}$
β (with 95% CI)	0.000624 (0.00601–0.000646)	0.0537 (0.0519–0.0555)	0.0103 (0.00772–0.0129)	0.00962 (0.00702–0.01221)
p (with 95% CI)			0.473 (0.435–0.511)	0.499 (0.446–0.552)
q (with 95% CI)			0.762 (0.719–0.806)	0.793 (0.730–0.855)
ϵ (with 95% CI)				0.00168 (−0.00092–0.00428)
AICc	−5.333	−104.711	**−291.142**	-291.125

For the stochastic model, we performed a linear regression of the sampled dZ/dt against the transmission function $f(S, Z)$ on which the stochastic model was based, then used the R^2 of this regression as our measure of goodness-of-fit. This is a reasonable measure since the information we want to gain from the data is how closely the rate of infection for any given pairing of S and Z ($\frac{dZ}{dt}$) matches the theoretical transmission function ($f(S, Z)$).

We repeated the schedule of model runs 50,000 times for each function, giving a histogram of R^2 values. Each of the four histograms demonstrates a high correlation between the models' $\frac{dZ}{dt}$ values and their respective transmission functions (see Figure 11.3) of around 0.8 to 0.9; this is expected since the only sources of error are stochasticity and the sampling process.

If two models are identical, any measure of the output of one should lie within the distribution of the same measure in another. Our goodness-of-fit value here is such a measure; the measure is calculated in the same way for the agent-based model and the stochastic model. Hence we can now generate a statistical null hypothesis H_0 that the agent-based model gives the same goodness-of-fit value as the stochastic model, or $H_0 : \mu_{\text{ABM}} = \mu_{\text{stoc}}$.

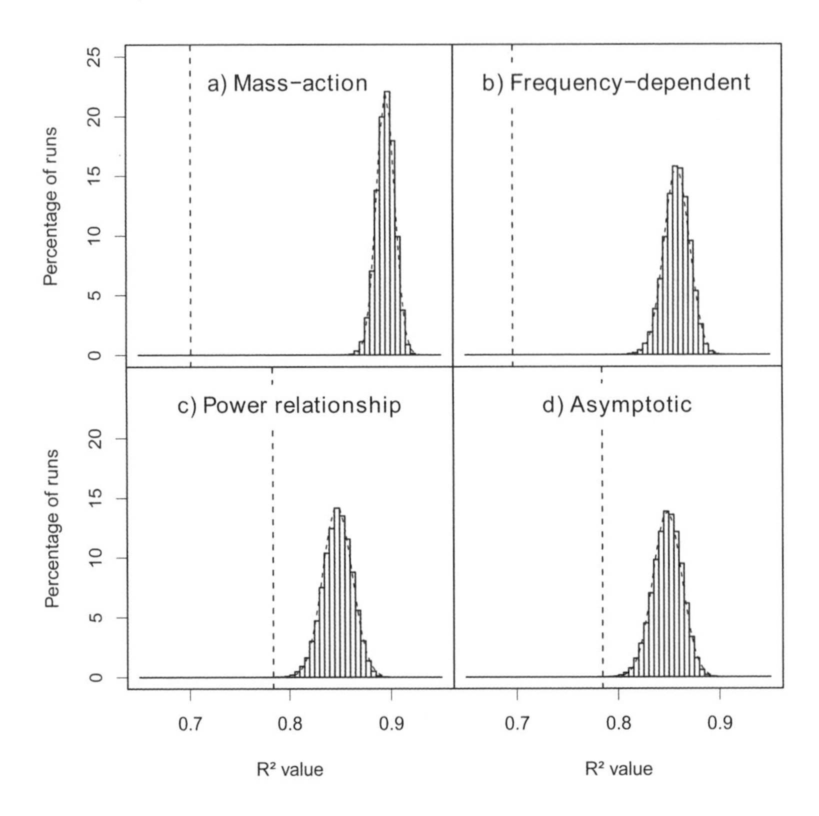

Figure 11.3: For each transmission function, a histogram of results from the stochastic model, a fitted normal curve to the histogram (dashed curve) and the value of R^2 gained from the agent-based model (vertical dashed line).

If the null hypothesis is rejected, it means that the stochastic model with a particular transmission function does not describe the data from the agent-based model, which suggests to us that the agent-based model is thus not governed by that transmission function. On the other hand, if the null hypothesis is not rejected, the stochastic model may describe this data. It is important to note that the null hypothesis not being rejected is a necessary but not a sufficient condition for accepting a particular transmission function. The goodness-of-fit value from the agent-based models against each of the four transmission functions is also described in Figure 11.3 by a vertical dotted line for the purposes of visual comparison.

From Table 11.2, we see that the normal curve underestimates the p-value of fit for the two cases where at least some of the stochastic runs were outperformed by the agent-based model, due to a left skew in the distribution. Even so, the probability of the transmission function being mass action or frequency dependent is vanishingly small. A power relationship, designed as it is for fitting less-easily-described transmission functions, provides a better estimate of transmission dynamics, and the introduction of an asymptotic contact function to it does nothing to improve its

Table 11.2: Estimates of the probability of the agent-based model being a better fit than any given run of the stochastic model using the 50,000 runs of the stochastic model and using a Gaussian estimate of the probability distribution of the R^2 value of the stochastic model.

Transmission function	p-value estimate from 50,000 runs	p-value estimate from normal curve
Mass-action	0	4.49×10^{-105}
Frequency-dependent	0	7.30×10^{-39}
Power relationship	0.00010	3.69×10^{-6}
Asymptotic contact	0.00008	3.52×10^{-6}

performance (a fact confirmed by the AICc results from Table 11.1). These results suggest that the complex interactions between humans and zombies, both spatially and dynamically, are not easily explainable by a simple random-motion-based model such as the stochastic model we have used. Therefore more complex models such as our agent-based model are useful when studying zombie epidemics.

11.6 Zombie Management

We attempted to estimate the impact of hunters on the infection process using simulations. We started each simulation using either fifty or one-hundred human agents, and we varied the number of hunters added to the population. The game map (see Figure 11.2) has an approximate area of 4000 m^2, so our two scenarios represent human densities of 12,500 and 25,000 people per km^2 respectively. As a reference point, Manila in the Philippines is the world's most densely populated city at 43,079 people per km^2 [29], so these numbers represent an urban area that is vulnerable to a zombie outbreak. We distributed the humans and hunters randomly across the map. To simulate a spontaneous outbreak, five zombies were randomly distributed among the population. The simulation was run until there were either no humans or no zombies left. This simulation was repeated ten times in each scenario. Table 11.3 summarizes the results.

As would be expected, whether we start with 50 or 100 humans, as more hunters are introduced, more scenarios occur in which some humans and hunters survive. Figure 11.4 shows more details of these results, not only the final outcome but the extent of survival of either zombies or humans and hunters.

Some noise is visible in the results from Figure 11.4, which is expected as we are only running ten simulations for each quantity of hunters. The trends, however, are quite clear.

Table 11.3: The tally (given in each row) of the final outcomes of each run of ten simulations, with varying initial numbers of hunters and humans in each row. Three results are possible, as described in the columns: the zombies take over entirely; some hunters manage to eradicate the zombies without any regular humans surviving; or both humans and hunters manage to survive.

50 Humans			
Number of hunters	Zombies win	Zombies lose	
		Hunters survive	Hunters and humans survive
1	10	0	0
2	10	0	0
3	8	2	0
4	3	5	2
5	4	1	5
6	2	1	7
7	0	0	10
10	0	0	10
25	0	0	10
100 Humans			
Number of hunters	Zombies win	Zombies lose	
		Hunters survive	Hunters and humans survive
1	10	0	0
2	10	0	0
3	10	0	0
4	10	0	0
5	6	1	3
6	2	2	6
7	2	0	8
10	0	0	10
25	0	0	10

The time taken for a zombie–human conflict to resolve itself has a maximum at around four zombie hunters for 50 humans, and six hunters for 100 humans (see Figure 11.4A). This makes sense, since if the army of hunters is much smaller than this, we see the zombies quickly take over. If it is much larger, the hunters make short work of the zombie plague.

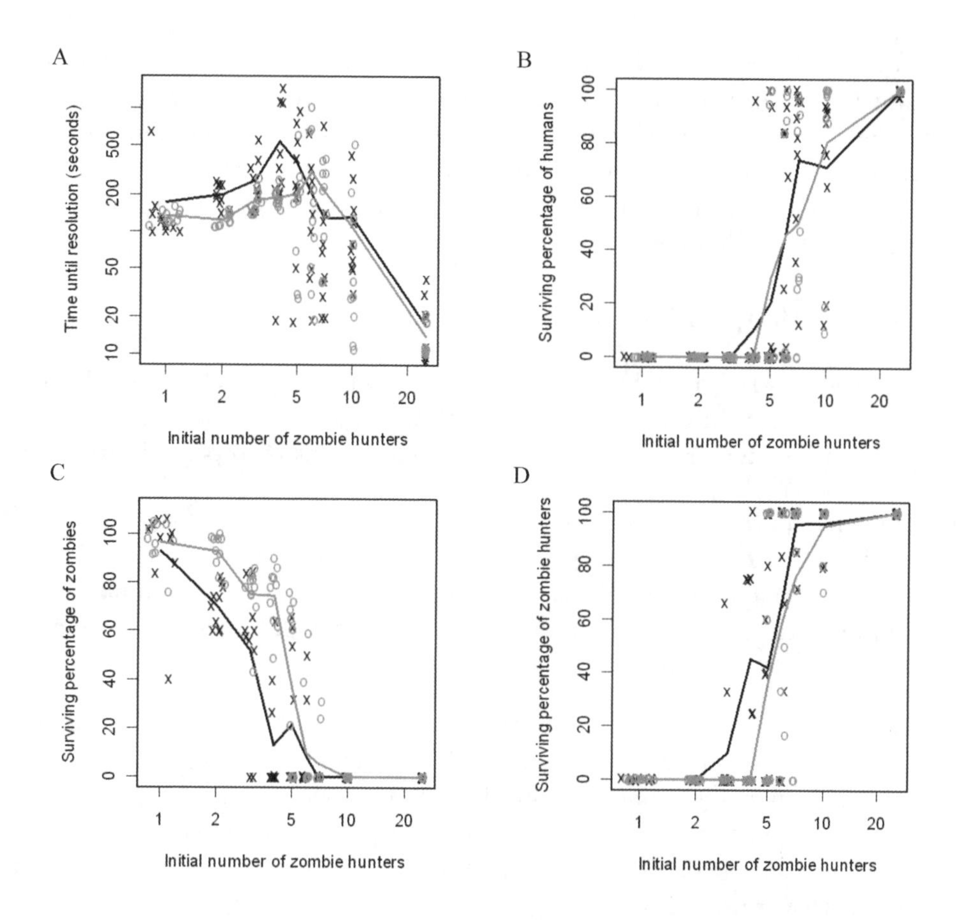

Figure 11.4: A: The effect of the number of hunters in the population on the amount of time the scenario took to resolve itself one way or the other. B: The surviving percentage of humans. C: The surviving percentage of zombies. D: The surviving percentage of hunters. For 50 humans, a black 'x' represents the result of a single simulation and a solid black line represents the mean of a set of 10 simulations. For 100 humans, a grey 'o' represents a single simulation and a grey solid line represents the mean of each set.

From Figure 11.4B and 11.4D, the proportion of surviving humans and hunters increases with the number of hunters; in either case, at least four or five hunters seem to be required for any humans to survive; with seven or more, we are very likely to keep at least some humans alive. Interestingly, how many humans that might be varies greatly from run to run, even with ten hunters. Additionally, there is a tendency for either nearly all or very few of the humans to survive (see Figure 11.4B); having about half the humans killed by a zombie encounter seems to be relatively unlikely. From observation, this is probably due to emergent behaviour in the agent-based model: either the zombies are eradicated quickly with few casualties, or they have a chance to establish themselves and cause many casualties, if not total annihilation. This hypothesis is supported by the results from the time of resolution (see Figure 11.4A), which also shows a bimodal tendency towards either rapid or slow resolution with relatively little middle ground around the five-hunter mark.

Zombie survival unsurprisingly decreases with an increasing number of hunters (see Figure 11.4C) though, as mentioned above, eradication cannot be guaranteed with any level of confidence until at least seven hunters are present for 50 humans, or around ten for 100 humans.

11.7 Observations on Behaviour

Though an agent-based model is no substitute for a live experimental trial, we can also gain some insight into zombie incursions by merely observing the simulations. A few emergent behaviours become qualitatively apparent by repeated observation:

- Zombies require a critical mass to be able to take down a hunter. Simulation runs in which each zombie begins near a hunter invariably end quickly with the hunters shooting the zombies; conversely, if zombies begin in an area free of hunters, this gives them the chance to infect enough humans to form a mob, which may then survive a hunter encounter.

- Zombies also require organization to take down a hunter. Even a large number of zombies who are wandering aimlessly a short distance apart are likely to be picked off one by one. From the point of view of a zombie, the ideal situation for encountering a hunter is while chasing humans, as they will be already clustered together to compete for a smaller number of humans, and can then swarm a hunter. Once a 'feeding frenzy' is over and there are no humans left alive in the immediate area, the zombies are immediately less threatening. This is an example of where spatial structuring of the map plays an important part.

- The initial random positioning of zombies, hunters and humans plays a vital part. In some simulations, all zombies are unlucky enough to be positioned near a hunter at the beginning and thus are destroyed without the chance for any infections. In other cases, only one zombie is well positioned to survive, but is near to enough humans to start an epidemic on its own before detection by the hunters, and eventually to wipe out the human population. The very high level of infectivity of the disease allows for this kind of scenario to be common. Indeed, in the absence of control measures, the basic reproduction number R_0 is effectively infinite due to zombies being . . . well, undead.

11.8 Discussion

We have chosen to replicate what has already proven to be the best type of prevention, a shotgun or another firearm. Our model has shown that an effective way to protect a population is to distribute trained hunters, armed with shotguns, among the population. With enough of these hunters randomly distributed throughout a community, a spontaneous zombie outbreak can be quickly quelled before gaining a foothold. Within our simulation framework, a ratio of seven hunters to 50 civilians is enough to prevent a major zombie outbreak in a sparsely populated community in each of our ten simulations. When the quantity (and hence the density) of humans is doubled, a lower ratio of ten hunters to 100 humans is sufficient to eradicate the zombie epidemic in all cases. This suggests that a similar proportion of armed humans would probably be feasible in a similar real-life setting, and that the number of hunters required is not directly proportional to the density of humans. More work would need to be done to demonstrate the nature of this relationship.

This result was unexpected; we believed that a dense population setting would accelerate the spread of zombie-ism. One probable explanation for the result is that the dense setting forces more of the population to stay within the 'protection zones' of the hunters.

It is clear that numerous important differences are likely to exist between the simulation scenario and a real-life zombie incursion; simulations are therefore unlikely to closely mirror an equivalent real-life situation. Further work would perform sensitivity analyses on all available parameters in order to gauge which are most important; at this point, ethically approved experimental trials would give useful estimates of these parameters with which to refine our modelling. However, obtaining informed consent from zombies has so far proven difficult.

According to a recent survey [30], 45 out of 178 countries already have enough civilian firearms to satisfy the 7:50 ratio that was required in our simulations. However, while the weapons may be available (if we assume that our simulations roughly represent reality), the training required to fight zombies is not. Governments will soon need to make tough decisions concerning the redistribution of firearms, and should also consider the creation of 'Zombie Survival Training Centres.' These centres may prevent instances of civilians cowering in fear and thus greatly increase their chance at survival.

Despite our best efforts, there is still no promising cure for the zombie virus. Although there has been some progress, no clinical trials have made it beyond phase III. The most effective strategy in curbing the spread of zombies is prevention. The most effective form of prevention is still in debate, with several moral questions to be addressed. New organizations such as PETZ (People for the Ethical Treatment of Zombies) believe in the "Zombies Were People Too" mantra, and push for isolation and control as the best and most morally acceptable prevention strategy [31]. This strategy has come under heavy criticism, not least from the zombies themselves. When inteviewed, they had this to say: "Braaaiiinnnsss!!!" (Contact with our interviewer was lost shortly afterwards.) Future modelling investigations that will be able to properly evaluate such a strategy are under way; however, for the time being, the authors prefer to keep their shotguns by their beds.

More research needs to be done in both simulation and collecting real-life experimental data to find out more about transmission of the zombie plague, and in turn allow for increasingly accurate simulation modelling. However, this pursuit of more accurate data can lead to some interesting ethical dilemmas. One of the benefits of modelling is that we are able to perform experiments in an environment (the computer) that has no impact on the real world. But a model is only as good as the data that it uses and the assumptions it is based upon. Some of the most interesting information provided by our brave data collection team came not from their structured experiments, but from their own survival stories.

Better information can be gained from a larger scale experiment, but organizing a zombie outbreak entirely to test our model is . . . somewhat unethical (more details outlined in [32]). Zombies are, first and foremost, the reanimated remains of deceased humans. The key consideration here is that a deceased human, to be used in a study, must first give written consent before death, and it is extremely difficult to find a sufficiently randomized group of humans who will all give consent for scientific experiments after death. Otherwise, using the bodies of the (living) dead can be considered ethically unviable.

To complicate matters further, zombies are also animate beings that react to the world around them. Their reactions show signs of some form of intelligence, on par with insects and daytime television viewers. As such, they could theoretically be considered to have a will of their own, allowing them to provide consent to experiments. However, attempts to collect consent from large numbers of zombies have proved . . . troublesome, and are therefore unlikely to be repeated (until we can get more students to replace the others). These ethical issues need to be resolved before further studies can be performed, and it must happen soon. Before it is too late for us all.

Appendices

A Acknowledgements

The authors would like to thank Acme™ Pty Ltd for funding and equipment (and the vacations to Southern France). We would also like to acknowledge the hard work and sacrifice made by our data collection team; may their brains provide troubling gastrointestinal problems for the zombies, or more likely malnutrition. A similar acknowledgement goes to friends and colleagues who have read various drafts of the manuscript.

Special acknowledgement goes to the creators of Looney Tunes whose characters and stories still provide inspiration, to Valve whose games keep the authors far too entertained and to Garry Newman for creating Garry's Mod.

Finally, we would like to thank Robert Smith? for introducing us to the world of zombie modelling, providing encouragement and advice, and allowing us to contribute this chapter.

B Glossary

Agent-based model. A model that takes into account the interactions between individuals within the scenario.

Akaike Information Criterion (corrected) (AICc). A measure of the relative goodness of fit of a statistical model, taking into account the number of its parameters: the lower the score, the better the fit.

Alternative hypothesis. The statistical hypothesis that is accepted if the null hypothesis is rejected.

Basic reproduction number (R_0). The expected number of secondary infections caused by a single infected agent during its lifetime; typically, only diseases with $R_0 > 1$ persist in the landscape.

Coefficient of determination (R^2). A measure of the proportion of variation in a measured variable that can be explained by its relationship with another measured variable.

Confidence interval (CI). The range that a parameter is expected to occupy with a certain level of confidence (usually 95%).

Emergent behaviour. Unexpected behaviour arising from inherent complexity in a system.

Exponential distribution. A probability distribution describing the time between discrete events occurring independently at a constant rate (otherwise known as a Poisson process).

Garry's Mod. A game modification written for Valve's Source engine enabling users to generate their own scenarios.

Goodness of fit. Any of a number of measures of how well a given model matches up with the measured reality.

Human. Moving source of zombie food.

Hunter. Human with a shotgun for killing zombies.

Likelihood-ratio test. A test comparing the goodness of fit of two related models by calculating the ratio of their likelihoods.

Maximum likelihood estimation. A method of estimating the parameters in a statistical model by determining which parameter values are most likely to describe a given dataset.

Nonlinear least-squares optimization. A (usually multi-dimensional) computational method of finding the minimum or maximum value of a function in a range.

Null hypothesis. A hypothesis that can be evaluated using statistical tests: a null hypothesis can be rejected based on a low probability (p-value) that it could have occurred by chance given the data, but it can never be accepted.

Ordinary differential equation (ODE). A mathematical model where quantities (here, population size or density) change continuously based on a known relationship with other quantities.

Predator–prey. A two-species system where one species preys on the other.

Probability distribution. A function that describes the probability of each possible outcome of some process.

p-value. See null hypothesis.

Regression. A statistical technique for estimating relationships between variables.

Source engine. A game engine written by Valve Software for running computer games.

Stochastic model. A model incorporating random (stochastic) processes that happen over time.

Transmission dynamics. The spread of a disease from an infectious host to a susceptible party.

Zombie. Undead creatures with a taste for brains.

References

[1] P. Munz, I. Hudea, J. Imad, R.J. Smith?. When zombies attack!: Mathematical modelling of an outbreak of zombie infection. In *Infectious Disease Modelling Research Progress*, J.M. Tchuenche and C. Chiyaka, eds., Nova Science, Happuage, NY 2009, pp. 133–156.

[2] D. O'Bannon. *The Return of the Living Dead*, directed by D. O'Bannon. Orion Pictures, 1985.

[3] A. de Ossorio. *Tombs of the Blind Dead*, directed by A. de Ossorio. Blue Underground, 1971.

[4] D. Tenney. *I Eat Your Skin*, directed by D. Tenney. Cinemation Industries, 1964.

[5] A. Garland, *28 Days Later*, directed by D.A. Boyle. 20th Century Fox, 2002.

[6] M. Brooks. *World War Z: An Oral History of the Zombie War.* Three Rivers Press, New York, NY, 2006.

[7] J. Gunn. *Dawn of the Dead*, directed by Z. Synder. Universal Pictures, 2004.

[8] M. Brooks. *The Zombie Survival Guide: Complete Protection from the Living Dead.* Three Rivers Press, New York, NY, 2003.

[9] G.A. Romero, J.A. Russo. *Night of the Living Dead*, directed by G.A. Romero. The Walter Reade Organization, 1968.

[10] S. Casey. *Zombies plan weekend invasion.* Brisbane Times, 19 October 2009. http://www.brisbanetimes.com.au/entertainment/your-brisbane/zombies-plan-weekend-invasion-20091019-h467.html

[11] A. Ramadge. Left 4 Dead 2 refused classification in Australia. news.com.au, September 17, 2009 http://www.news.com.au/technology/left-4-dead-2-refused-classification-in-australia/story-e6frfro0-1225775873481

[12] C. Faliszek. *Left 4 Dead 2*, directed by E. Johnson. Valve, 2009.

[13] S. Hung, Y. Ling. *Encounters of the Spooky Kind*, directed by S. Hung. Golden Harvest, 1980.

[14] R. Kitamura, Y. Yamaguchi. *Versus*, directed by R. Kitamura. Tokyo Shock, 2000.

[15] M. Chow, S. Man-Sing, W. Yip. *Biozombie*, directed by W. Yip. Media Blasters, 1998.

[16] J. Kalangis. *The Mad*, directed by J. Kalangis. Peach Arch, 2007.

[17] C. Watson, A.J. Rausch. *Zombiegeddon*, directed by C. Watson. Troma Films, 2003.

[18] I.M. Mort. Natural Selection Under the Evolutionary Pressure of a Zombie Horde. *Int. J. Zombie Dynamics* (in press).

[19] I.C. Dedpeeple. To mmrrr is zombie: behavioural observations of the zombie horde. *Undead Anthropol.* (in press).

[20] G. Newmann. Garry's Mod, software version 10. 2006. www.garrysmod.com

[21] Source game engine. Valve Software, 2004. www.valvesoftware.com

[22] Marc Laidlaw. *Half-Life 2*, Valve Software, 2004. www.valvesoftware.com

[23] B. Brister. *Shotgunning: The Art and the Science.* Winchester Press, New York, NY, 1976.

[24] W.E. Coyote, W. Runner. Acme™ Weapons Testing 2009. Acme™ Weapons Division, 2009.

[25] P.W.S. Anderson. *Resident Evil,* directed by P.W.S. Anderson. Constantin Film, 2002.

[26] M.A. Gibson, J. Bruck. Efficient exact stochastic simulation of chemical systems with many species and many channels. *J. Phys. Chem. A*, 104(9):1876–1889, 2000.

[27] H. McCallum, N. Barlow, J. Hone. How should pathogen transmission be modelled? *Trends Ecol. Evol.*, 16(6):295–300, 2001.

[28] K.P. Burnham, D.R. Anderson. *Model Selection and Multimodel Inference: A Practical Information-Theoretic Approach.* Springer-Verlag, Berlin, 2002.

[29] Population of cities of the Phillipines (2007 census). http://www.census.gov.ph/tags/2007-census

[30] *Small Arms Survey 2007: Guns and the City.* Graduate Institute of International Studies, Geneva, Switzerland, 2007.

[31] E. Wright, S. Pegg. *Shawn of the Dead,* directed by E. Wright. Universal Pictures, 2004.

[32] M.M.M. Brains. Killing Humans to save Humans? An analysis of the Jack Bauer approach. *Exitus Acta. Probat.* (in press).

ZOMBIES IN THE CITY: A NETLOGO MODEL

Jennifer Badham and Judy-anne Osborn

Abstract

We present a NetLogo model of an infestation of zombies in a city environment. Our model offers hope to those who may have been dejected by previous mathematical modelling of zombie attacks, which painted a bleak picture for the survival of humanity under most scenarios. We show that factors alluded to by the authors of a genre-defining chapter [1], but not incorporated into their mathematical model, are of essential importance in survival scenarios for humans in the face of zombie infiltration. This is particularly true of humans' higher speed and capacity to gain skill through experience. Thus, in our model, we observe the full triad of possibilities: zombie-win, human-win and stalemate. This demonstrates the usefulness of agent-based models to complement analytical approaches.

12.1 Introduction

THE entry of 'zombie-ism' into the disease-modelling literature came with the 2009 publication of "When Zombies Attack!: Mathematical Modelling of an Outbreak of Zombie Infection" [1]. We refer to this original chapter for a pre-modern and modern mythology of zombies. We list the characteristic of zombies given in [1] that are essential to our treatment:

- Slow-moving
- Unintelligent (this was implied by a discussion of brain damage, rather than stated)
- Prone to attacks on humans that, if successful, transform humans into zombies
- Capable of being removed from the infectious population ('killed') by decapitation.

The original model of Munz, Hudea, Imad and Smith?, henceforth referred to as MHIS, is an SIR model of infection, comprising a population made up of three classes: Susceptibles (humans), Infected (zombies) and Removed (dead bodies awaiting possible reanimation as zombies). It shares many characteristics of the basic epidemic model [2], which applies to diseases such as measles where recovery confers immunity. The MHIS model differs from the basic epidemic model and its extensions in its consideration of partial immunity. In epidemiology, partial immunity is modelled as a transition from Removed to Susceptible; but in the MHIS zombie model, reanimation is modelled as a transition from Removed to Infected (zombie). The chapter also considers the effects of latency, quarantine and treatment, which parallel epidemiological concepts and models [3], modified as required for the reanimation process.

The narrative structure of MHIS is one of ever-deepening despair for the reader who feels an affiliation with humans in the modelled human–zombie struggle. The authors of [1] begin with a basic differential equation model that includes human birth and natural death. They argue that this model leads to a 'doomsday' scenario in which an increasing supply of bodies leads to an increasing dominance of zombies in the population. They then shift their attention to short time spans in which natural birth and death are excluded. They look for equilibrium solutions of the systems of differential equations and analyze them for stability. In their basic model, they find only two equilibrium states, one consisting of all humans, which they find to be unstable, and one consisting of all zombies, which they find to be stable. Various additional assumptions are then added to the model, including a latency in the infection period, the potential for quarantine, the possibility of effective treatment and finally the use of large-scale impulsive eradication attacks on zombies by humans. Several cases are solved, both analytically and numerically, and graphed for selected parameter choices. The authors report that only the eradication attacks allow for the survival of a substantial human population.

Different models have different purposes and highlight different aspects of the real-world system [4, 5]. As seen previously, variations on the original model have produced more complex outcomes, although in each case the zombies tend to win. In this chapter, we choose to explore aspects of the zombie-attack scenario that were only briefly mentioned in [1]. In particular, the authors of [1] state that their aim is to model 'classical zombies': slow-moving and not as smart as certain contemporary varieties. However, speed and intelligence are encoded in the MHIS model only through selection of the parameter values for the probability of transmission and the probability that a human will kill a zombie during an encounter.

Using an agent-based model (ABM), we are able to explore some effects of these human attributes. In addition, we can model the spatial implications of movement, instead of being restricted to the mathematically convenient assumption of random contact based solely on population densities. In the previous chapter,

ABMs were used to investigate the inclusion of zombie hunters. Here we forego zombie hunters and concentrate instead on what we as mere mortals can do in the face of an undead uprising.

12.2 Methodology

12.2.1 Why Use ABMs and NetLogo?

Our rationale for building a model is our expectation that factors such as human speed and skill are likely to be important in the outcome of an attack by zombies upon humans. The features of ABMs that make them particularly suitable for this investigation of zombie infestation are:

- capacity for heterogeneous agents, so that the skill of each human depends on that human's experience in fighting zombies;
- representation of the environment, so that humans and zombies can be spatially aware and decide where to move using information about where other humans and zombies are located; and
- agent interaction rules, so that fights occur only when humans and zombies are close to each other and the result of the fight is determined by the characteristics of the specific humans and zombies who are fighting.

ABM uses computer programs to simulate simplified models of real-world situations, particularly in the social sciences [6]. In ABMs, agents (in this case, humans and zombies) are simulated individually and interact in some environment according to specific rules. The model comprises the computer code necessary to define the agents, environment and interaction rules.

While ABMs can be programmed from scratch, with any object-oriented language forming a natural fit, several purpose-built tool kits make model construction relatively simple (see [6] for a comparison). NetLogo [7] is our choice of modelling environment because of its ease of model construction and support for systematic experimentation through parameter variation. In addition, free access to the software and the simplicity of the graphical user interface provide secondary educational benefits, since interested readers can run the model and perform their own experiments without any previous knowledge of ABMs or NetLogo.

For all the advantages we gain by using ABMs and NetLogo, there are some costs. Because our model is based purely on simulation, we lose the capacity to formally prove anything, such as calculating where the world remains stable, as was done in [1]. We also lose the potential to mathematically analyze a continuous range of values for some parameter of interest, as is possible in many mathematical models. While we can identify likely boundaries between regions with qualitatively different behaviours, we cannot formally calculate the boundaries between such regions.[1] Also, there are some computer processing costs associated with using NetLogo as compared to a purpose-built program in a suitable object-oriented language. On the bright side, the intuition about a model gained from a NetLogo implementation may lead to more efficient implementation later in another programming language or to more sophisticated or useful analytical mathematical models.

12.2.2 Our NetLogo Model

Our NetLogo model should not be taken as a realistic representation of probable outcomes in the event of a zombie attack! Instead, we aim to explore the importance of parameters representing human 'speed,' 'skill' and 'ability to improve' in the event of zombie attack. The reader is likely to be able to see many potential adjustments and improvements to the basic model presented, and is encouraged to experiment with implementing these. The NetLogo model is available for download from [9]. NetLogo itself is available for free download from [7].

The spatial structure for our NetLogo zombie model is a 21×21 grid of 'patches' (NetLogo terminology for physical environment units), wrapped so that the top and bottom boundaries are connected (an agent moving north through the top edge will reappear on the bottom edge, directly south of its previous position); similarly, the left and right boundaries are connected. The patches can be thought of as 'street intersections' in a city laid out on a grid structure. That is, streets run north-south or east-west, and moving from one patch to another represents a distance of one city block between the intersections. The size is arbitrary, and was selected so as to be large enough for different humans and zombies to have different local situations, but small enough for reasonable processing times over thousands of simulations.

Zombies and humans are initially located at random (x, y) coordinates within the grid, in numbers determined by the parameters. Each patch may be occupied with one or more agents (which may be humans or zombies). A zombie density of '1' corresponds to an average of one zombie per city block; similarly for humans. See Figure 12.1.

Zombies are all assumed to have fixed speed of one patch (city block) per unit time. In contrast, humans are potentially faster and can move multiple blocks in the same unit time, such as when running away from zombies. This is set by the human-speed parameter.

Humans also have some skill level, interpreted as the probability (or certainty if greater than 1) of defeating a zombie in single combat. The initial-skill parameter sets an identical starting skill level for each human. The skill increment is the increase in skill gained by a human upon defeating a zombie in combat and accrues to the specific human who defeated the zombie. Thus, over time, the various humans

[1] Interested readers should consult a text such as [8] for more information about 'critical points' and 'phase transitions.'

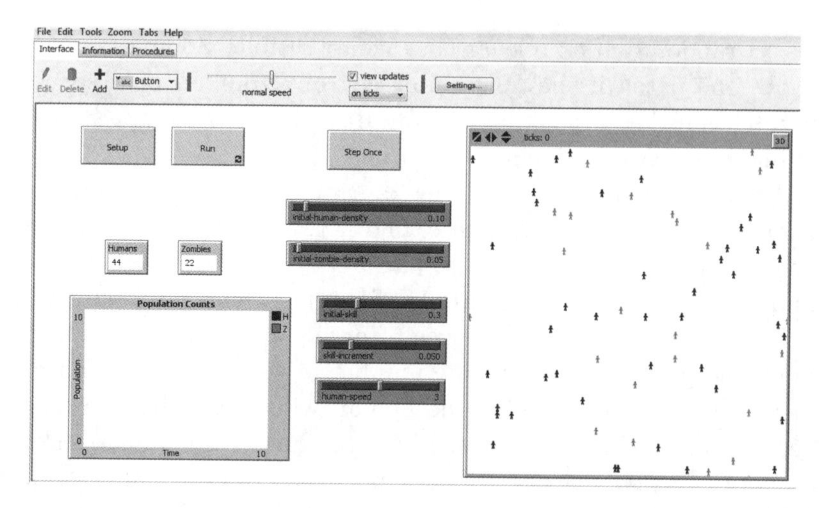

Figure 12.1: A screenshot of the NetLogo Zombie model interface, as it looks when a user has manipulated the sliders to choose parameters and pressed the Setup button to generate an initial configuration of humans and zombies, but the simulation has not yet begun.

have different skill values. There are thus five parameters of special interest that can be varied in our NetLogo model:

- Initial zombie density
- Initial human density
- Initial human skill
- Skill increment
- Human speed

The agent interaction rules for the NetLogo model fall into two groups: i) movement based on perception of the environment and ii) combat. The model operates synchronously. That is, all zombies and humans decide their movements, then they all move. After the movement phase, combat occurs between all humans and zombies that are in the same location.

Conceptually, in our model, zombies and humans can see three blocks along the street in any direction, but the city buildings block diagonal views. This is implemented in the NetLogo model as zombies and humans 'seeing' a distance of three patches, but only able to see along vertical and horizontal lines. Humans look for zombies and orient themselves away from the largest number of visible zombies (in preparation for running). Zombies look for humans and orient themselves toward the largest number of visible humans (in preparation for chasing). If there are multiple directions with the same number of 'looked for' agents, then direction is set by the first tested of the possibilities (in the order north, south, east, west).

Direction is set randomly if no relevant agents are visible. Zombies move one patch. Humans move one patch if they have not seen any zombies or a specified higher number of patches if they are running away. If humans and zombies cross the same locations while moving, they ignore each other.

Once all movement has been completed, humans and zombies that are at the same location (on the same patch) fight each other. All pairs of humans and zombies on the same location engage in one-to-one fights (for example, if there are two humans and three zombies, there are six fights). There is a skill multiplier to recognize the effects of the relative sizes of each force, calculated as the number of humans on the patch divided by the number of zombies on the patch (0.67 for the same example). For each fight, the human wins if a uniform pseudo-random number from the interval [0,1) is less than the product of the human's skill and skill multiplier. Note that a sufficiently skilled human will always win unless substantially outnumbered. If the human wins, the human's skill immediately increases by the specified skill increment and the zombie is marked for death. If the zombie wins, the human is marked for zombification. The zombie deaths and human zombifications are not performed until after all fights are completed. That is, there is a small latency for change of status from human to zombie and from zombie to removed.

These interaction rules provide all the characteristics of the classic *weak notion of agency* [10, p.116]:

- Autonomy: each agent determines its own movement;
- Social ability: each agent interacts with other agents through combat;
- Reactivity: each agent responds to its local environment by choosing a direction to move based on the presence of other agents; and
- Proactivity: each agent pursues its goals; zombies chase humans and humans evade zombies.

12.2.3 Experimental Design

The model requires five input parameters, which were systematically varied over the following ranges:

- Initial zombie density (30 values): varied from 0 to 0.05 in increments of 0.005, and from 0 to 1 in increments of 0.05;
- Initial human density (21 values): varied from 0 to 1 in increments of 0.05;
- Human speed (5 values): varied from 1 to 5 in increments of 1;
- Initial skill (11 values): varied from 0 to 1 in increments of 0.1; and
- Skill increment (5 values): varied from 0 to 0.2 in increments of 0.05.

Note that the range boundaries are arbitrary. For example, there is no reason why initial densities cannot be greater than 1. However, these ranges were chosen to cover the range of interesting behaviour, based on some preliminary simulations.

Instead of performing multiple simulations for all 173,250 parameter combinations, we ran three sets of simulations that varied some of the parameters and selected specific values for the others. For each tested parameter combination, we ran twenty-five simulations. The run stopped when either the humans or the zombies died out, or at a predetermined time step. For each run, the information recorded was final number of zombies, final number of humans and the time step at which the simulation stopped.

The key results are the number of humans and zombies in the final population, averaged over the twenty-five simulations for each parameter combination. Twenty-five might seem like a small number of simulations if the objective is to predict the outcome given certain inputs. However, over changes in parameter values in experimental sets with twenty-five simulations each, we can observe trends in general behaviour and hence multiple parameter combinations.

The first set of simulations focused on the effect of skill. For this set, we used all initial skill and skill increment values, selected speed values and both population densities. Runs were stopped at 1000 time steps if not already completed. In total, we ran 33,000 simulations—twenty-five simulations for each of 1320 parameter combinations:

- Initial zombie density (four values): 0.005, 0.15, 0.5 and 0.85;
- Initial human density (three values): 0.15, 0.5 and 0.85;
- Human speed (two values): 1 and 3;
- Initial skill (eleven values): varied from 0 to 1 by 0.1; and
- Skill increment (five values): varied from 0 to 0.2 by 0.05.

The second set of simulations focused on the effects of population densities and speed. For this set (run in four subsets), we used all speed values, all initial human densities and a full range of initial zombie densities, with selected values for initial skill and skill increment. If not already completed, runs were stopped at 250 time steps for speed 1 and no skill, at 1000 time steps for speed 1 with skill and at 2500 time steps for runs with higher human speeds. In total, we ran 110,250 simulations—twenty-five simulations for each of 4410 parameter combinations:

- Initial zombie density (twenty-one values): varied from 0 to 1 by 0.05;
- Initial human density (twenty-one values): varied from 0 to 1 by 0.05;
- Human speed (five values): varied from 1 to 5 by 1; and

- Initial skill and skill increment (two values): either 0 and 0 (no skill) or 0.3 and 0.05.

The third set of simulations focused on the effects of low initial zombie population density, to provide a finer resolution in an area of varying behaviour. This set is similar to the second set, except that a different range is used for the zombie density and only selected human speeds are run. If not already completed, runs were stopped at 2500 time steps. In total, we ran 23,100 simulations—twenty-five simulations for each of 924 parameter combinations:

- Initial zombie density (eleven values): varied from 0 to 0.05 by 0.005;
- Initial human density (twenty-one values): varied from 0 to 1 by 0.05;
- Human speed (two values): 1 and 3; and
- Initial skill and skill increment (two values): either 0 (no skill) or 0.3 and 0.05.

12.3 Generic Behaviours Exhibited by the Model

Here we next describe the kinds of behaviours that unfold in the world we have created, under various choices of parameters. It turns out that we find instances of each of the following three generic possibilities:

- Zombies win;
- Humans win;
- Stalemate.

We give examples of parameter sets for each of these behaviours.

12.3.1 Zombies Win

In the 'zombies win' scenario, humans become extinct. Figure 12.2A, showing the population over time, is typical. The graph shows the grey line (representing zombie population) climbing and crossing the black line (representing human population), which drops swiftly to zero. Our observation is that when zombies win, they win quickly. Our interpretation is that in these scenarios, the humans' zombie-fighting skill stays low across the human population for the duration of the run.

In this particular run, the initial population was 66 agents in total. The final population was 52 zombies. Thus we see that humans collectively only killed fourteen zombies between them; evidently not enough for any individual human to become sufficiently skilled to survive subsequent attacks.

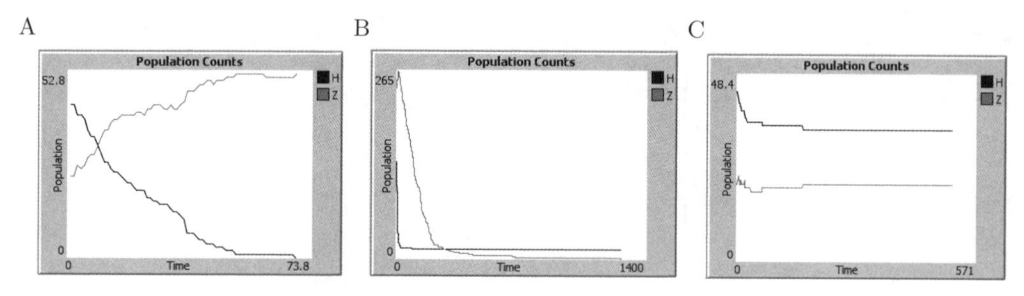

Figure 12.2: Three possible outcomes. A: Zombies win. Parameters of this run: initial human density = 0.10, initial zombie density = 0.05, initial human skill = 0.3, skill increment = 0.05, human speed = 3. B: Humans win. Parameters of this run: initial human density = 0.40, initial zombie density = 0.50, initial human skill = 0.3, skill increment = 0.05, human speed = 3. C: Stalemate. Parameters of this run: initial human density = 0.10, initial zombie density = 0.05, initial human skill = 0.3, skill increment = 0.05, human speed = 1.

12.3.2 Humans Win

In the 'humans win' scenario, zombies become extinct. Figure 12.2B, showing population over time, is typical. Our interpretation is that initially, zombies making contact with humans tend to zombify them, so that the zombie population (shown in grey) grows, while the human population (shown in black) drops dramatically. However, after a while, the remaining humans have become very skilled, and humans start to win their fights with zombies. At this point, the black line representing the human population stabilizes, while the grey line representing the zombie population plummets, eventually dropping below the black line to zero.

In this particular run, fourteen humans survived in the end. Humans as a group win, but it takes a long time to extinguish the last remaining zombies, since humans never chase zombies, and must bump into them by accident to kill them.

12.3.3 Stalemate

In a 'stalemate' situation, both zombies and humans survive: we have coexistence. In some ways, this is the most interesting phenomenon. Humans and zombies survive indefinitely because they arrange themselves into 'packs' of 'chasers' and 'chased.'

We illustrate, in Figure 12.3, the spatial configuration of zombies and humans at various time steps (or 'ticks') throughout the simulation. Their population counts are graphed in Figure 12.2C.

Consider the leftmost frame of Figure 12.3, corresponding to 511 ticks. The bottom pack shows a group of zombies (grey) chasing a group of humans (black) south. Because the boundaries wrap in our model, this pack disappears off the bottom edge of the screen and reappears at the top (middle frame), and continues

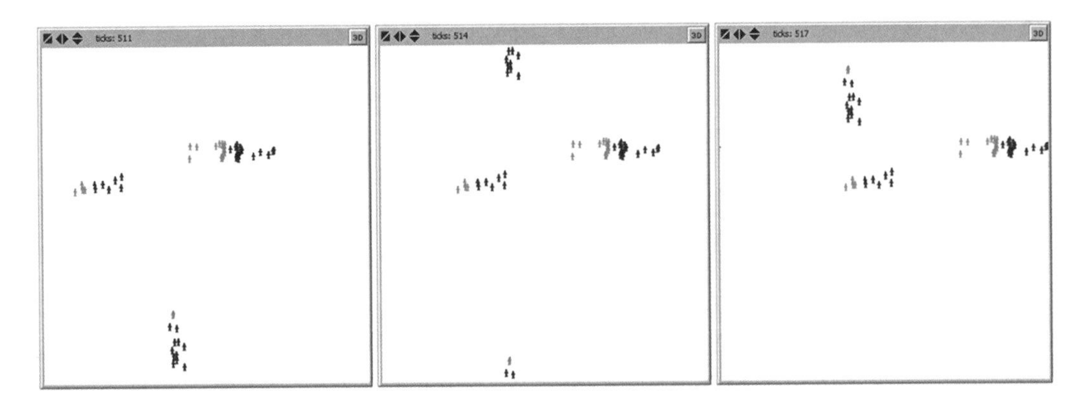

Figure 12.3: Example of chasing groups. Parameters of this run are as in Figure 12.2C. The zombies and humans are occurring in 'chasing packs.' Snapshots of spatial configurations at times: 511 ticks, 514 ticks and 517 ticks are illustrated.

to repeat this behaviour. The other two packs are moving east, and will disappear off the right of the screen and reappear at the left, ad infinitum.[2]

12.4 Effects of Speed and Skill

Speed and skill of humans turn out to be important in determining whether zombies win, humans win or stalemate occurs, for given parameter choices. To see this, we now systematically vary input parameters over the experimental ranges. We study how variation in initial human and zombie densities modifies the effects of speed and skill. The randomness in the model is addressed by averaging over multiple runs. The results are graphed using shading. Pale grey indicates a region of parameter space in which zombies dominate the final population, black indicates a region of parameter space in which humans dominate the final population, and there is continuous variation in between. The grey-scale is with respect to the proportion of zombies in the final population, not the initial population, so that a black area could correspond to a single human (and no zombies) surviving.

12.4.1 Humans Slow and Unskilled

We begin by considering the simplest case, in which humans have speed identical to that of zombies and no skill. This is reminiscent of the situation described by the basic MHIS model, except in this case humans have no chance of defeating a zombie in combat, whereas in MHIS's model, a human had a fixed chance of defeating a

[2]The phenomenon of cellular automata that reproduce themselves as translations moving across the screen is well known [11]. In that context, the reproducing configurations are known as 'spaceships.' Yes, really.

zombie opponent. The small world-size of our NetLogo model—and hence limited population—restricts our capacity to emulate extremely low probabilities.

The chart in Figure 12.4A looks entirely pale, except for a vertical band of black at initial zombie density equal to zero. An interpretation is that when no zombies are initially present, the human population remains intact, but otherwise the humans that interact with them always get zombified, hence zombies win overall.

A subtle feature of the chart in Figure 12.4A reveals a more complicated story: very close to the black band there are some patches of medium grey, indicating that some simulations with low initial zombie densities end without complete zombie domination. We investigate further by considering zombie densities over a smaller range: 0.000–0.050, in much smaller increments of 0.005. The results are shown in the horizontally stretched-out view of the left-most twentieth of Figure 12.4A, pictured in Figure 12.4B.

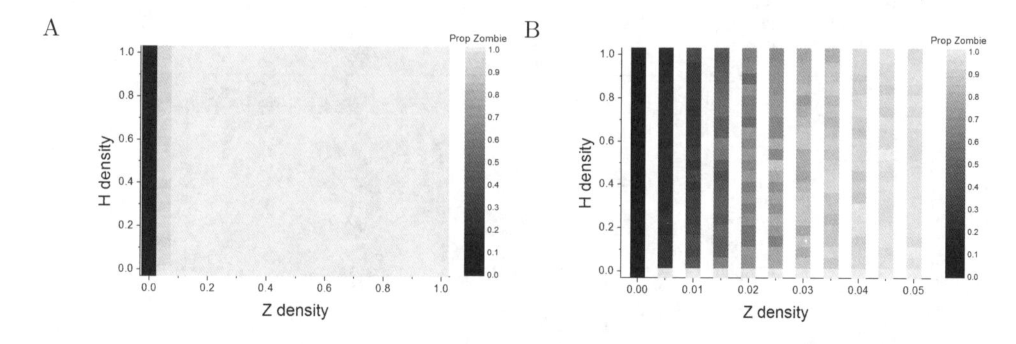

Figure 12.4: Proportions of zombies and humans in final populations under varying initial zombie (Z) and human (H) densities, subject to initial human skill = 0, skill increment = 0.05, human speed = 1. Initial zombie densities were simulated in steps of 0.05. A: Zombies on a scale of 0–1.0. B: Zombies on a scale of 0–0.05.

Now we see a range of shades. It appears that under very low initial zombie densities, some humans are surviving despite not being able to kill zombies. There could be two reasons for this phenomenon. It might be a rounding error resulting from truncating the runs after 2500 steps: zombies simply haven't had time to zombify all the humans. Or humans could be surviving some other way. In the next subsection, we address this issue.

The importance of the spatial structure of the model

Spatial structure, spatial awareness and direction-setting intelligence turn out to be crucial to human survival in a low initial-zombie-density situation in our model, even in the absence of speed or skill. In the MHIS model, a population of humans and zombies in which zombies are sparse will continue to interact with some probability under the mass-action assumption and hence eventually run into each other.

By contrast in our NetLogo model, humans are smart enough to run away. Zombies chase the humans, but if the humans are lucky enough in the original spatial arrangement, they will be able to flee forever.

We looked at individual runs of the NetLogo model to see if we could observe the phenomenon of human survival through early formation of 'chase packs.' We found what we were looking for: an example is pictured in Figure 12.5. The phenomenon occurs as a consequence of the details of the behaviour rules for the agents. Each agent can see a distance of three patches in any direction. Zombies move towards the largest number of humans. Humans move away from largest number of zombies. Because humans and zombies are moving at the same speed, this leads to groups of zombies chasing groups of humans, in perpetuity, provided that all agents have formed clusters.

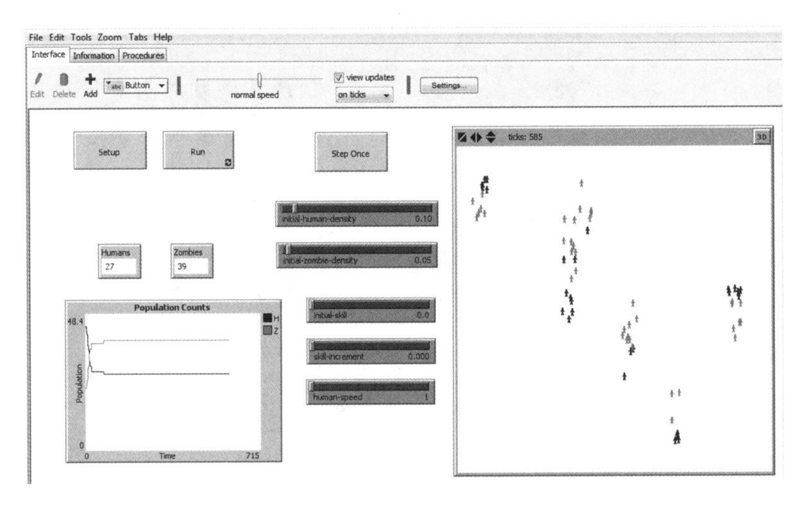

Figure 12.5: A parameter set and run such that zombies and humans arrange themselves into 'chase groups'; hence large numbers of humans survive without ever engaging in combat.

We conclude that spatial configuration and spatial awareness, not accounted for in the MHIS model, can be critical to human survival under our set of assumptions. Our 'chasing stalemate' occurs with parameters that are of the order of the parameters in MHIS. It would be interesting to further develop the NetLogo model to take more realism into account. For instance, we would expect real humans to eventually get tired and need to rest or sleep. If these features were added, would zombies always win even when they had a very low initial densities as predicted by MHIS's original model? The appeal of an agent-based model for experimentation is demonstrated here. It would be possible to model a 'sleep cycle' for humans, as well as adaptations such as 'human cooperation.' It is not feasible to incorporate this level of complexity into an OED-based model such as that used in [1].

12.4.2 Humans Fast and Unskilled

Ironically, speed does not help unskilled humans in our model. Our fast humans, in running away from one group of zombies, tend to run into another: out of the frying pan, into the fire.

We plotted final population density charts similar to that in Figure 12.4A (speed = 1) for each of the faster human speeds '2,' '3,' '4' and '5.' The charts were all similar, with a vertical band of black at initial-zombie-density = 0, and otherwise pale grey. However there was a qualitative difference between the speed = 1 case and the collective speed $\in \{2, 3, 4, 5\}$ cases.

The qualitative difference between the speed = 1 case and the 'fast' case was the absence of stalemate, perpetual chasing loops. We observed no analogue of the grey-banded Figure 12.4B for fast unskilled humans given very low initial zombie densities. All the individual runs that we observed in this case terminated with zombies winning.

Our interpretation of the reason that speed is a hindrance rather than a help to fast unskilled humans in our model lies in the detailed behaviour rules for agents. Consider the representative case of speed = 3. Suppose a zombie is within sight range of a human. Then, unless there are more humans in sight range in some other direction, the zombie will step towards the human. The human will also see the zombie, and unless there are more zombies in sight range in some other direction, then the human will run away, moving three patches at once. Now the distance between the zombie and human has increased by 2. The two agents may still be in sight range, but after another iteration of the zombie chasing and the human running away, the human and the zombie will be out of sight range each other. Hence all agents will resume moving in random directions at speed 1 until they sight other agents. As a result, 'chase packs' find it difficult to form and soon break up if they do form. The unfortunate humans are more likely to run into other zombies by accident, and thus meet their demise.

There is a circumstance in which 'chase packs' could conceivably form and be maintained for fast humans and slow zombies, which we call 'oscillation.'[3] We have not observed oscillation in the unskilled human case, so we will delay discussion of it to Subsection 12.4.4 below, in which this curious form of chasing pattern emerges.

12.4.3 Humans Skilled and Slow

Non-zero values of skill and skill increment make a big change to the structure of final population dominance. A chart of final zombie density is plotted in Figure 12.6A for an interesting selection of parameters: initial skill = 0.3, skill increment = 0.05. We chose these parameters because they are low in the range of values of skill and skill increment and yet yield results quite different from the no-skill scenario. We further discuss this observation below.

---

[3]This is consistent with terminology for similar behaviour in cellular automata [11].

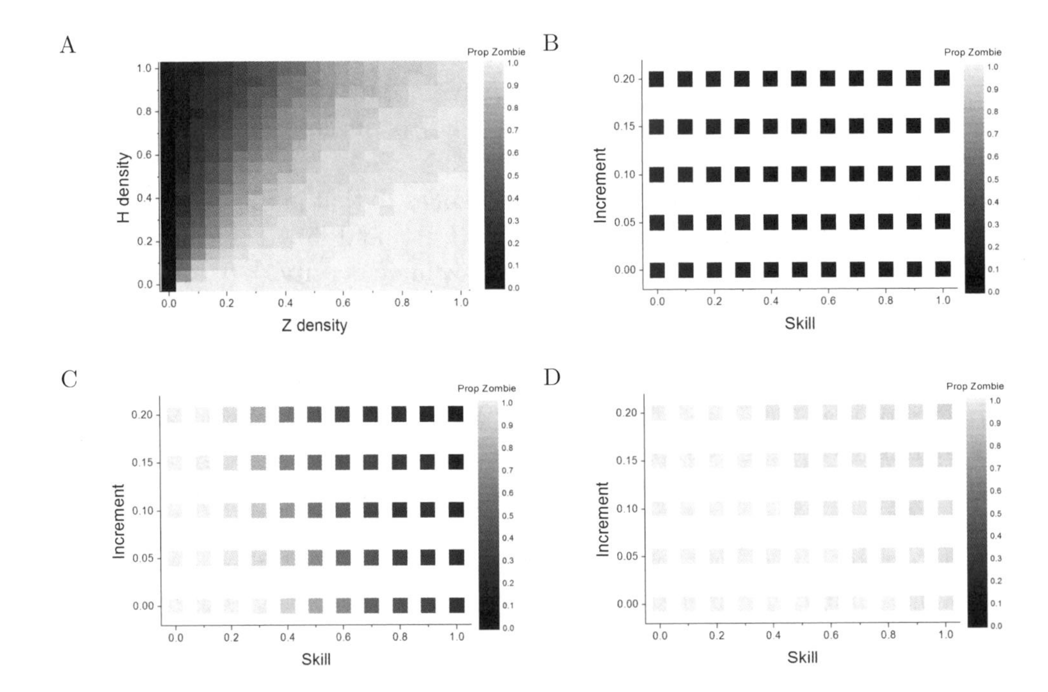

Figure 12.6: Final proportions of zombies under varying skill and skill-increments values. A: Initial human skill = 0.3, skill increment = 0.05, human speed = 1. B: Initial zombie density of 0.005 and moderately low initial human density of 0.15, subject to speed = 1. C: Initial human and zombies densities both equal to 0.5, subject to speed = 1. D: Initial human density of 0.15 and initial zombie density of 0.85, subject to speed = 1.

We observe that, in Figure 12.6A, there is a series of gradient lines through the origin, corresponding to shifting from black in the upper left to pale grey in the lower right of the figure. Thus, in the lower right of the diagram, where zombies outnumber humans initially, zombies tend to win outright. In the upper left of the diagram, zombies no longer win outright. Instead we have chasing behaviour (distinguished from the other possibility of a mix of human and zombie wins by observation of the underlying data), except of course where there are no zombies to start with and the original human population remains intact.

We now present three charts (Figures 12.6B–12.6D) in which initial skill and skill increment are varied, for selected fixed initial zombie and human densities. Our three sets of initial densities are chosen so that (B) zombies are very sparse, (C) humans and zombies are in equal proportions, and (D) zombies dominate. As usual, our principal interest is in which group wins, or whether we have stalemate; thus the shading in the plots is designed to emphasize the final proportion of zombies present, averaged over twenty-five simulations.

The shading in Figure 12.6B is strikingly dark. A very low initial density of zombies tends to mean a human win or at worst a stalemate. Also, the higher the human skill, the more likely that humans win, as indicated by dark shading. We observe that increased skill increment does not help very much (as indicated by the top of the chart not being particularly darker than the bottom), probably because an individual human does not meet many zombies upon which to practise their skill in this sparse-zombie environment.

Most epidemics start with a very small proportion of the population infected, so Figure 12.6B may be quite realistic in this sense. The scenario of this chart is therefore comparable to the original MHIS model, since MHIS also assumed an initial low proportion of zombies (in keeping with the spirit of a zombie infestation or attack on an existing human population). However, our results are more optimistic than MHIS, with spatial properties of the model helping even the unskilled humans, as explained above in Subsection 12.4.2, and non-trivial initial skill level helping even more. In our model, very highly skilled individual humans defeat zombies in all cases except possibly when badly outnumbered. An initially highly skilled human population defeats a sparse infiltration of zombies decisively. This possibility never arises in the MHIS model. Indeed, where the final conclusion of [1] is that only 'impulsive eradication' (which sounds to us rather like weapons of mass destruction) is likely to be effective in dealing with a zombie threat, in our model humans would achieve effectiveness in eradicating zombies by each individually learning sword skills.

Suppose that the world woke up one morning, and every second person was a zombie. This 50:50 scenario is out of the range typically considered in epidemiology. We indicate our model's behaviour in Figure 12.6C. We see a range of shades from pale grey where skill is low, up to a dark grey where both skill and skill increment are highest. We also observe fairly steeply diagonal (top left to bottom right) lines separating darker from lighter shades. We interpret this as meaning that *skill increment* now helps humans, as there are enough zombies for humans to get practice in fighting them. Observe for instance that an initial skill of 0.4 with increment of 0.2 is similar in shade to an initial skill of 0.6 with increment of 0.0.

For low initial human skill, zombies win decisively. Once initial skill level exceeds about 0.3, humans stand a fighting chance, particularly if they are quick learners. For initial skill greater than about 0.6, the final population is often composed of more than half humans (being chased). When initial skill and skill increment is highest, outright human wins are quite often observed in individual simulation runs.

Suppose zombies initially outnumber humans by almost a 6:1 ratio. This circumstance is pictured in Figure 12.6D. Not surprisingly, the situation looks bad for humanity. The chart is entirely pale grey, indicating zombie dominance for all skill and skill-increment ranges. The only faint sign of hope is the slightly darker grey in the top right corner. This indicates that perpetual chase scenarios can occur allowing some humans to survive even when strongly outnumbered, for humans who are initially very highly skilled.

Summarizing Figures 12.6B–12.6D, a very high level of skill always helps humans, but this effect is moderated strongly by initial population densities. Very high skill will generate clear human victory for low initial zombie density, and will allow humans to survive in stalemate for high initial zombie density.

When zombie densities are low to middling, even low to moderate initial skill significantly shifts outcomes in favour of humans. When initial zombie densities are low, initial human skill matters more than skill increment, because human fighters have few opportunities to practise. When initial zombie densities are moderate to high, skill increment plays a more important role because there is more chance that some humans will meet and kill several zombies, thus becoming highly skilled and less vulnerable themselves.

Note that in selecting Figures 12.6B–12.6D, we are trying to capture the key outcomes in a parameter space of four dimensions: skill, skill increment, initial zombie density and initial human density. More charts are available on our website [9].

12.4.4 Humans Faster Than Zombies and Skilled

Skill and speed interact in our model to produce pictures of parameter space that are different from any that we have seen before, with qualitatively different regions of zombie dominance. In the presence of skill, it turns out that speeds in the 'fast' set $\{2, 3, 4, 5\}$ produce collectively similar behaviours, which are different from those produced by unitary speed. In many but not all respects, speed '3' is typical of the high speeds and we will use it as a representative example.

A chart of final zombie density is plotted in Figure 12.7 for the parameters: initial skill $= 0.3$, skill increment $= 0.05$ and human speed $= 3$. This should be compared with Figure 12.6A, which uses the same parameters except for unitary speed. Unlike Figure 12.6A, Figure 12.7 shows a large swath of black in a curved region abutting the right-hand side of the upper half of the chart. This is a substantial region of human wins. The lower half of the chart is largely pale, indicating zombie dominance, with a curved rough border of medium greys.

We interpret the general shape of the boundary in Figure 12.7, which turns out to be similar for all 'fast' speeds, as follows:

i. The vertical line of black at initial-zombie-density $= 0$ is because there are no zombies, so that all original humans survive.

ii. The narrow stretch of medium greys in the upper left of the chart arises because when there are few initial zombies, humans have few chances to become skilled. These unskilled humans, while running away from some zombies, often accidentally run into some other zombies and get killed.

iii. The large black patch in the upper half of the chart arises because when there are sufficiently many humans, even though many may die, some of them survive a large number of fights and become highly skilled, so they survive to kill off the remaining zombies.

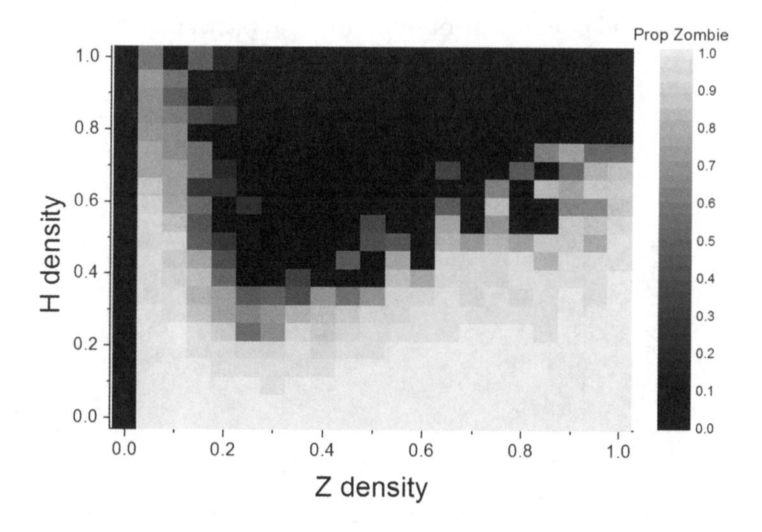

Figure 12.7: Final proportions of zombies under varying initial densities of zombies and humans, subject to initial human skill = 0.3, skill increment = 0.05, human speed = 3.

iv. The pale region shows situations in which zombies so outnumber humans that they zombify them all quickly, though there may be occasions where a human survives and could win, given sufficient time.

A word of caution is necessary in interpretation. The runs that generated this data were truncated at 2500 ticks if they had not finished naturally by then. However, experience with individual runs shows that when humans are fast, they can take a long time to win. Thus some of the shades in the border regions might have been darker if the runs had not been truncated. More or longer runs might allow us to more accurately locate the boundary between dark and pale regions.

The boundary between black (all runs have only humans surviving) and pale grey (all runs have only zombies surviving) is made up of four distinct situations, of which more than one can occur for any boundary point:

- Humans win in some simulations and zombies win in others;
- Some simulations fall into cyclic behaviour, equivalent to the chasing that occurs with a speed of 1;
- Some simulations would have resulted in humans winning but were truncated before this could occur (as would have occurred for the simulation shown in Figure 12.8 if it had been truncated); and

- Some simulations would have resulted in zombies winning but were truncated before this could occur (not observed in the particular simulations conducted, but theoretically possible).

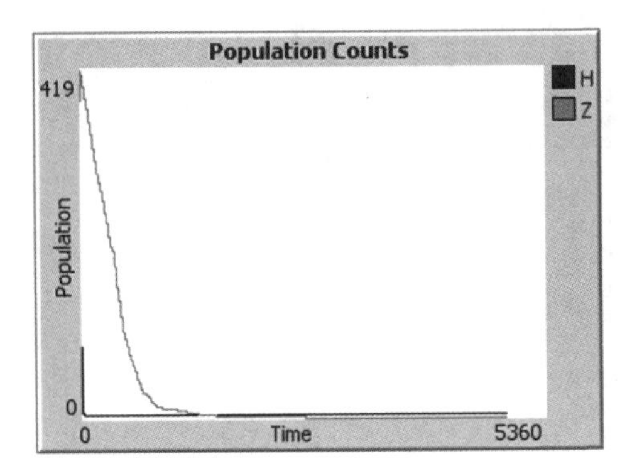

Figure 12.8: Example with long completion time. When this run terminated after 4916 ticks, four humans survived. Parameters for this run were initial human density = 0.30, initial zombie density = 0.80, initial skill = 0.30, skill increment = 0.05 and speed = 3.

We note that, for speeds 2 and 3, we have never observed chasing behaviour in an individual run of a simulation. This is consistent with humans being able to run fast enough to get out of sight of zombies. Hence the intermediate shaded regions of the chart correspond to parameters that led to some runs terminating with human wins and other runs with zombie wins. Darker greys indicate that humans were more often triumphant, and paler greys indicate that more often zombies were the victors.

For speeds 4 and 5, we occasionally observe a special kind of chasing behaviour that was initially puzzling. We show an example of this oscillatory behaviour in Figure 12.9, for a human speed of 5. Each zombie sees humans adjacent and takes a step toward the humans. Each human sees a pack of zombies nearby and runs five steps. Having done this, in the next tick, the zombie sees a new human in the opposite direction, so it steps back to its original positions. Similarly, the humans each see a pack of zombies approaching from the opposite direction, so they run five steps back to their original position. The chasers jumps back and forth endlessly, in a loop analogous to the state known as 'perpetual check' in a game of chess. This interesting behaviour may be a result of the myopia for distances greater than 3 that is hard-coded into our model combined with specific speeds and the world size used for the simulations.

Returning to the speed 3 case, we choose particular initial zombie and human densities corresponding to medium grey points in Figure 12.7, and investigate how details of the skill and skill-increment parameters affect zombie/human dominance at these points. This story is encapsulated in Figures 12.10A–12.10C.

We see that, for initial zombie densities that are very low, as in Figure 12.10A, humans dominate after skill is above about 0.3. The line between pale grey on the

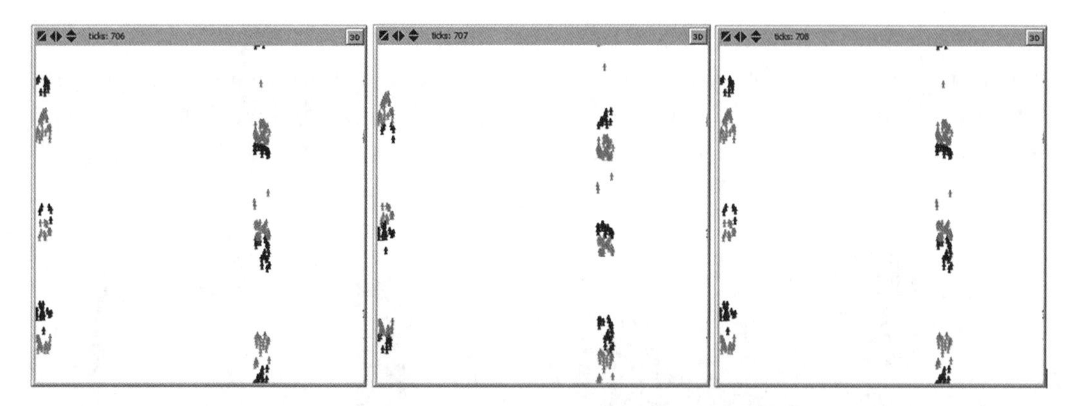

Figure 12.9: An example of oscillation: successive screenshots for a run with initial human density = 0.90, initial zombie density = 0.10, initial skill = 0.30, skill increment = 0.05, speed = 5.

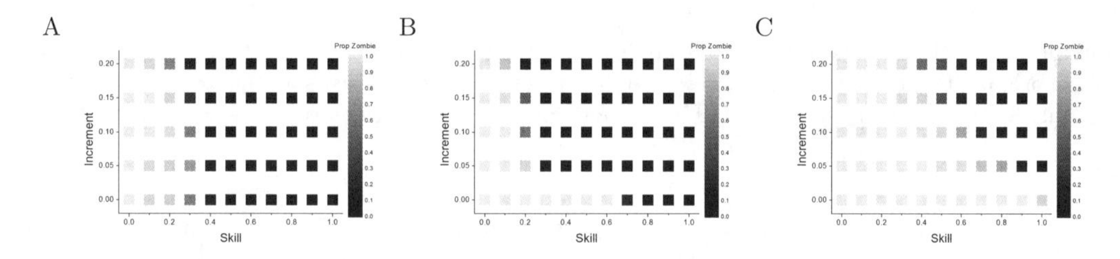

Figure 12.10: Final proportions of zombies under varying skill and skill-increments values. A: Initial zombie density of 0.005, moderately low initial human density of 0.15 and human speed = 3. B: Initial human and zombies densities both equal to 0.5, and human speed = 3. C: Initial human density of 0.15, initial zombie density of 0.85, and human speed = 3.

left and black on the right is almost vertical, so skill increment has little effect, since humans have few zombies on which to practise. For moderate zombie densities as in Figure 12.10B, fast humans still win more often than not, but now the skill-increment factor matters more. For very high zombie densities as in Figure 12.10C, zombies win more often than humans, but when humans have both high skill and skill increment, they are able to win.

Increasing zombie density generally tends to shift the charts to the right, decreasing the benefits to humans of skill and increasing zombie dominance overall. Higher human density increases the benefits to humans of skill and decreases final zombie dominance. These general observations may be made more easily by referring to the complete set of charts available on our website [9].

Proportion of humans surviving

Our optimistic note must drop a tone or two here. Figure 12.11A shows the final number of humans (or humans at 2500 ticks if truncated) as a proportion of the original number of humans. The conclusion must be: 'victory is hollow.' For the upper region of the chart corresponding to the human winning black section for Figure 12.7, most of the human population is dead (or zombified if there are still zombies 'alive') throughout most of parameter space.

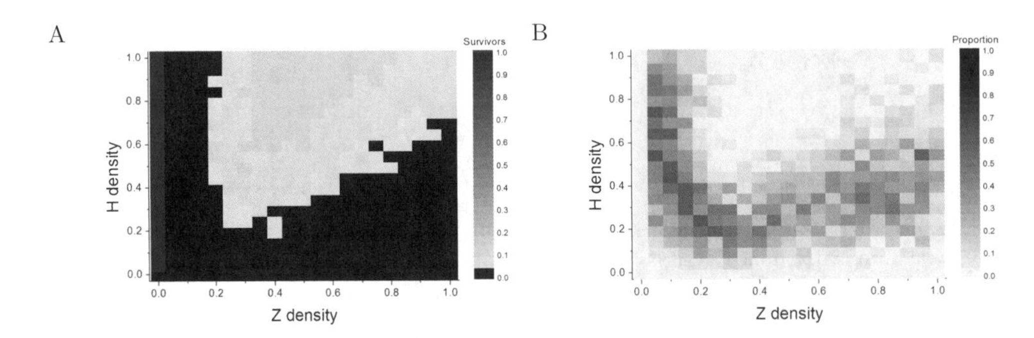

Figure 12.11: Initial human skill = 0.3, skill increment = 0.05, human speed = 3. A: The proportion of the original human population that survives. Note that we have used solid black for 'proportion 0' to make the boundary clearer. B: The proportion of the simulations truncated at 2500 ticks.

The intriguing boundary that was present in Figure 12.7 also occurs in Figure 12.11A since it marks the turning-point where humans start winning so at least some survive. Furthermore, near this boundary, the model takes longer to resolve and truncation is more likely to occur because either group could win.[4] Figure 12.11B shows the proportion of simulations truncated for the same parameter sets.

The consequence for our model is that dark regions of Figure 12.11B may correspond to populations of zombies that take a long time to get killed by humans but would eventually resolve as a small number of humans surviving the zombie attack. We investigated this possibility in Table 12.1, which we constructed by examining our stored data for the original runs. Indeed, in a number of cases, we see runs that were truncated at 2500 ticks and that were probably not looping, indicating that without truncation more zombies (or potentially humans) would die.

For higher speeds, truncation occurs both for simulations with slow resolution and those with oscillatory behaviour illustrated in Figure 12.9. On examination of the raw data, some runs for 'speed 5' appear to be looping (see Table 12.1).

Clearly it would be useful to have an automated way of identifying looping behaviour in our model, whether it be the perpetual chase in a single direction around the screen as commonly seen with speed 1, or the oscillatory behaviour exhibited by speeds 4 and 5. The technical difficulty in implementing this lies in

[4]This is indicative of a phenomenon called 'critical slowing down.' This terminology is used in statistical mechanics to indicate the slow rate of convergence that can occur near a phase transition boundary. For a readable explanation of this phenomenon in a slightly different context, see [8, pp.7–8].

Table 12.1: Best estimates of how many runs (out of twenty-five) were probably exhibiting looping behaviour, and how many runs would probably terminate naturally if they had been allowed to keep running beyond the truncation point of 2500 ticks.

Initial Zombie Density	**Initial Human Density**	**Number of Runs Expected to End Naturally**	**Looping Behaviour**	**Total Number of Simulations Truncated**
		Speed=3		
0.1	0.9	8	0	8
0.1	0.5	14	0	14
0.3	0.3	12	0	12
0.55	0.95	0	0	0
0.8	0.4	9	0	9
		Speed=5		
0.1	0.9	6	5	11
0.1	0.5	9	1	10
0.3	0.3	11	0	11
0.55	0.95	0	4	4
0.8	0.4	4	0	4

the large memory requirements for 'remembering' former states. To implement a 'short term memory' would be easier and sufficient to recognize oscillation.

The clear message of Figure 12.11A, independently of any subtleties of truncation, is that the price of speed and victory for humans is death for most of the original population, consistent with MHIS.

12.5 Implications for Education and Research Training

A key advantage in using NetLogo simulations is the potential to motivate and train students and researchers. The visual display provides an interactive environment in which to immediately see the effects of parameter changes on the behaviour of agents. This stimulates novel *What-if* questions as well as assisting in understanding why the behaviour occurs. In addition, models are particularly easy to construct in NetLogo. Apart from fine tuning, the (admittedly simple) zombie model was built very quickly by one of the authors, based only on a workshop attended over a year earlier [12] and the documentation and sample models provided with NetLogo.

MHIS extended their model to consider issues of latency, quarantine and treatment. The NetLogo model could be extended relatively simply to introduce equivalent effects and investigate whether skill and speed also enable humans to win more easily in these situations. The model could also be extended to implement more realistic interpretations of skill and intelligence. For example, humans could move toward each other to provide protection, or sufficiently skilled humans could attack isolated zombies.

12.6 Conclusion

The ABM and analytical approaches are able to investigate different questions and should be considered complementary. ABMs can incorporate spatial structure and differences between individuals; they can provide greater understanding of the issues being investigated and the factors that may be important for further modelling. However, ABMs cannot provide the rigour of an analytical mathematical model.

We have been able to show that the grim conclusions for the future of the human race in the event of zombie attack as drawn by the authors of [1] were unnecessarily pessimistic, if human traits such as skill and ability to learn are allowed for. In our model, for very low initial zombie densities (parallel to epidemiological levels) and slow-moving humans, large proportions of the human population commonly survive. For this sparse-zombie, slow-human context, only when human skill is very low do zombies stand a chance of winning, and only when human skill is very high do humans have a reasonable chance of victory; for the in-between state of moderate skill, a substantial proportion of humans tend to survive, albeit in packs that are being forever chased by zombies. However, these packs are not maintained for faster humans, and survival proportions for fast humans drop dramatically.

For higher zombie densities in our model, there are conditions where human skill and speed allow humans to eliminate all zombies. For very high initial skill levels, humans generally win and the majority survive. At moderate skills, however, the majority of humans are zombified in the early phase of the zombie invasion (and then eventually killed) because humans require experience before acquiring sufficient skill to generally win.

We have also shown that agent-based models and NetLogo are useful tools for investigating features that would make an ordinary mathematical SIR model of infection too difficult to solve. We have demonstrated the benefits of NetLogo as a tool for experimentation and theory development, and for training students and researchers.

Appendices

A Acknowledgements

Financial support from the Australian Research Council through its support for the Centre of Excellence for Mathematics and Statistics of Complex Systems is gratefully acknowledged by the second author.

The second author would also like to thank the first author for introducing her to the zombie-modelling question and to NetLogo. The first author would like to thank the second author for drafting her into an interesting project. Thanks also to Robert Smith? and co-authors for writing the first mathematical model of a zombie epidemic, and for the opportunity to participate in this project.

B Glossary

Agent-based model. Computer simulation where each individual (such as a human) takes independent action based on its own characteristics (such as skill), situation (are zombies visible?) and rules (run away!).

NetLogo. Software for developing and running agent-based models, available from http://ccl.northwestern.edu/netlogo/.

ODE (Ordinary differential equation). A common mathematical approach to *SIR* modelling. Differential equations describe the rate at which the proportion of the population in each disease state changes.

Patch. The NetLogo term for spatial units. For this model, a patch simulates locations where humans and zombies can move.

SIR model. A mathematical approach to modelling the spread of a disease in a population. People are assumed to be in one of three states: susceptible (S), infected (I) or removed (R). Susceptible individuals become infected if exposed to the disease; infected individuals recover after some period; and removed individuals are immune from further infection.

Skill. A human characteristic that describes the chance that a particular human will kill a zombie in a one-to-one fight.

Speed. A human characteristic that describes the number of patches by which a human can move in a single simulation step.

Zombification. The process whereby a human becomes a zombie.

References

[1] P. Munz, I. Hudea, J. Imad, R.J. Smith?. When zombies attack!: Mathematical modelling of an outbreak of zombie infection. In *Infectious Disease Modelling Research Progress*, J.M. Tchuenche and C. Chiyaka, eds., Nova Science, Happuage, NY 2009, pp. 133–156.

[2] W.O. Kermack, A.G. McKendrick. Contributions to the mathematical theory of epidemics - I. *Proc. R. Soc.* 115A:711–721, 1927.

[3] F. Brauer, P. van den Driessche. J. Wu. *Mathematical Epidemiology.* Springer-Verlag, Berlin, 2008.

[4] J.D. Sterman. A Skeptic's Guide to Computer Models. In *Managing a Nation: The Microcomputer Software Catalog*, G.O. Barney, W.B. Kreutzer and M.J. Garrett, eds., Westview Press, 1991, pp. 209–229.

[5] N. Gilbert, K.G. Troitzsch. *Simulation for the Social Scientist.* Open University Press, Maidenhead, UK, 2005.

[6] N. Gilbert. *Agent-Based Models (Quantitative Applications in the Social Sciences).* Sage Publications, Los Angeles, 2007.

[7] U. Wilensky. NetLogo, Center for Connected Learning and Computer-Based Modeling, 1999 http://ccl.northwestern.edu/netlogo/

[8] D. Stauffer, A. Aharony. *Introduction to Percolation Theory.* Taylor and Francis, London, UK, 1992.

[9] J. Badham, Zombies in the City: Supplementary Materials http://www.research.criticalconnections.com.au/zombies/zombies.htm

[10] M. Wooldridge, N.R. Jennings. Intelligent Agents: Theory and practice. *Knowledge Eng. Rev.* 10, 115–152, 1995.

[11] D. Griffeath, C. Moore. *New Constructions in Cellular Automata.* Oxford University Press, New York, 2003.

[12] M.J. Berryman, S.D. Angus. Tutorials on Agent-based modelling with NetLogo and Network Analysis with Pajek. In *Complex Physical, Biophysical and Econophysical Systems*, R.L. Dewar, F. Detering, eds., World Scientific, Hackensack, NJ, 2010, pp. 351–375.

AN EVOLVABLE LINEAR REPRESENTATION FOR SIMULATING GOVERNMENT POLICY IN ZOMBIE OUTBREAKS

Daniel Ashlock, Joseph Alexander Brown and Clinton Innes

Abstract

During a zombie outbreak, the issue of government policy—as well as the population's awareness of the threat—can change the effectiveness of the response to that threat. During emergencies, governments have the ability to take over the local media, call citizens to form a militia, enforce curfews, vaccinate the population and call in the military. The modelling of the course of action and the timing of these actions is critical to making an informed policy choice. Governmental influence is also potentially variable in its effect; a vaccine's effectiveness, for example, is not known until after it is developed. Furthermore, it takes time for policy to be legislated and implemented. This study models the government's policy as a sequence of actions that allows for randomness in the effects of the actions upon a model of a zombie outbreak. One action from the list is applied in each model time step, and the state variables of the model are updated based on that action. An evolutionary algorithm is used to optimize the sequence of actions to minimize the impact of the zombie outbreak.

13.1 Introduction

THE issue of formulating effective government policy for dealing with a zombie outbreak has never been more critical than since the zombie wars began at the start of this millennium. Government intervention has proven to be quite ineffectual during many outbreaks. Command and control breaks down in sectors with outbreaks, while civilians are uninformed or misinformed of locations for evacuations and medical supplies [1]. Furthermore, Brooks [2] gives a highly detailed account of the impacts policy initiatives have after humanity has been overrun. One horrific example is the Redeker Plan. In this plan, a phased retreat was made into largely gated areas and naturally defensible areas allowing for lines of defence. This plan, while effective for those within the enclosure, doomed a large section of the public that had not been preselected. The selection criteria were survivability, a cross-section of useful skills and the protection of military and political officials. The policy also provided for military forces and civilians to be deliberately misinformed as to their areas of withdrawal in order to lure groups of zombies away from the real withdrawal, feeding the opposition while the main body escaped. While this policy did have the effect of saving a survivable portion of the populations of various countries, it was substantially mismanaged. Having no planning for a zombie outbreak in place before the outbreak allowed the implementation of the reprehensible ideas realized in the Redeker Plan. The establishment of a government plan during the early stages of outbreak is therefore of primary importance.

Previous chapters have illustrated the importance of decision-making in a zombie epidemic, even if only partial information is available. Decisions may be made on a broad scale, such as country- or state-wide, or may be required at the local level, such as a hospital or university campus. Future planning already exists for forthcoming events such as pandemics, earthquakes and tsunamis, despite the fact that the nature and timing of such events cannot be fully comprehensible in advance.

The number of differing zombie types that might arise, and hence require modelling, is quite large. Sacchchetto [3] enumerates eleven main classes of zombies, giving preferred means of destruction. Of these classes, those caused by a form of zombie virus are the most prolific in terms of spread, with alien-controlled zombies perhaps the second most. The classes that have a viral transmission he denotes as the *common zombies*. We will focus in this study on common zombies not only because of their prevalence, but also because of their use of the virus as a means of transmission. Policies for rarer types of zombies are not covered here but will be examined in later phases of our research.

The typical zombie virus is highly infectious. It can be contracted through bites, scratches or any contact with bodily fluids of an infected person. A single drop of contagious liquid landing on the surface of the cornea has been shown to lead to zombification [4], so goggles are advised when dealing with the undead. Brooks [5] traces it back to a disease known as Solanum and gives a full accounting of its properties. There is also the hypothesis of an airborne strain where natural death acts as a catalyst for zombification [3]. This could also be the case in a weaponized form of zombie infection [6]. However, Brooks has noted that field reports are frequently exaggerated [2] and the virus cannot infect dead flesh [5].

Furthermore, given that the speed of transformation to the active zombie form is highly dependent on both the location of infection and the severity of the exposure to the virus (a small amount of zombie saliva is not as infectious as a pint of zombie blood), it would not be impossible to hide an infection or to be unaware of it. The occurrence of an asymptomatic death can make it seem as if natural death resulted from an apparent airborne strain. Erring on the side of caution, some form of safe disposal of the dead is advisable during even the smallest of outbreaks; burning of all necrotic issue is advised [7]. This is just one policy a government could initiate in order to ameliorate or prevent the outbreak; we will look at others.

Government planning for zombie outbreaks can be improved by using computer models such as those provided in this study. Simulation allows us to view various outcomes based on the sequence of actions selected. Multiple models can be run to assess the typical and extreme impact of stochastic events to allow for more accurate planning. We use an evolutionary algorithm (EA) in order to search the space of possible government responses. EAs are a global optimization technique that can efficiently search complex, discrete model spaces. Based on the principles of natural evolution, an EA operates on a population of possible zombie outbreak policies, examining and comparing tens of thousands of policies in each modelling run.

13.2 The Discrete-Time Model

The model of zombie outbreaks treated in this study is similar to those in [8]. This model differs from previous work in that it uses discrete time steps rather than continuous, differential models. The exact formulation of how zombies form from the human population also differs.

Common zombies do not rise from the grave, so there is not a constant flow of zombies into the system. This model incorporates only deaths caused by zombie infection because death from other causes is at a far lower rate during a zombie outbreak. We also assume that all survivors killed will be resurrected as zombies. We make the assumption at the start of an outbreak, before the population becomes fully aware of the situation, that ignorance of the virus makes those infected unlikely to commit suicide or for others to carry out mercy killings; for example, see [9].

The model assumes that infection has a fixed incubation period. This means that the incubation period is some fixed number of time steps of the model. This is a simplification of the actual state of affairs, in which incubation time is best modelled as a random variable based on the initial dose of the zombie virus and the method of death. We have used this simplification because no measurement has been made of the distribution of actual zombie resurrections as of this date. The model we use can be easily changed when such studies are available.

This is a compromise for the sake of simplicity of analysis. It has been established that incubation times follow a distribution based on the location of the bite and the means of death. This distribution could be taken into account if we used a finer gain mathematical model of a breakout situation based upon the behaviour of the virus. The model retains a well-mixed population, making it equally likely for any zombie to meet any survivor. The state equations for the model are as follows:

$$Z_t = Z_{t-1} + I_{t-d} - R_t \tag{13.1}$$
$$S_t = S_{t-1} - I_t \tag{13.2}$$
$$I_t = \beta Z_{t-1} S_{t-1} \tag{13.3}$$
$$R_t = \alpha Z_{t-1} S_{t-1}, \tag{13.4}$$

where S_t is the number of humans, Z_t is the number of zombies, I_t is the number infected (but not yet zombies), and R_t are the number killed or removed in time step t. The virus infects the population of survivors in any one time step based upon the percentage of immunity to the virus, γ, the aggressiveness of the zombies, δ, and the chance that an encounter with a zombie will cause infection, ω. The β value used in other models is therefore the conjunction of these factors: $\beta = (1 - \gamma)\delta\omega$. R_t is the number of zombies killed, based on the α value, which quantifies human agression. A zombie is killed by a human depending upon weaponry and training ability, ρ, and the chance of them encountering zombies while at an advantage, ϕ. This gives $\alpha = \rho\phi$. The population of survivors and zombies is therefore dependent upon these two values.

This model also differs from those previously in that the zombies do not rise immediately from the dead. They are instead delayed for a number of rounds, d. This represents the infected proportion of the population taken out of action due to their sickness and rising back to join the undead after the incubation period.

Government policies of various types change the infection rate β through application of health polices and the kill rate α through military action and education. The state variable S_t is changed by the addition of personnel in military forces. There are three other factors in the simulation: the level of scientific progress towards the understanding of the virus, *vax*; the threat and political awareness variable, *aware*; and the number of troops dispatched in a call-up of military personnel, *army*. The creation of a vaccine would be based upon the *vax* variable. The *aware* variable is more thoroughly explained in Subsection 13.3.2.

Examining the state equations, we see two stable equilibria: no survivors and no zombies. If both survivor and zombie variables have a positive value, then there is a probability that a zombie will infect a survivor or that survivors will kill a zombie. Since people and zombies interact, this is an intrinsic feature of the simulation.

An effective policy must somehow steer the state variables away from the pure-zombie equilibrium by manipulating the α and β values. It is desirable to have $\beta < \alpha$ as this means the humans are killing zombies at a faster rate than humans are being killed. The more quickly this system is steered toward a large value of $\frac{\alpha}{\beta}$, the more rapidly the system approaches the pure-survivor equilibrium. This provides a basis for assessing the effectiveness of a policy.

13.3 Representation of Government Policies

This study uses a *string representation*. This is an evolvable linear representation that can be used to model sequential actions taken to steer a state-conditioned discrete model. String representations have been used previously as models for the design of starfighters and election strategies [10].

The actions in a sequential policy are applied to the state variables per time step. A time step represents a few days of the zombie outbreak. We limit the actions taken by the government to one per time step, representing the intrinsic inefficiencies of implementing government policy in a democracy. The costs that limit the government to one action per time step include the expenditure of money, time, effort and logistical difficulties. The cost of taking an action, which takes one round of the simulation, only represents initial investment in a policy and the effects persist for the remainder of the simulation, even when other actions are chosen. Actions thus represent shifts in the pattern of governmental efforts, with the state variables of the simulation representing inertia that limits the impact of those variables.

13.3.1 Overview of Policies

Table 13.1 lists the policies available to the government. The Effect column specifies the way that an action updates the state variables. *Quarantine* includes collecting and burning bodies and working to reduce the ability of zombies to infect the population. This can be accomplished via curfew and martial law [2, 11, 9, 12]. No specific group of zombies is quarantined; rather, the action represents the isolation of infectious materials.

Armed citizens in the form of a militia have demonstrated their usefulness during outbreaks to act in conjunction with local military forces [2, 9, 1, 12]. Further, citizens can take it upon themselves to arm themselves and defend their homes. This behaviour is evidenced in perhaps every known outbreak, via the looting of sources of weaponry. Two prominent examples: the 'trip' to the Monroeville Mall gun store [2] and Andy the gun-store owner [11]. Other provisions will also affect the defence ability of a survivor: food, water, fuel, medicine and radios. These inform the *army* and *kill* actions, which represent directives by the government to call in military forces or to arm citizens respectively.

TV and radio broadcasts are followed by survivors to allow dissemination of information by the government [11, 9, 1, 12]. Furthermore, the DVD extras in [11] provide a day-by-day accounting of the coverage provided by one such emergency broadcast in an hour-by-hour breakdown of the outbreak. Broadcasts provide actionable intelligence (the video clip of the militia force showing how to properly dispatch a zombie [11] and accurate news reports [9, 12]), misinformation (locations of shelters that were full or not operational [1]) or pure fearmongering (recommendations to feed the zombies or the call for nuclear strikes on population centres [1]).

Table 13.1: Actions that the government can take in a time step; **rand** produces a random number from a uniform distribution in the range [0,1].

Action	Symbol	Effect
Quarantine	Q	$\omega = (2\omega + 0)/3$, $Aware = (2Aware + 0)/3$
Army	A	$\rho = (\rho + 1)/2$ if ($\mathbf{rand} < Aware$), $S_t = S_t + 100$, $Aware = (Aware + 0)/2$ if $(Z_{t-1} > S_{t-1})$
Kill	K	$\rho = (\rho + 1)/2$, $\omega = (2\omega + 1)/3$
Warn	W	$\rho = (3\rho + 1)/4$, $Aware = (2Aware + 1)/3$
Fearmonger	F	$\rho = (2\rho + 1)/3$, $Aware = (3Aware + 1)/4$
Science	S	$Vax = (Vax + 1)/2$ if ($\mathbf{rand} < Aware$)
Vaccinate	V	$\gamma = (3\gamma + 1/4)$ if ($\mathbf{rand} < Vax$) $Aware = (Aware + 0)/2$ if $(Z_{t-1} < S_{t-1})$

This informs the *warn* and *fearmonger* actions where the government appeals to the public to follow its policies and gives information about the threat at hand.

During the crisis, scientific agencies can be called in to evaluate the threat [13]. This leads to the development of techniques to slow the threat by learning more about the mechanisms that the virus uses for transmission. This informs the *science* and *vax* actions, where the government forms a policy to press for scientific progress and the development and administration of a vaccination.

13.3.2 Public Response to Government Policies

Many of the actions are stochastic in their effect, either due to the action being random in its effect—such as a vaccine—or that they require the populace to give approval, a declaration of war on the zombies. The tenement building cleared at the start of *Dawn of the Dead* [1] gives an example of the populace resisting martial law and quarantine. In this engagement, the two survivors were part of a SWAT[1] team.

[1]Special Weapons and Tactics, a police force that has been established in many large cities in the United States. The officers have specialized training for dealing with known armed offenders, high-risk warrants, hostage situations and any situation that the regular police force would have difficulty containing. The Emergency Task Force (ETF) is the Canadian equivalent.

Not only did they find armed resistance from the building occupants, but the occupants had been dumping the bodies of the dead into the basement of the tenement, leading to a large reservoir of zombies, in common parlance a *swarm*. Throughout the building, there was also a smaller number of zombies generated by recent or hidden infections. Without SWAT intervention, the tenement would probably have been overrun and the infection would have spread. The officers speculated that the dumping of the dead bodies was due to a 'respect for the dead' which the health initiative that was put forth by the government would not have allowed. The scientists' opinions on the zombies given in news reports were also the subject of public scorn.

In *Day of the Dead* [13], once again a government policy was put to the test when a group of military personnel had been tasked with looking after a group of scientists who were searching for either a cure or a method to control the zombies. The team was operating inside a military bunker. The military overseeing the scientists were not pleased with the progress and decided to kill as many of the zombies as possible instead of using them as research materials for a deranged experimenter. This division about which policy to follow led to a breakdown in trust in the civilian-scientist-military coalitions and the eventual collapse of the organization within the bunker.

In our model, the *aware* factor is built into all of the actions except *kill*. The *kill* action is in many ways a default action. The citizens, noticing a zombie outbreak, do their best to arm themselves and fight back. Therefore public approval will not change from its current levels. Approval of the government by the populace factors into the majority of other actions. For example, the scientist will only be listened to if the population is aware of the threat and has been convinced of its reality. The military will always be called in, but the use of their weaponry will only be effective if they are fighting the zombies alongside the civilians and not just acting as crowd control against an angry armed citizenry. This makes an increase in the weaponry rate dependent upon the degree of public awareness. If the number of zombies is substantially smaller than the human population, there is also a loss of awareness, as the public finds the very idea of zombies hard to believe. The reverse happens in an attempt to deploy a vaccine. The public will only have sympathy towards a vaccination effort when they outnumber the zombies; the military wants direct action as they see humans being overwhelmed and the cost of protecting the scientific effort is seen as wasted [13]. The change or cost of awareness is independent of the effect to the system; while the representatives have been allowed to institute a policy, the population may still object to the implementation.

13.4 The Evolutionary Algorithm

EAs are a branch of computational intelligence derived from the theory of evolution [10, 14]. They can be used as general-purpose optimizers. They are most effective for optimizing discontinuous, stochastic or anisotropic problems. Discrete problems, like the government policy model used in this study, are natural targets for optimization with an EA. This study provides a overview of our system sufficient for implementation.

An EA operates on a *population* of *chromosomes* that are subject to *fitness evaluation*. The fitness evaluation generates a figure of merit for each chromosome—in this case, the string of actions specifying a government policy—by running them through a simulation of a zombie outbreak. The fitness function scores or ranks government policies on their ability to solve the problem. Once fitness evaluation has taken place, a *selection operator* selects *parent chromosomes* with a bias toward high fitness values. These parent chromosomes are copied and then probabilistically subjected to the genetic operations of *crossover* and *mutation*. The resulting new structures are called *children*. Some of the population may be selected and transferred into the population without crossover or mutation; this is called *elitism*. This process creates a new generation of chromosomes. The process of creating a new population continues until some stopping condition is met, usually a predefined number of iterations of the breeding step.

In our simulations, the chromosome is a string of actions representing a government policy. Each policy lasts for ten time steps and so requires a string of ten symbols. Each position in a policy is known as a *locus*. The fitness of a string is first calculated by running the simulation of the outbreak, as defined in Section 13.3. The number of human survivors, minus military personnel called in to deal with the threat, is the fitness of a policy. The goal is to find a policy that reduces the number of human deaths during the outbreak. By penalizing the population by the number of troops called up, we prevent the government policy from just selecting the *army* action repeatedly to 'throw bodies at the problem' in order to receive a better score. Such a policy is retrograde to the intention of the policy optimization. This was just the kind of policy action seen in World War II, Vietnam and the Redeker Plan. The value of a human life, either in the armed services or civilian, is of equal weight. They must be fairly scored against each other. This method of scoring rewards the overall saving of lives based on the initial population of the area. An army unit ordered forward by a government using this system will do the most good in terms of their losses and the population losses. Tossing soldiers at a doomed sector without gain is avoided.

The selection operation used is 'size four single tournament selection.' In this method of selection, four policies are selected at random. Children are generated by crossover and mutation applied to copies of the two best policies and these children replace the worst two policies. The genetic operators are a two-point crossover operation and a single point mutation. In a two-point crossover, two loci on the chromosome are selected at random and the actions between them are swapped between the parents. Single point mutations change a single locus, selected from the parent at random. The selected locus is changed to a new action selected uniformly at random. Examples of crossover and mutation are given in Figure 13.1.

A

parent 1	W	W	Q	Q	S	V	...
parent 2	F	F	K	K	A	K	...
child 1	W	W	Q	**K**	**A**	V	...
child 2	F	F	K	**Q**	**S**	K	...

B

parent	W	W	Q	Q	S	V	...
mutant	W	W	Q	**K**	S	V	...

Figure 13.1: A: Two-point crossover in the fourth and fifth loci (in bold). B: Point mutation in the fourth locus (in bold) from a *quarantine* to a *kill* action.

Thirty runs with differing random-number seeds were run for 10,000 population updates. This allows the experiments to permit characterization of the stochasticity of outcomes and provide a sampling of the set of possible policies.

13.5 Experiments

Our experiments investigate three topics: the type of zombies chosen, the latency period of the virus and the initial government approval. The city is relatively small with an initial population of 100,000 survivors and 10,000 zombies, representing an outbreak that has progressed to one zombie per ten survivors.

The human aggression is set to $\phi = 0.0001$, with an initial weaponry level of $\rho = 0.1$, resulting in a removal rate of $\alpha = 0.00001$. The population of this city is therefore not well-trained in the use of firearms and other combat, at least initially. The value of immunity begins at $\gamma = 0.01$ and the viral effect is $\omega = 1.0$, so that all who are bitten are infected. This places the initial β values within 1% of the zombie aggressiveness, δ. In other words, the number of humans infected initially depends only upon the aggressiveness of the zombies. The aggressiveness of the zombies is set to a value selected from $\delta \in \{\frac{1}{2}\alpha, \alpha, 2\alpha, 3\alpha, 4\alpha, 5\alpha, 10\alpha\}$. This allows for a swift comparison of types of zombies which are killing humans from a half to ten times the rate at which the humans are destroying them. Allowing zombie aggressiveness to vary, while maintaining viral nature and initial immunity, allows us to study different zombies that use this type of infection vector. It shows, for example, the difference between shambling zombies and fast zombies. If we see a group of survivors lasting through the simulation where $\beta > \alpha$ initially, then we have shown the value of optimizing the governmental planning.

We varied the latency using values of 1, 3 or 5 time steps in order to show the role played by latently infected zombies. These zombies are not killed during latency because they appear to be normal humans. We would expect to see a greater impact from holding back the spread of the infection than from killing zombies. Furthermore, increases in latency extend the outbreak, making it harder to contain.

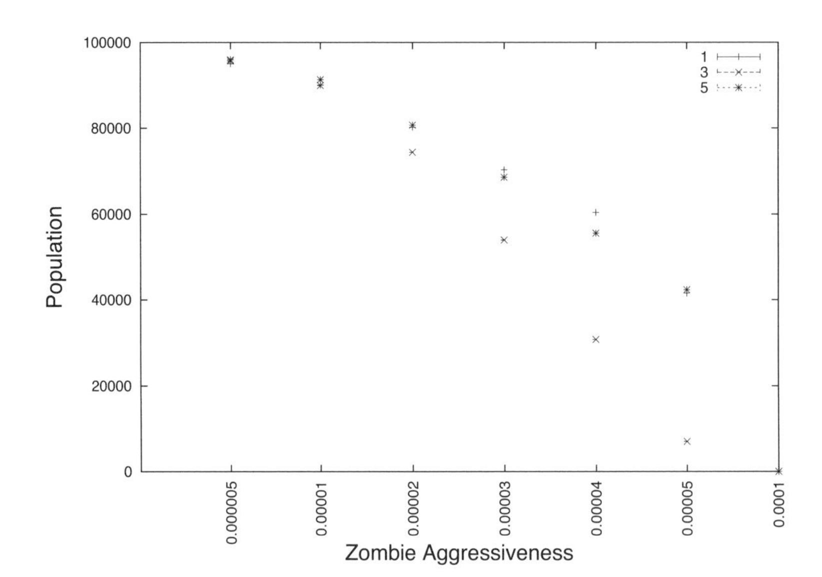

Figure 13.2: Difference in survivor population levels as a function of zombie latency with 95% confidence interval about the mean.

The initial approval of the government is also varied using the values drawn from $aware \in \{0.1, 0.5, 0.8\}$. These are hatred, no-strong-opinion, and popular government. A popular government, by starting with public opinion on its side, should have to spend less time working on improving public perception.

To compare the frequencies of actions with the final set of policies, we use a normal approximation to the binomial distribution to estimate the range of each of these actions. The 95% confidence interval of a probability of an observed action is

$$Z_{95} = 1.96\sqrt{\frac{P(\text{action})\,[1 - P(\text{action})]}{N}}.$$

13.6 Results and Discussion

The latency rate was first set to 1, representing immediate return from the dead. The government's optimized policy, no matter what public perception of the government, was found to be successful in driving the zombies to extinction in all chosen δ levels other than ten times α. When δ was ten times α, there was no chance of human survival; the zombies kill humans at such a high rate that humans are all removed before the policies can be enacted effectively. However, the final number of zombies differed depending on the strategy chosen. While these would not be optimal strategies for lowering the zombie population, government policy nevertheless affects the outcome.

As zombie aggressiveness increases, effective policies change from *killing* to *quarantine*. *Killing* is risky, as it exposes humans to the virus, more so when the zombies

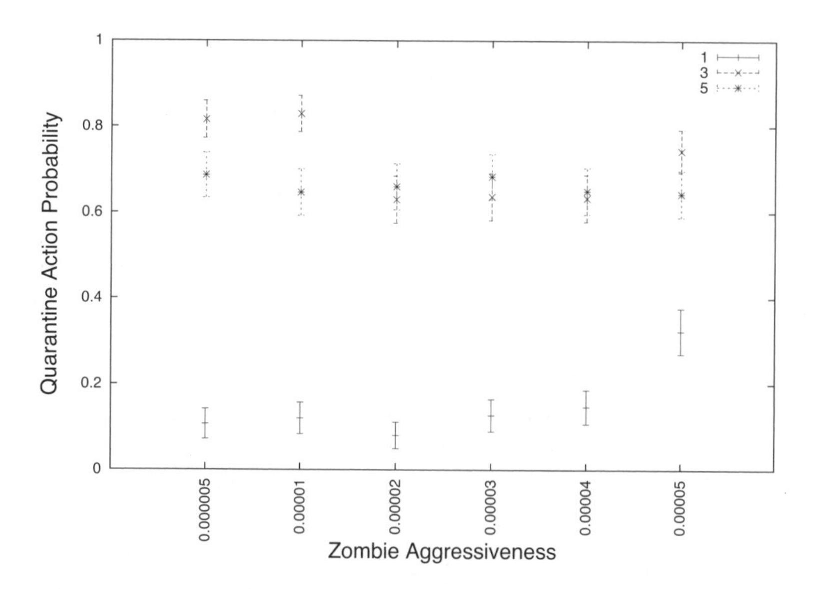

Figure 13.3: Difference in *quarantine* action probability as a function of zombie latency with 95% confidence interval about the mean.

are more likely to bite. *Quarantine* actions prevent the spread as the humans are kept indoors and away from the aggressive zombies. The increased level of quarantine becomes statistically significant when $\delta = 5\alpha$. This confirms some of the conclusions of [8]; that is, when the rate of aggressiveness of zombies, δ, is sufficiently small compared to the human ability, α, to kill zombies, increasing the kill rate will produce results. Otherwise, an aggressive quarantine is the preferred solution. The influence of the government allows for a larger difference between α and β before quarantine is necessary.

The latency rate of the zombies leads to a reduction in the surviving human population when the aggressiveness increases (Figure 13.2). The greatest change, however, is in the actions. The actions selected in an effective policy change from working on the viral factors and became less about killing zombies. This is due to the reduction in the number of zombies that can be killed during a time step. The most interesting result is that humans facing zombies with a short latency or a long latency may have a survival rate longer than humans faced with zombies who have a medium level of latency.

This is also reflected in the probability of the quarantine action (Figure 13.3). The first rising zombies happen later in the simulation, meaning there are fewer zombies available to be removed by killing actions. This makes fighting the virus a much more attractive option; however, it takes many rounds for the quarantine action to have a large impact upon the viral spread. When extra latency is added, this gives time for the effectiveness of the zombie bites to be reduced enough to keep the population alive, and the threat of waiting is removed. This is interesting because, in *Dawn of the Dead*, a remark is made by one of the survivors that he has never seen anyone last more than three days after being bitten [1]. According to our simulations, these zombies would be the most likely to cause just the kind of outbreak that Romero documents.

The role of the government's initial awareness apparently has no effect on the final population of survivors. In hated-government scenarios, the *warn*, *kill* and *army* actions are more frequent; in the popular-government scenarios, these actions are exchanged for *fearmonger*. These differences are not, however, statistically significant for actions taken on the entire string. We can say that a combination of *kill* and *army* is more common for the less popular governments. At the start of a zombie outbreak, killing seems to be a favoured method, since the large number of zombies renders this action more effective. Once the population of zombies is sufficiently reduced, the removal of the infection becomes more favourable. The actions for reducing the infection require a higher level of awareness. Furthermore, as actions like *army* depend on the state of public opinion, a government that is less popular will need to spend more time legislating (using more probabilistic actions) in order to call in the military.

When the zombie aggressiveness increases, stopping the spread of infection is a more effective way to save the population. Therefore quarantine events are found to occur more often, no matter which government type is used. A popular government, however, incurs less damage. We see the number of scientist actions reduced in popular governments, though not by a statistically significant amount. Once again, given that it is a probabilistic action based on *aware*, the scientist action is more likely to work for the popular government and so it has to use violent actions less frequently.

Evolution produces a counterintuitive effect when probabilistic actions are considered. Where the common reaction to having a better chance of positive outcomes would be to take more probabilistic actions, evolution takes the minimum number of probabilistic actions required to increase the survivor population. The fitness of these strings is only evaluated once; therefore the policies with high mean fitness based on probabilistic actions would have to be 'lucky' in a single evaluation and so children created by them might well not share in such luck. There is an evolutionary cost in taking a probabilistic action because a single evaluation does not accurately measure expected fitness.

The effect of evaluation of the fitness function has also been shown to sometimes hide the differences between government types. As we are looking at a single saved evaluation of the policy, we evaluate the policy once rather than multiple times, and average to see the effects of the random events. Thus the random events are shown at their luckiest. A different evaluation when the fitness function takes the average, or each member is re-evaluated every time it is used for breeding, would perhaps allow for a more critical look at the differences in the government types. The findings in the differences of the government types may be affected by this issue of randomness by not making multiple evaluations. Both of the types of governments can meet the same population limits by using differing actions (stochastic versus deterministic). This finding should be stronger when the fitness is evaluated as an average of multiple trials.

13.7 Conclusions and Future Directions

Governments have a role to play in the effort to slow down the spread of a zombie virus. Through actions such as quarantines and arming the citizens, the government extends the life of the population when facing a variety of different zombie types. In many cases, effective government policy can lead to saving a population from what would be its doom when the number of humans infected for every zombie killed is greatly unbalanced in favour of the walking dead. Governments must take into account the wishes of the populace, and this will affect the opportunities open to them.

During a real outbreak, the system can even be used in an 'online' capacity. That is, after an action is taken and a result is known, these new values could be used as the initial point for another iteration of the EA, allowing for a plan to be amended if necessary. This would be especially useful if the plan included stochastic events. Furthermore, changes to the tentative policies could be made to allow for the actions to be time dependent. Other effects could be changed to have time limits for the effects. For example, a vaccine might only have a short period of effectiveness as the virus mutates.

Multi-objective methods could also be used for fitness evaluation. As we have seen, there are perhaps three goals in the policy where we have only looked at one, maximizing the number of survivors. Other objectives include reduction in the numbers of zombies, and maintaining popular opinion in favour of the government. Multi-objective EAs optimize conflicting goals as part of fitness evaluation using techniques such as Pareto optimality.

We have seen that, in this first system of testing for government policy, the zombie outbreaks can be reduced in effect. Evolution allows for the system to quickly search policies, evaluate them and provide a set of actions that can be put forward by governments. It can be used to prevent the needless expense of life and resources. Outbreaks in the future will not see the deplorable use of initiatives like the Redeker Plan.

Appendix

A Glossary

Chromosome. A data structure that encodes a complete solution or a process for creating a complete solution. Furthermore, the structure should be robust: a small change will still be a valid solution.

Crossover. A binary operator that acts on the chromosomes to trade segments of the data structure between the chromosomes. This trade should create valid candidate solutions or have a process for changing an invalid structure into a valid structure.

Elitism. Sometimes, we are interested in always keeping a best solution. The elite are those chromosomes of high fitness that are transferred into the next generation without change. Any evolutionary algorithm having this property can be said to be elitist.

Evolutionary algorithm. An algorithm that uses the principles of evolutionary thought (Darwinian, Lamarkian, Baldwinian, etc.) in order to solve a problem. Generally, it involves a population of candidate solutions that undergo fitness evaluation, selection and breeding cycle. Common evolutionary algorithms include genetic algorithms, genetic programming, evolutionary programming and evolutionary strategies.

Fitness evaluation. Biologically, the evolutionary fitness of an individual is the number of offspring able to reach maturity and have offspring themselves. For an evolutionary algorithm, this concept is used to represent the ability of an individual to solve a problem and informs the selection mechanism as to its chances of being able to breed.

Mutation. A unary evolutionary operator that acts on a chromosome to provide a small change, usually at random, in the current structure of the chromosome. This change should create another valid structure or have a process for changing an invalid stucture into a valid structure.

Selection. Evolutionary algorithms are guided by the ideas of natural selection. The selection operation decides which of the individuals in the population should be allowed to breed based upon their fitness evaluations. Common mechanisms are fitness proportional selection, where each individual is given a probability to breed equal to its share in the total fitness score of all members of the populations; rank selection, in which rank replaces fitness; and tournaments, where a number of members of the population are randomly selected and the best are allowed to breed.

References

[1] G.A. Romero. *Dawn of the Dead*, directed by G.A. Romero. United Film Distribution Company, 1978.

[2] M. Brooks. *World War Z: An Oral History of the Zombie War*, Three Rivers Press, New York, NY, 2006.

[3] R. Sacchetto. *The Zombie Handbook: How to Identify the Living Dead and Survive the Coming Zombie Apocalypse*, Ulysses Press, Berkeley, CA, 2009.

[4] R. Ma. *The Zombie Combat Manual: A Guide to Fighting the Living Dead*, Berkley Books, New York, NY, 2010.

[5] M. Brooks *The Zombie Survival Guide: Complete Protection from the Living Dead.* Three Rivers Press, New York, NY, 2003.

[6] R. Rodriguez. *Robert Rodriguez's Planet Terror*, directed by R. Rodriguez. Dimension Films, 2007.

[7] T. Contento. Classification and causation of zombification, and guidelines for risk reduction and management. In *Braaaiiinnnsss!: From Academics to Zombies*, Robert Smith?, ed., University of Ottawa Press, Ottawa, ON, 2011.

[8] P. Munz, I. Hudea, J. Imad, R.J. Smith?. When zombies attack!: Mathematical modelling of an outbreak of zombie infection. In *Infectious Disease Modelling Research Progress* J.M. Tchuenche and C. Chiyaka, eds., Nova Science, Happuage, NY 2009, pp. 133–156.

[9] E. Wright, S. Pegg. *Shawn of the Dead*, directed by E. Wright. Universal Pictures, 2004.

[10] D. Ashlock. *Evolutionary Computation for Modeling and Optimization*, Springer, Berlin, Germany, 2006.

[11] J. Gunn. *Dawn of the Dead*, directed by Z. Synder. Universal Pictures, 2004.

[12] G.A. Romero, J.A. Russo. *Night of the Living Dead*, directed by G.A. Romero. The Walter Reade Organization, 1968.

[13] G.A. Romero. *Day of the Dead*, directed by G.A. Romero. United Film Distribution Company, 1985.

[14] M. Mitchell. *An introduction to genetic algorithms.* MIT Press, 1996.

BANELING DYNAMICS IN *LEGEND OF THE SEEKER*

Gergely Röst

Abstract

We propose a model of baneling dynamics, following the description in the television series *Legend of the Seeker*. A deceased human being may return to life as a baneling, allowed to remain in the world so long as it has killed at least one human in the previous day. We consider two intervention strategies, one in which humans have an effective way of killing banelings (wizard's fire), and one in which a cure is available but in very limited quantities (shadow water). With our model, we predict the future baneling dynamics and evaluate the two intervention strategies.

14.1 Introduction

PREVIOUS chapters have dealt directly with the zombie threat, with aspects including, but not limited to, spatial diffusion, government decision-making, the estimation of parameters, etc. However, in order to understand the zombie epidemic from all perspectives, it is instructive to examine the effects of similar undead creatures on humanity. Although banelings are mere fiction—unlike the zombies we all live in fear of—they are still helpful in showcasing the rise of an undead species.

In the *Legend of the Seeker* television series [1, 2], banelings are those who have died and been offered a second chance at life in service to the Keeper. The Keeper is the lord of the Underworld and his hatred for life is endless. The boundary between the world of the dead and the world of the living is called the veil, which prevents the Keeper from being loosed on the world of life. Due to some unfortunate events, the veil has been breached, allowing the Keeper to send the dead back to the world of the living. When a person dies, the Keeper may grant him a second chance in life in exchange for the person becoming a baneling. This will not stop until the veil is repaired.

Banelings are required to have killed at least one person sometime in the preceding twenty-four hours in order to stay in the world of the living. The clock is reset after each kill. Therefore killing multiple people at once will not result in multiple days of life for the baneling. Banelings are removed from the world of the living if they are unable to kill within any time interval of length one day. Humans have no effective way of killing the banelings. The only way to prevent banelings from remaining in the world of the living is burning their physical bodies.

The purpose of this paper is to create a dynamical model that takes into account the essential features of the baneling phenomenon to predict the future of the human population. Though banelings are fictional characters, our model resembles some real-life situations. These are discussed in the concluding section.

Banelings are undead corpses, just like zombies. However, previous models of zombie outbreaks [3] are not applicable to banelings. First, a person bitten by a zombie will become a zombie inevitably, while any dead person may become a baneling by accepting the evil offer of the Keeper regardless of the cause of death; i.e., anyone who has died, by natural or other causes, may become a baneling as a result of the choice made by that person. The other major difference that characterizes baneling dynamics compared to zombie outbreak models is the removal term, which has to reflect the requirement that a baneling must kill someone every day.

There are major differences in the appearance and behavior of zombies and banelings. Though these are not relevant to the mathematical modelling, for the sake of clear distinction and to provide adequate context to the nonexpert reader, we briefly outline the major similarities and differences between the characteristics of banelings and standard contemporary zombies.

Zombies display visible signs of desiccation, decay and emaciation on their face and body, and have blank, expressionless faces; on the other hand, banelings look like normal people. Thus, while zombies are easily recognizable by visual inspection after their incubation period, one can never be sure whether someone is a baneling or not. Banelings show symptoms such as rotting and deterioration of flesh only when they are near the end of their one-day limit without having killed anyone. When a zombie attacks, it will tear out organs from the living belly and feast upon them, while making moaning and guttural sounds. Banelings, on the contrary, prefer to slash the throat quickly, quietly and effectively, because their purpose is to stay in the world of the living as long as they can. Zombies have limited intelligence and they are mostly dangerous when gathered in relentless hordes, but a single baneling can outsmart and trick his opponent. Zombies stay in their pitiable state indefinitely (however, some speculate that eventually a decaying zombie will break down to the point where it cannot function anymore), and it is very difficult to kill them, since they are already dead; they are vulnerable only to attacks that chop the head off or crush the brain. By contrast, banelings will perish if they do not manage to kill, though, like zombies, they can only be killed with great difficulty and through specific methods (in this case, burning). For further details on zombies, banelings and the world they belong to, we refer to [4], [5], [6] and [1].

14.2 Model Derivation

In our model, we use the following state variables:

- $H(t)$ denotes the number of humans at time t,
- $B(t)$ denotes the number of banelings at time t,
- $P(t)$ denotes the number of permanently deceased humans at time t.

Since the events are happening in a relatively short time period, natural human demographics are neglected: there are no births and all deaths are attributed to the killing spree of the banelings. Let D be the rate humans are being killed by banelings, while $q \in [0, 1]$ is the fraction of killed persons who accept the Keeper's evil offer and become banelings. Those who refuse the Keeper's evil offer move into the class of permanently deceased humans and remain there for all time. Let R be the baneling removal rate: those banelings who were not able to kill in the last day must return to the Underworld, so they move into the class of permanently deceased humans and remain there. In this initial model, the humans have no proactive way of removing banelings, so the only negative effect on the baneling population arises from their inability to kill a human in the last twenty-four hours.

Then the basic model has the structure

$$\begin{aligned} H'(t) &= -D, \\ B'(t) &= qD - R, \\ P'(t) &= (1-q)D + R. \end{aligned}$$

Next we determine how D and R depend on the population sizes. The number of encounters between banelings and people is assumed to be proportional to both the human and baneling densities with a searching-efficiency parameter α. Let κ express the chance that such an encounter results in the killing of the human by the baneling. Then we have $D = \kappa\alpha H(t)B(t)$.

Given a baneling, it has a time window of one day to kill, or else it is removed from the baneling population. For the removal term R, we need to determine the fraction of banelings who were unsuccessful and unable to kill in the time interval $[t-1, t]$. The number of encounters of the baneling with humans is assumed to follow a Poisson distribution, and the expected number of encounters is proportional to the average human population size in this time interval and is equals to $\alpha\bar{H}_t$, where

$$\bar{H}_t \equiv \int_{t-1}^{t} H(u)du.$$

Hence the parameter of the Poisson distribution is $\lambda = \alpha\bar{H}_t$. The probability that a baneling had k encounters with humans in the time interval $[t-1, t]$ is then

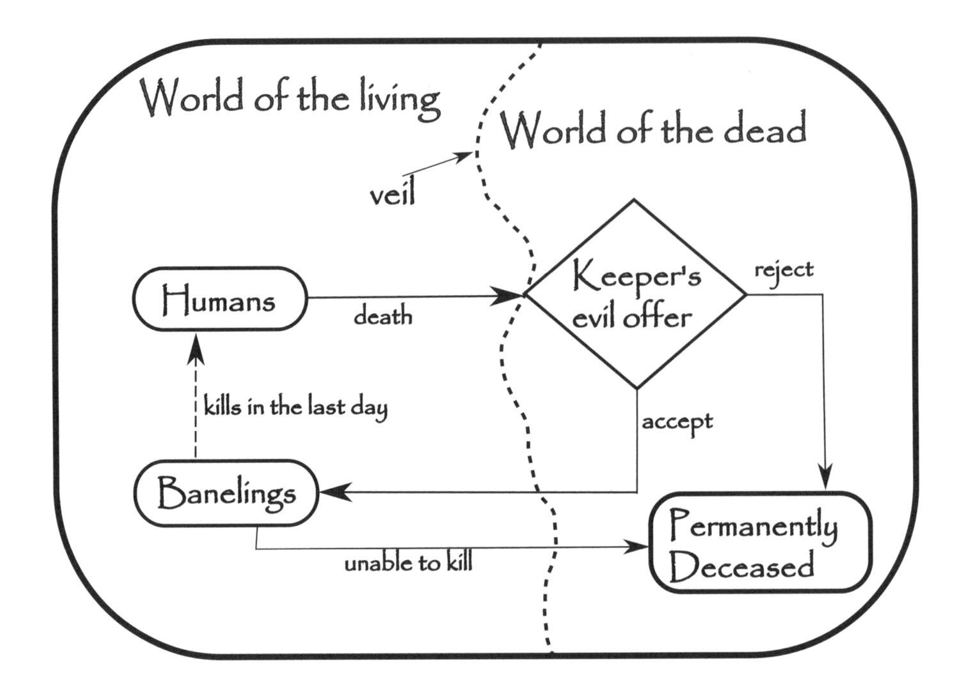

Figure 14.1: Baneling dynamics diagram. There are three population groups: humans, banelings and the permanently deceased. A dead human becomes a baneling (by accepting the Keeper's evil offer) or remains deceased permanently (by refusing the offer). Banelings stay in the world of the living as long as they have killed a human in the last day. If they fail to do so, then they move to the permanently deceased class.

$p_k = \frac{e^{-\lambda}\lambda^k}{k!}$ and the probability that all of these encounters were unsuccessful (from the baneling's perspective) is $(1-\kappa)^k$. The chance that a baneling was unable to kill is

$$\sum_{k=0}^{\infty}(1-\kappa)^k p_k = e^{-\lambda}e^{\lambda(1-\kappa)} = e^{-\lambda\kappa} = e^{-\alpha\bar{H}_t\kappa}.$$

Thus our baneling model becomes

$$H'(t) = -\kappa\alpha H(t)B(t) \tag{14.1}$$

$$B'(t) = q\kappa\alpha H(t)B(t) - B(t)\exp\left[-\alpha\kappa\int_{t-1}^{t} H(u)du\right] \tag{14.2}$$

$$P'(t) = (1-q)\kappa\alpha H(t)B(t) + B(t)\exp\left[-\alpha\kappa\int_{t-1}^{t} H(u)du\right]. \tag{14.3}$$

Since $H(t)+$ $B(t)+$ $P(t)$ is constant, and the equation for $P'(t)$ decouples, we omit the variable $P(t)$ from the further analysis.

14.3 Analysis of a Simplified Baneling Model

We first consider a simplification of our model by approximating $\int_{t-1}^{t} H(u)du$ with $H(t)$. Since $H(t)$ is decreasing, this estimation results in a smaller value of $B'(t)$, so that the model will underestimate the number of banelings compared to the model (14.1)-(14.2). The simplified model takes the form

$$H'(t) = -\kappa\alpha H(t)B(t) \tag{14.4}$$
$$B'(t) = q\kappa\alpha H(t)B(t) - B(t)\exp[-\alpha\kappa H(t)]. \tag{14.5}$$

Note that this model is similar to some of the classical predator–prey, host–parasite and disease-spread models [7], but not identical to any of them. Given the biological meaning of the model, we consider solutions only with nonnegative initial values (H_0, B_0). Notice that the system has a continuum of baneling-free equilibria (BFE) given by $(H_0, 0) \in \mathcal{R}^2$ (since we are assuming that there are no natural deaths). We summarize the basic properties of system (14.4)-(14.5) in the following theorem.

Theorem 1 *The solutions of (14.4)-(14.5) remain positive and $H(t)$ is monotone decreasing. The limits $H_\infty \equiv \lim_{t\to\infty} H(t)$ and $B_\infty \equiv \lim_{t\to\infty} B(t)$ exist with $B_\infty = 0$.*

Proof. From (14.4)-(14.5), we can express $H(t)$ and $B(t)$ as

$$H(t) = H_0 \exp\left[-\int_0^t \kappa\alpha B(s)ds\right]$$

and

$$B(t) = B_0 \exp\left[\int_0^t q\kappa\alpha H(s) - \exp[-\alpha\kappa H(s)]ds\right].$$

Hence they remain positive for all $t > 0$ if the initial values are positive. It follows that $H(t)$ is monotone decreasing and thus converges. Note that

$$\frac{d}{dt}[H(t) + B(t)] = -(1-q)\kappa\alpha H(t)B(t) - B(t)\exp[-\alpha\kappa H(t)] \leq 0.$$

It follows that the combined human-baneling population decreases during a baneling outbreak in our model. Considering

$$\frac{d}{dt}[qH(t) + B(t)] = -B(t)\exp[-\alpha\kappa H(t)] \leq -B(t)\exp[-\alpha\kappa H_0],$$

we conclude that $B(t)$ converges to zero since otherwise $qH(t)+B(t)$ would become negative, which is not possible. □

Now we define the baneling reproduction number $\mathcal{B}_0$: in the initial phase when banelings first appear, the number of humans is approximately H_0. The sojourn time of one single baneling in the world of the living is $1/\exp[-\alpha\kappa H_0]$ on average and it kills $\kappa\alpha H_0$ humans per day, resulting a total number of

$$\mathcal{B}_0 \equiv q\kappa\alpha H_0 \exp[\alpha\kappa H_0]$$

new banelings. We say that a baneling outbreak is occurring when the number of banelings starts to increase rapidly. By preventing the outbreak, we mean that (possibly because of human intervention) the number of banelings cannot increase from the very beginning and the baneling population drops to zero with minimal human deaths.

Theorem 2 *System (14.4)-(14.5) has threshold parameter $\mathcal{B}_0$: there is an outbreak of banelings if and only if the baneling reproduction number $\mathcal{B}_0$ is greater than* 1.

Proof. One can see that $B(t)$ is increasing if and only if

$$q\kappa\alpha H(t) - \exp[-\alpha\kappa H(t)] > 0.$$

It is easy to see that $f(z) \equiv q\kappa\alpha z - \exp[-\alpha\kappa z]$ is a monotone increasing function of z with a unique positive zero z^*. Since $H(t)$ is decreasing, $f(H(t))$ is also decreasing as time elapses, so we can conclude that there are threshold dynamics. There is no baneling outbreak if

$$f(H_0) = q\kappa\alpha H_0 - \exp[-\alpha\kappa H_0] < 0,$$

or equivalently $\mathcal{B}_0 < 1$. In the opposite case, the number of banelings will initially increase, until $H(t)$ drops below z^*. In the special situation $\mathcal{B}_0 = 1$, $B'(0) = 0$, but, for any $h > 0$, the monotonicity of $H(t)$ and $f(z)$ imply that

$$q\kappa\alpha H(h) - \exp[-\alpha\kappa H(h)] < 0$$

so $B'(h) < 0$. Thus we observe an outbreak of banelings if and only if the baneling reproduction number satisfies $\mathcal{B}_0 > 1$. □

Since $\mathcal{B}_0 \equiv q\kappa\alpha H_0 \exp[\alpha\kappa H_0]$, one can see that if $\kappa\alpha H_0 \exp[\alpha\kappa H_0] < 1$, then $\mathcal{B}_0 < 1$ regardless of the value of q; i.e., it does not matter how many people accept the Keeper's evil offer, there will be no outbreak. For example, if $\alpha\kappa = 0.01$ and the population size is $H_0 = 400$, then the critical value of q is approximately 0.005 (that is, only one out of every 200 deceased humans becomes a baneling). If q is larger than that, then there will be a baneling outbreak. Besides q, humans may decrease the searching parameter α (by hiding from the banelings) or the chance κ of being killed upon an encounter with a baneling (by improving their defence skills). For example, if $q = 0.2$, then the humans can prevent the baneling outbreak by reducing $\alpha\kappa$ below 0.003.

Theorem 3 *The fraction of the human population that survives the baneling outbreak can be determined from the final-size relation*

$$a\kappa qH_0 - Ei(-a\kappa H_0) = a\kappa qH_\infty - Ei(-a\kappa H_\infty),$$

where $Ei(x) = \int_{-\infty}^{x} \frac{e^t}{t} dt$ *is the first exponential integral.*

Proof. For the properties of the first exponential integral $Ei(x)$, see [8]. We have $Ei'(x) = \exp[x]/x$. Consider the function

$$W(H, B) = B + qH - Ei(-\alpha\kappa H)/\alpha\kappa.$$

Then

$$\begin{aligned}\frac{dW}{dt} &= B'(t) + qH'(t) - \big(\exp[-\alpha\kappa H(t)]/\alpha\kappa H(t)\big)H'(t)\\ &= q\kappa\alpha H(t)B(t) - B(t)\exp[-\alpha\kappa H(t)] - q\kappa\alpha H(t)B(t)\\ &\quad - \big(\exp[-\alpha\kappa H(t)]/\alpha\kappa H(t)\big)(-\kappa\alpha H(t)B(t))\\ &= 0.\end{aligned}$$

We can conclude that W is a first integral of our system.

Using $W(H_0, B_0) = W(H_\infty, 0)$, we obtain the final-size relation

$$\alpha\kappa qH_0 - Ei(-\alpha\kappa H_0) = \alpha\kappa qH_\infty - Ei(-\alpha\kappa H_\infty).$$

The final-size relation is graphically represented in Figure 14.2. □

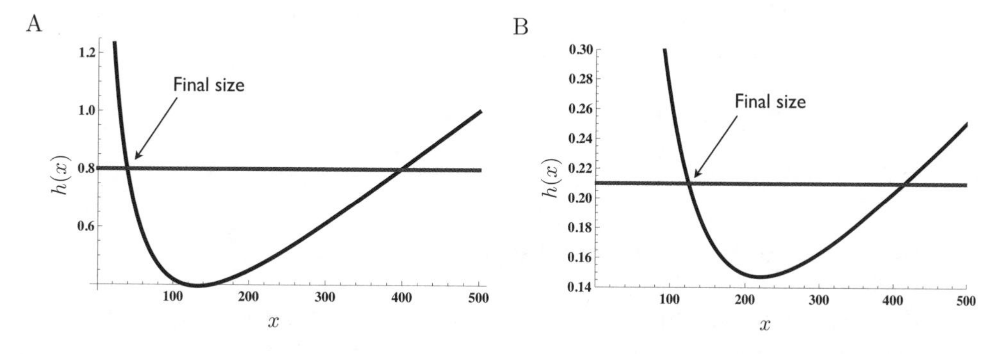

Figure 14.2: Plot of the function $h(x) = q\alpha\kappa x - Ei(-\alpha\kappa x)$. The final size is determined by the relation $h(H_0) = h(H_\infty)$. A: Parameter values are $q = 0.2$ and $\alpha\kappa = 0.01$. For an initial population of $H_0 = 400$, we have $h(H_0) \approx 0.8$, so the final size is given by the other intersection of $h(x)$ with the horizontal line; i.e., by a value x_* for which $h(x_*) = 0.8$, so $H_\infty \approx 50$. B: Parameter values are $q = 0.05$ and $\alpha\kappa = 0.01$. For an initial population of 400, the final size is 130.

14.4 Intervention 1: Wizard's Fire

To be removed from the world of the living, a baneling's body must be burned. In the television series, a particularly potent method of burning large quantities of bodies is the so-called wizard's fire [1], which is a possible intervention strategy. To incorporate the effect of the wizard's fire into our model in a simplified way, we assume that banelings can be killed and burned at rate w. Thus our model takes the form

$$H'(t) = -\kappa\alpha H(t)B(t) \tag{14.6}$$

$$B'(t) = q\kappa\alpha H(t)B(t) - B(t)\exp[-\alpha\kappa H(t)] - wB(t). \tag{14.7}$$

This modified model can be treated analogously to system (14.4)-(14.5). Define the wizard-baneling reproduction number

$$\mathcal{B}_w \equiv \frac{q\kappa\alpha H_0}{\exp[-\alpha\kappa H_0] + w},$$

which expresses the expected number of banelings generated by a single baneling in the early phase of the outbreak, when wizard's fire intervention is present.

Theorem 4 *The solutions of (14.6)-(14.7) remain positive and $H(t)$ is monotone decreasing. The limits $H_\infty \equiv \lim_{t\to\infty} H(t)$ and $B_\infty \equiv \lim_{t\to\infty} B(t)$ exist with $B_\infty = 0$. System (14.6)-(14.7) has threshold parameter $\mathcal{B}_w$: there is an outbreak of banelings if and only if the wizard-baneling reproduction number satisfies $\mathcal{B}_w > 1$.*

Proof. This can be shown analogously to Theorems 1 and 2; the only difference is that, instead of $f(z)$, we use $g(z) \equiv q\kappa\alpha z - \exp[-\alpha\kappa z] - w = f(z) - w$. □

Theorem 4 implies that banelings can be controlled by sufficiently intense wizard's fire. If w is sufficiently large; or, more precisely, if

$$w > f(H_0) = q\kappa\alpha H_0 - \exp[-\alpha\kappa H_0],$$

then $g(H_0) < 0$ and $\mathcal{B}_w < 1$, so the baneling outbreak is contained.

14.5 Intervention 2: Treatment by Shadow Water

In the television series, banelings can be cured (turned back into humans) by shadow water, but this resource is extremely scarce. In episode 34 ("Hunger," Episode 12 of Season 2; see [9] and [2]) a source of shadow water is discovered and used to cure banelings, but the Keeper finds out and demolishes the source. Taking that episode as a guide, we will suppose that, once the treatment by shadow water starts, the Keeper will learn the location of the source very quickly and destroy it. Therefore the cure will be available only for a limited amount of time. For the modelling of this very brief window of opportunity for initiating treatment, we assume that all available cures are administered at once at some time T, healing a fraction σ of the banelings, who then become normal humans.

Thus our model can be formulated as an impulsive system with one impulse at $t = T$. More precisely, for $t \neq T$, we have the basic model

$$H'(t) = -\kappa\alpha H(t)B(t) \tag{14.8}$$
$$B'(t) = q\kappa\alpha H(t)B(t) - B(t)\exp[-\alpha\kappa H(t)], \tag{14.9}$$

while, for $t = T$, we have the discontinuity

$$H(T+) = H(T) + \sigma B(t) \tag{14.10}$$
$$B(T+) = (1 - \sigma)B(T), \tag{14.11}$$

where $0 < \sigma < 1$.

For further details on the theory and simulations of impulsive systems in life sciences, we refer to [10]. Note, however, that cured banelings are susceptible to becoming banelings in the future, at the same rate as the rest of the human population. The shadow treatment impulse decreases the number of banelings and increases the number of humans, but the long-term effect of such a healing impulse is not immediately obvious. The following theorem provides an encouraging result.

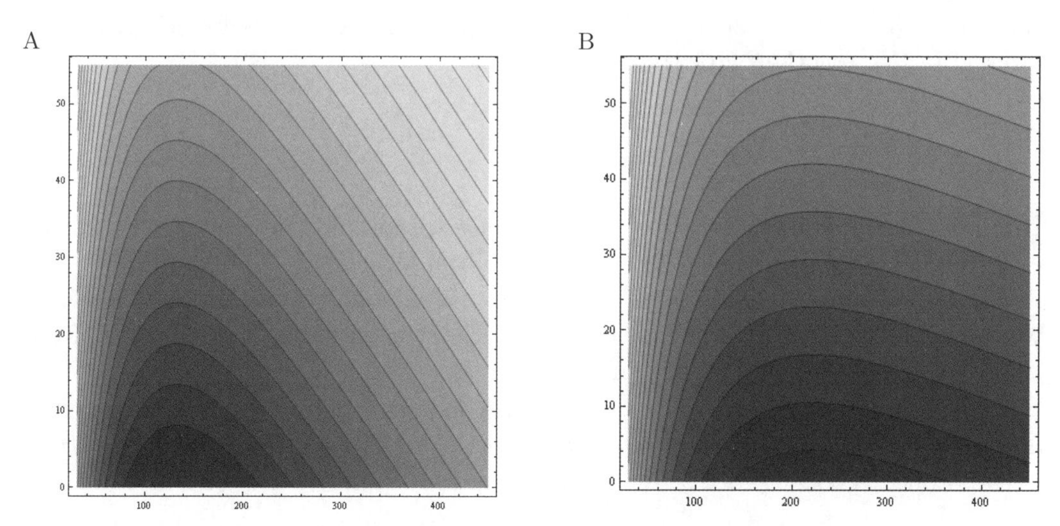

Figure 14.3: Level curves of the first integral $W(H, B) = B + qH - Ei(-\alpha\kappa H)/\alpha\kappa$ on the (H, B)-plane. The phase curves of the solutions move left along the level curves, since $H(t)$ is decreasing. A: Parameter values are $q = 0.2$ and $\alpha\kappa = 0.01$. B: $q = 0.05$ and $\alpha\kappa = 0.01$.

Theorem 5 *Even a single impulse of shadow water healing increases H_∞, the final size of the human population.*

Proof. The downward-pointing normal vector of the level curve of the first integral W (see Figure 14.3) is:

$$\mathbf{n} = (-W_H, -W_B) = \left(-q + \frac{\exp[-\alpha\kappa H]}{\alpha\kappa H}, -1\right).$$

An impulse at time T pushes the state of the planar system from (H, B) to $(H + \sigma B, (1 - \sigma)B)$ by the shadow-water treatment vector $\mathbf{s} = (\sigma B, -\sigma B)$. We can see that the scalar product $\mathbf{n} \cdot \mathbf{s} = 1 - q + \frac{\exp[-\alpha\kappa H]}{\alpha\kappa H} > 0$, which means that, after the impulse, the state of the system jumps to a level set that corresponds to a lower value of W and thus a larger final population size. □

Besides increasing the final human population size, shadow-water impulses have an additional potential benefit: the baneling peak can be delayed. See, for example, the numerical simulations in Section 7.

14.6 The Time-Delay Baneling Model

Our original model (14.1)-(14.3) is a delay differential system with distributed delay, because the second equation involves the history of $H(t)$. The analysis of such systems is fairly technical [11]. To define a dynamical system, we use the phase space $C \times R$, where C denotes the Banach space of continuous functions on the interval $[-1, 0]$ together with the norm

$$||\phi|| = \sup_{t\in[-1,0]} |\phi(t)|, \quad \phi \in C.$$

It is possible to define a semiflow on our phase space, and the standard existence and uniqueness results hold, for specified initial values $H_0 \in C$ and $B_0 \in R$. Although the detailed analysis of this system is beyond the scope of this study, we can nevertheless draw some conclusions without using the advanced tools of the theory of delay differential equations.

The positivity of solutions can be shown for this system as well, completely analogously to the case of the simplified system of ODEs. Then it follows that $H(t)$ is monotone decreasing for $t \geq 0$ and therefore converges; we also obtain

$$H(t-1) \geq \int_{t-1}^{t} H(u)du \geq H(t).$$

This implies

$$\exp[-\alpha\kappa H(t)] \geq \exp\left[-\alpha\kappa \int_{t-1}^{t} H(u)du\right] \geq \exp\left[-\alpha\kappa H(t-1)\right].$$

To prove that $B(t)$ tends to zero, consider

$$\begin{aligned}\frac{d}{dt}[qH(t)+B(t)] &= -B(t)\exp\left[-\alpha\kappa\int_{t-1}^{t} H(u)du\right] \\ &\leq -B(t)\exp[-\alpha\kappa H(t-1)] \\ &\leq -B(t)\exp[-\alpha\kappa H_0],\end{aligned}$$

for $t \geq 1$; hence $B(t)$ must converge to zero, or else the left-hand side becomes negative. Since

$$\begin{aligned}B'(t) &= q\kappa\alpha H(t)B(t) - B(t)\exp\left[-\alpha\kappa\int_{t-1}^{t} H(u)du\right] \\ &\geq q\kappa\alpha H(t)B(t) - B(t)\exp[-\alpha\kappa H(t)]\end{aligned}$$

for $t \geq 1$, we can use standard comparison arguments to deduce that the simplified baneling model predicts a lower number of banelings than the model with distributed delay.

If we approximate $\int_{t-1}^{t} H(u)du$ by $H(t-1)$, then we obtain the system

$$H'(t) = -\kappa\alpha H(t)B(t), \tag{14.12}$$

$$B'(t) = q\kappa\alpha H(t)B(t) - B(t)\exp[-\alpha\kappa H(t-1)] \tag{14.13}$$

with a single constant delay. The basic properties of this system can be deduced similarly to Theorem 2, but the further analysis—for example, finding an appropriate invariant that replaces the first integral—is again beyond the scope of this study. Nevertheless, from the monotonicity of $H(t)$ and standard comparison arguments, we can conclude that the single-delay model overestimates the baneling population.

14.7 Numerical Simulations

We have run computer simulations for various parameter values. For the sake of comparison, we consider two outbreaks of different severity. For both scenarios, we also examine the effect of interventions. We choose our population size to be 400. In the first case, let $q = 0.2$ and $\alpha\kappa = 0.01$. These parameter values appear in Figures 14.2 and 14.3 as well. In the more severe case, we assume that half of the deceased accept the Keeper's evil offer and that the banelings are three times more effective; i.e., $q = 0.5$ and $\alpha\kappa = 0.03$. The parameters were chosen because they make it easy to highlight the interesting features of the model.

The time course of these two baneling outbreaks can be seen in Figure 14.4. We can see that, in the first case, a small number of people survive, while in the more serious case the banelings reduce the population to a negligible size.

Figure 14.5 shows the application of wizard's fire with $w = 0.6$. We can see that the wizard's fire was very effective in the first case and prevented almost half

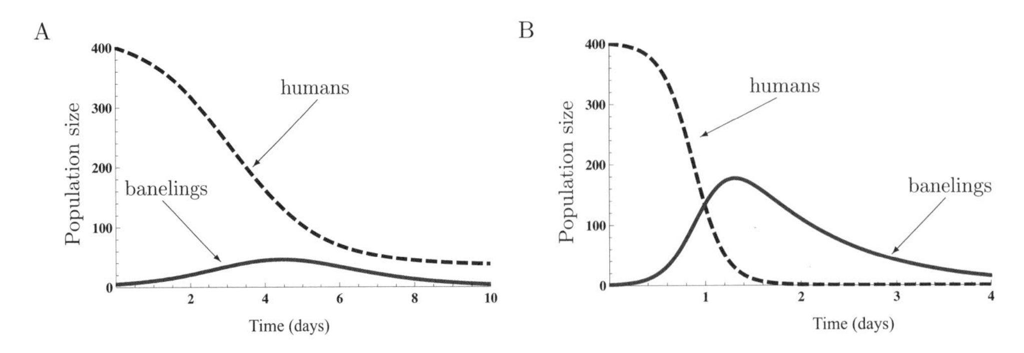

Figure 14.4: Time course of two baneling outbreaks without intervention in a population of 400. A: $q = 0.2$ and $\alpha\kappa = 0.01$. B: $q = 0.5$ and $\alpha\kappa = 0.03$.

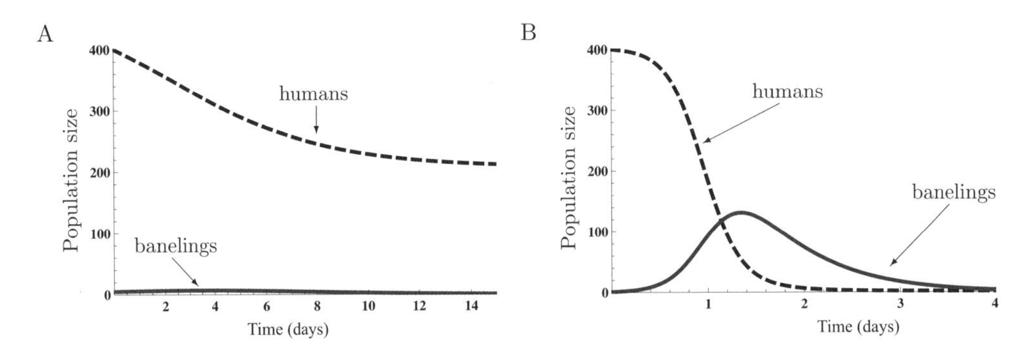

Figure 14.5: Time course of two baneling outbreaks in a population of 400, with wizard's fire $w = 0.6$. A: $q = 0.2$ and $\alpha\kappa = 0.01$. B: $q = 0.5$ and $\alpha\kappa = 0.03$.

of the baneling murders, but it only slightly mitigated the damage in the second case. To prevent the outbreak (i.e., to ensure that $\mathcal{B}_w < 1$), the necessary intensity of wizard's fire must be at least $w = q\kappa\alpha H(0) - \exp[-\alpha\kappa H(0)] = 0.78$ in the first case, and $w = 6$ in the second case.

Figures 14.6 and 14.7 demonstrate the effect of shadow-water treatment in various situations. We can see that the the impact of the impulsive healing depends crucially on the timing of the cure. It seems that, for maximal benefit, shadow-water treatments should be administered at the peak time of the baneling outbreak. However, this conjecture is still to be proven. Conversely, if the outbreak is too severe, the impulsive intervention has no long-term benefit.

14.8 Summary and Conclusions

We have developed a dynamic population model to analyze the time course of a baneling outbreak, following the description in the television series *Legend of the Seeker*. Our model allowed us to make predictions for various values of the key parameters, such as a) the efficiency of banelings, b) the relative number of people

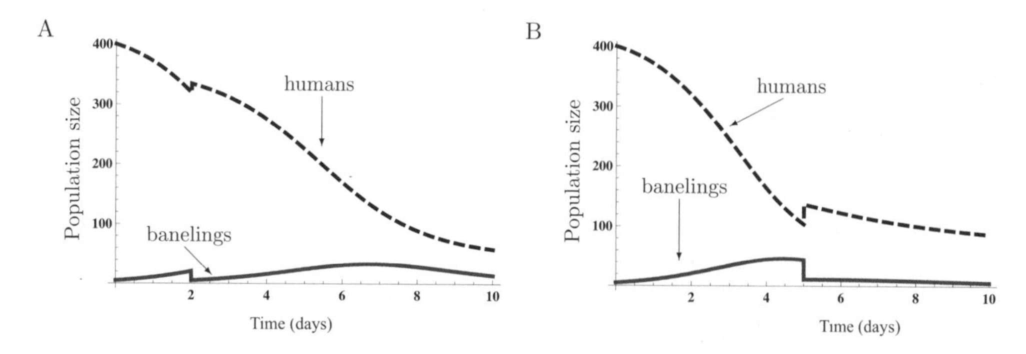

Figure 14.6: The effect of shadow water treatment impulse with $\sigma = 0.75$ in a population of 400, when $q = 0.2$ and $\alpha\kappa = 0.01$. A: Healings occurred at $T = 2$. B: Healings occurred at $T = 4.7$.

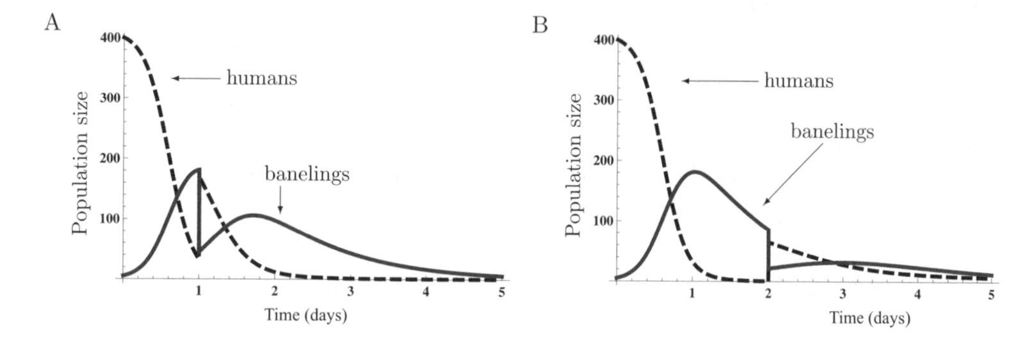

Figure 14.7: The effect of shadow water treatment impulse with $\sigma = 0.75$ in a population of 400, when $q = 0.5$ and $\alpha\kappa = 0.03$. A: Healings occurred at $T = 1$. B: Healings occurred at $T = 2$.

who accept the Keepers evil offer and return as banelings upon their deaths, and c) the human population density.

The major conclusion of our research is that an outbreak of banelings occurs if and only if the baneling reproduction number is greater than one. The baneling reproduction number expresses the average number of new banelings produced by the deeds of one single baneling in the initial phase of the outbreak. This number depends on how effectively the banelings are able to find and kill humans, and how many humans accept the Keeper's evil offer and become banelings themselves. Taking all these paremeters into account, our model is able to predict whether there will be an outbreak of banelings or their numbers will drop quickly without causing significant damage to the human population.

In any case, banelings will eventually disappear. This makes sense, since they perish if humans can defend themselves effectively; but they also perish if they are so successful that they manage to kill most humans. In this latter case, there remain no available victims to fulfill the requirement of the Keeper to kill someone every day.

The most important question is how many people survive the baneling attacks. We derived a formula that gives a precise answer to this question. Our results show that banelings can never completely eradicate the human population, since the final population size is always positive. However, this positive final size can be too small to be viable.

Humans can increase the odds of surviving a baneling encounter by training and improving their defence skills. The number of such encounters can be reduced by hiding from the banelings. These are reflected in the model parameters; by changing those, we can calculate how the final population size changes. We have also examined two possible intervention strategies: the application of wizard's fire and shadow water. Wizard's fire is an effective way of removing banelings. The outbreak can be controlled by sufficiently intensive baneling killing combined with wizard's fire. Shadow-water treatment effectively cures banelings, but cannot be applied continuously because this resource is very scarce and the Keeper destroys the resources shortly after treatments start. However, even one impulse of shadow water healing delays the baneling peak, and also increases the final size of the human population.

On a demographic time scale, if the veil cannot be repaired and the humans are capable of sustaining a viable population, we can expect a constant baneling supply by natural deaths, so the banelings are expected to become endemic.

Though banelings are fictional characters, our model resembles some real-life situations. For example, one can think of a severe disease, where a patient who would otherwise perish needs care every day, but there is a chance that the caregivers die or get infected. Then wizard's fire plays a similar role to permanent isolation of the sick, and shadow-water treatment corresponds to a cure that is available in very limited quantities.

Another analogue from real life is the rapid spread of a fashion, political agenda or evangelization. Then banelings are activists who spread some ideas. If they cannot recruit new followers in a time interval of some length, then they become disappointed, lose enthusiasm and give up the idea themselves. Unsuccessfully approached individuals will be dismissive of the idea forever. In this context, wizard's fire corresponds to some counter-propaganda which constantly reduces the number of followers, while shadow-water treatment can be some particular event that reflects bad light on the given idea and makes some followers turn away from it (but later they can be convinced to rejoin).

The television series *Legend of the Seeker* is based on *The Sword of Truth* fantasy novel series by Terry Goodkind [5]. In the books, banelings are the agents of the Keeper, doing his bidding. In this chapter, we have chosen to follow the representation of the TV series, where banelings are undead people who are required to kill a human every day. And finally, an obvious real-life application of the work herein is the very zombie epidemic we are all facing. Good luck.

Appendices

A Acknowledgements

The author is grateful to Tabrett Bethel (Cara) and Craig Parker (Darken Rahl) for the inspiration. The author was partially supported by the following grants: OTKA K75517, TÁMOP-4.2.2/08/1/2008-0008 and ERC StG Nr. 259559. The author is thankful for the enormous work of the anonymous referees.

Competing interests. The manuscript recommends the application of wizard's fire and shadow water. The author has never received any funds from the Magic Guild or the shadow water industry and has no potential conflict of interest (except his personal admiration for one specific Mord-Sith).

B Glossary

Baneling. A fictional creatures in the TV series *Legend of The Seeker*. They are deceased persons who accepted the offer of the evil ruler of the underworld so that they can return to the world of the living and remain there as long as they kill at least one person each day.

Baneling reproduction number. In epidemiology, the basic reproduction number is a key concept, expressing the expected number of secondary infections generated by a single infected individual introduced into a susceptible population. Analogously, the baneling reproduction number represents how many banelings will appear due to the activities of a single initial baneling.

Final-size relation. The final size relation is an equation that relates the initial population sizes, the model parameters and the final population sizes. Having known the former ones, one can use this equation to predict the final population size.

First integral. A function of the model variables, which is a conserved quantity (which in many physical systems can be interpreted as energy); i.e., it does not change as time elapses, although the variables themselves change in time. The first integral tells you very important information about the behaviour of solutions.

Impulse. A discontinuous jump.

Impulsive system. A set of differential equations with discontinuities representing interventions that happen in a short time with a large effect. Impulses generate discontinuous solutions in models where the state changes smoothly between impulses.

Legend of the Seeker. An epic fantasy television series, based on the series of novels *The Sword of Truth* by US fantasy writer Terry Goodkind. The show premiered in November 2008 and ran for two seasons.

Shadow water. A scarce substance that can cure banelings; i.e., turn them into normal human beings again.

Wizard's fire. Wizard's fire is a particularly strong fire, shrieking through the air as if shot from the fingers of high-level wizards, which can effectively burn to ashes a large number of bodies, thus preventing them from returning as banelings.

References

[1] Sword of Truth Wiki. http://sot.wikia.com/wiki/Main_Page

[2] Wikipedia, Legend of the Seeker. http://en.wikipedia.org/wiki/Legend_of_the_Seeker

[3] P. Munz, I. Hudea, J. Imad, R.J. Smith?. When zombies attack!: Mathematical modelling of an outbreak of zombie infection. In *Infectious Disease Modelling Research Progress*, J.M. Tchuenche and C. Chiyaka, eds., Nova Science, Happuage, NY 2009, pp. 133–156.

[4] A.T. Blumberg, A. Hershberger. *Zombiemania.* Telos Publishing, Tolworth, UK, 2006.

[5] T. Goodkind. *The Sword of Truth.* Tor Fantasy, New York, NY, 1994–2007

[6] Monstrous http://zombies.monstrous.com/

[7] L. Edelstein-Keshet. *Mathematical Models in Biology.* SIAM, Philadelphia, PA, 2005.

[8] N.M. Temme. Exponential, Logarithmic, Sine, and Cosine Integrals. In *NIST Handbook of Mathematical Functions* F.W.J. Olver, D.W. Lozier, R.F. Boisvert, C.W. Clark, eds., Cambridge University Press, Cambridge, UK, 2010, pp. 149–158.

[9] Internet Movie Database, *Legend of the Seeker: Hunger* (Season 2, Episode 12). http://www.imdb.com/title/tt1562516/combined

[10] J. Karsai. *Models of Impulsive Phenomena: Mathematica Experiments.* Typotex, Budapest, 2002

[11] Y. Kuang. *Delay Differential Equations with Applications in Population Dynamics.* Academic Press, Boston, MA, 1993

THE ZOMBIE SWARM: EPIDEMICS IN THE PRESENCE OF SOCIAL ATTRACTION AND REPULSION

Evelyn Sander and Chad M. Topaz

Abstract

We develop a spatiotemporal epidemic model incorporating attractive-repulsive social interactions similar to those of swarming biological organisms. The swarming elements of the model describe the ability of distinct classes of individuals to sense each other over finite distances and react accordingly. Our model builds on the 2009 non-spatial SZR model of Munz *et al.* modelling a zombie attack. This case is interesting from the modelling standpoint, as zombie epidemics are particularly virile, albeit restricted to a theatre near you and certain parts of San Francisco, Washington DC, and other major metropolitan areas [http://www.zombiewalk.com]. Spatial effects not only enhance entertainment value (as a cinematic portrayal of a spatially invariant zombie attack would be somewhat lacking in thrill), but are critical to understanding the epidemic. We show that, in the absence of a cure for zombie-ism, the alert human population will eventually be annihilated, but at a slower rate than in the non-spatial model. The extra time to extinction might allow the development of a cure. We also show that, without any assumption of collusion, the system self-organizes into transient travelling pulse solutions consisting of a swarm of zombies in pursuit of a swarm of alert humans. In the presence of a zombie cure, the travelling solutions exist and are stable for all time.

15.1 Introduction

THE notion of a zombie has its roots in the African Kingdom of Kongo, from whence the African diaspora brought it to Haiti. In Haitian voodoo, a sorcerer (or boko) could force someone (either living or dead) to become a zombie, while the boko maintained control of the soul. In contrast, the more modern view of zombie-ism is that it does not result from a sorcerer's control, but rather is an uncontrolled and unmodulated epidemiological effect, spread by contact like a disease. This 'viral' zombie-ism is depicted the classic 1968 movie *Night of the Living Dead*, in Michael Jackson's 1993 music video *Thriller*, in the computer game *Plants versus Zombies* and at public events such as zombie walks and the game *Humans vs. Zombies* played on many college campuses.

On the scientific front, in 2009, Munz et al. [1] first quantified the epidemiological view of zombies. This work (of which we later provide a detailed discussion) adapted classical epidemiological models to predict the course and ultimate outcome of zombie attacks. This important, foundational zombie modelling effort focused on the zombification of a population of healthy alert humans, but did not consider any geographic or spatial information pertaining to zombie attacks. As seen in many previous chapters, spatial information is often critical for understanding how to effectively treat an epidemic. For example, consider the citizens of Copenhagen in 1349, during the bubonic plague. A year later, the epidemic arrived in the city and spread explosively. Knowledge that the epidemic front was moving steadily northwards in an east-west band from the south [2] would have been much more useful for Copenhangen's citizens than only knowing that the epidemic was adversely effecting the total population of Europe.

In this chapter, we develop an epidemiological model of zombies that addresses the need for spatial information. A previous chapter incorporated spatial movement using spatial diffusion, which assumes that individuals move at random and each interaction with an infected individual results in a potential spread of the infection. Other chapters developed agent-based models (ABM), which ascribes a similarly random component to the movement of zombies. However, random motion is not an appropriate assumption in the case of zombies, who by nature seek out healthy humans as prey. Furthermore, ABMs by their very nature are restricted to computer simulations and cannot be generalized as theoretical models can. Thus our modelling methods will necessarily be different.

One aspect of the standard epidemiological models that we do adopt here is the *mean field hypothesis*, which says that the behaviour of individuals can be understood by considering a continuous function describing average behaviour. From observations of fluids, one finds it perfectly plausible that their flow and motion can be modelled (for most applications) by such averages, rather than by keeping track of the motion of individual molecules. One might wonder if this is a reasonable modelling assumption to make about the macroscopic behaviour of populations of complex biological organisms. In fact, such an assumption does prove to be quite successful in many situations, and the mean field approach in biology is supported by a substantial literature tracing its roots (at least) back to work such as Keller and Segel [3] and Okubo [4]. As a visual example, consider the 'human wave' crowd phenomenon in a stadium. Although each individual initiates an individual action to stand and sit, the macroscopic effect resembles the motion of a fluid wave.

We now return to the question of how to model zombies, keeping in mind that their motion is mostly directed, rather than purely random. Our goal is to address the observation that zombies can behave in an organized fashion, such as pursuing humans in packs. In a modern zombie epidemic, there is no external controlling force. A possible explanation is collusion; that is, the agreement of a master plan among the zombies (or perhaps as assigned by a zombie leader). We discard this possibility as well, due to the minimal brain power possessed by zombies. They may have the brain power of a fish or a bird, but certainly not sufficient mental capacity to design any sophisticated strategy of directed attack. Consider, however, that both birds and fish can behave in organized schools and swarms, and that these behaviours arise without control or collusion. A body of mathematical research has successfully modelled such swarming and flocking behaviour. The key model element is that each individual is able to sense the behaviour of others within reasonable proximity and to instinctively react accordingly. The result of the reaction to local behaviour is that the population spontaneously organizes itself into a structure such as a swirling school of fish, a directed flock of birds or a travelling wave of sitting-and-standing humans in a sports arena.

Based on the successes of the swarm modelling approach, the spatial aspects of the epidemiological model we will present are adapted from the swarming literature, with the assumption that each individual zombie is attracted to populations of alert humans, while each alert human is repelled by zombies. Our primary result is that the population self-organizes into swarms of zombies pursuing humans. If there is no cure for zombie-ism, the alert human population eventually dies out, but not as quickly as predicted by a non-spatial epidemic model. However, in the presence of a cure, it is possible for a population of zombies to pursue a population of alert humans for all time: a travelling wave. In addition, we find many other configurations of zombies in pursuit, with humans on the run. Much can be gained by full knowledge of these different moving population configurations, and the likelihood of successful persistence of the alert human population. Specifically, an understanding of the types of behaviour possible and how to recognize their onset is critical for the purposes of planning and preparation for large-scale epidemiological events.

The remainder of this introduction consists of a detailed and more technical overview of the body of scientific literature on epidemiological models and swarming models, and a description of how our modelling fits into this framework.

Epidemiological modelling in perspective. Though the history of epidemic modelling reaches back to studies of smallpox in the mid-eighteenth century [5, 6], the incorporation of spatial effects has roots in the last hundred years [7, 8, 9]. One major class of spatial models treats the population as a continuum in space and tracks the population density in space and time (using the mean field assumption discussed above). Within this realm of continuum models, the most basic studies of spatial epidemics modify compartmental models, based on non-spatial differential equations, which divide a population into different epidemiological classes such as susceptible, infected and removed. The basic spatial models add simple diffusion as the macroscopic description of the underlying random motion of individuals within a population [10, 11], and focus on the propagation of epidemic fronts from a clustered initial condition of infected individuals. A generalization of these models is the distributed infectives model [12, 13], which incorporates longer-range dispersal than that described by diffusion. Another class of models uses spatial terms to describe not the movement of the individuals, who are now assumed to be stationary, but rather the transmission of the disease itself. For instance, in the distributed-contacts model [14, 15], the infectivity at each point in space is computed from a weighted spatial average of the field of infected individuals over long ranges.

Most short-range spatial effects can be modelled with mathematical terms that involve spatial derivatives, which compute rates of change of a quantity over an infinitesimal spatial range. One intuitive example is that of chemotaxis, in which a biological organism such as a bacterium senses a chemical field (perhaps a nutrient field) immediately surrounding it and moves in the direction of greatest increase [16]. In contrast, long-range effects such as those in the distributed-infective and distributed-contact models cannot be accurately described with spatial derivatives. Instead, one typically uses so-called nonlocal operators such as integral operators, which compute quantities over finite (or potentially infinite) ranges.

Swarming models. We introduce a spatial model for epidemics in which nonlocal terms capture a completely different effect from those previously examined. In particular, we will model the directed motion of individuals due to social forces in an epidemiological context. By social forces, we mean forces such as attraction to and repulsion from other members of the population. These forces have received substantial attention in studies of swarming groups such as fish schools, bird flocks and insect plagues [17, 18, 19]. In the traditional biological swarming context, attraction and repulsion operate simultaneously between organisms of the same species. Some fish, for example, are evolutionarily preprogrammed with attraction (which provides safety and group cohesion) and repulsion (which helps avoid collisions). The balance between the two forces typically leads to a preferred separation distance between neighbouring organisms in a group. Social forces may arise directly from organisms' sensing over finite distances through sound, sight, smell or touch, or indirectly through chemical or other signals.

We now turn our attention to zombie epidemics in the presence of social forces. The recent work of Munz *et al.* [1] was the first to develop a mathematical model for zombie epidemics. It concluded that, in the absence of a cure or impulsive attacks, the only possible outcome of a zombie invasion is complete annihilation of the healthy, alert human population. However, as we have mentioned, by neglecting spatial features, the original model leaves open a key question of zombie epidemics; namely, that of how a swarm of zombies self-organizes to pursue and attack a swarm of self-organized healthy humans without any collusion on the part of either the zombie or the alert human populations. We will probe this phenomenon by generalizing the Munz model [1] to include spatial variation by incorporating social interaction terms. These terms account for zombies being socially attracted to healthy humans and healthy humans being socially repulsed from zombies.

Organization of the chapter. The rest of this chapter is organized as follows. In Section 15.2, we construct our model. In Section 15.3, we examine some of its basic properties. The model conserves total population, but does not conserve centre of mass. We then consider mass-balanced states and show that spatial coexistence of healthy humans and zombies is impossible. Finally, we show that the only possible steady states are homogeneous in space and we compute their stability. In Section 15.4, we run numerical simulations of our model, focusing on the effect that the spatial terms have vis-a-vis the non-spatial model. Though the spatial and non-spatial models reach the same doomsday equilibrium in which zombies extinguish healthy humans, the approach to the equilibrium can be slower in the spatial case. Furthermore, in the spatial case, the population can self-organize into a travelling pulse of susceptible humans who are pursued by a group of zombies. Though the pulse is merely transient, it is possible that the added time necessary to reach the doomsday equilibrium would afford the opportunity to develop a cure for zombie-ism. In Section 15.5, we consider an extension of the model in which such a cure exists, again examining equilibria and their stability. In Section 15.6, we perform numerical simulations for this case of a cure and show that travelling solutions can persist. Finally, we conclude in Section 15.7 with some open questions for the zombie research community.

The model we will develop and the mathematical tools we will use to study it draw on partial differential equations, dynamical systems and analysis. However, we strive to keep mathematical details to a minimum in the body of our paper. For the material presented in Sections 15.2 through 15.6, we focus our presentation on the main ideas; the reader who is especially mathematically inclined will find details and derivations in the appendices.

15.2 Model Construction

We begin by reviewing the following spatially homogeneous SZR Munz model [1]:

$$\dot{S} = \Pi - \beta SZ - \delta S \tag{15.1a}$$

$$\dot{Z} = \beta SZ + \zeta R - \alpha SZ \tag{15.1b}$$

$$\dot{R} = \delta S + \alpha SZ - \zeta R. \tag{15.1c}$$

The total population consists of three classes of individuals: active alert humans, denoted by S; zombies, who are deceased reanimated humans, denoted by Z; and those who are recently deceased but not reanimated, and thus are removed from either alert human or zombie populations, denoted by R. Each over-dot in (15.1) represents a time derivative. The right-hand sides represent rules for the instantaneous rates of change of the number in each population class.

Members of the healthy human population can become zombies as a direct result of an encounter with a zombie, with a zombie bite as the usual mechanism of transmission. Alert humans are recruited to the zombie population through mass-action kinetics with transmission parameter β. The mass-action kinetics assumption means, for instance, that the rate at which alert humans are bitten by zombies is proportional to the rate of contact between the two classes. This is in turn assumed to be proportional to the product of the number of individuals within each class, giving rise to the product SZ in (15.1). Though the model uses the law of mass action, in future studies one could modify this to be the standard incidence model, which assumes that the contact rate is not dependent on the total population size [7].

The reanimation of the removed population R is independent of social interaction since the members of this class are deceased and inanimate. It is assumed to be directly proportional to R with proportionality constant ζ, the natural undeath rate for reanimation of recently deceased humans. In addition to the two routes into the zombie class, it is possible for zombies to join the removed class as a result of an altercation between a zombie and a healthy, alert human in which the latter triumphs. Once again, this route from Z to R occurs through mass action, with attack rate α.

The model above incorporates the fact that, independent of zombie interaction, the susceptible class has natural birth and death rates, denoted respectively by Π and δ. However, these effects are negligible compared to the time scale of a zombie epidemic and one can set $\Pi = \delta = 0$ as in Munz *et al.* [1]. Therefore the total population size $N = S + Z + R$ is fixed, and the equations become

$$\dot{S} = -\beta SZ \tag{15.2a}$$

$$\dot{Z} = \beta SZ + \zeta R - \alpha SZ \tag{15.2b}$$

$$\dot{R} = \alpha SZ - \zeta R. \tag{15.2c}$$

Governing equations. The model (15.2) neglects the spatial structure of the zombie epidemic. We now construct a spatiotemporal version of this model. As discussed in Section 15.1, the most basic spatial models of epidemics incorporate linear diffusion, modelling the random movement of individuals [10, 11]. However, such an assumption is insufficient to model zombie epidemics in that it does not include any directed motion. Crucial to the understanding of interactions between zombies and alert humans is the fact that zombies tend to be attracted to alert humans, and alert humans seek to avoid zombies. Furthermore, the attraction and repulsion are not local effects. As with many socially interacting populations, both zombies and humans can sense population densities at finite (rather than infinitesimal) distances. Hence we will need to incorporate into our model swarming-type social interaction terms such as those discussed in Section 15.1 and used commonly in the biological swarming literature [17, 18].

In mathematical models of swarms, one commonly assumes that social interactions take place in a pairwise, linear manner, so that to compute the total social force on a given organism, one sums the interaction force between it and each other organism. Continuum swarming models, then, typically involve convolution-type integral terms of the form

$$\int \mathbf{K}(\mathbf{x}-\mathbf{y})\rho(\mathbf{y},t)\,d\mathbf{y} \equiv \mathbf{K} * \rho,$$

where $\rho(\mathbf{x},t)$ is population density and $\mathbf{K}$ is a kernel in which are embedded the rules for social interaction as it arises from sensing; that is, $\mathbf{K}$ gives the effect that organisms at location $\mathbf{y}$ have on those at location $\mathbf{x}$. Some methods of sensing, such as sound, are essentially omnidirectional. Others, such as sight, are more unidirectional. Because many organisms process a combination of communication signals, one often assumes that communication is omnidirectional [20, 19]. Another common assumption is that distinct organisms exert equal and opposite social forces on each other.

The simplest continuum swarming model assumes that the total population is conserved (no births or death, for example) and neglects inertia. In this case, the population is governed by the conservation equation

$$\dot{\rho} + \nabla \cdot (\rho \mathbf{v}) = 0, \quad \mathbf{v} = \mathbf{K} * \rho. \tag{15.3}$$

In a one-dimensional domain, the aforementioned assumptions of omnidirectional communication and equal-and-opposite forces cause the social kernel $\mathbf{K} = K$ to be odd. Negative values of $\mathrm{sgn}(x)K(x)$ correspond to attraction and positive values to repulsion. Recently, Leverentz *et al.* [18] explored the manner in which the asymptotic dynamics of (15.3) in a one-dimensional domain depend on properties of K.

Using these ideas from biological swarming, we now modify the non-spatial zombie epidemic model (15.1). We only include spatial terms in the equations for S and Z, since the individuals in the removed class are deceased and not animated, and are thus completely immobile. Our model is

$$\dot{S} + \nabla \cdot (v_S S) = D_S \Delta S - \beta S Z \tag{15.4a}$$

$$\dot{Z} + \nabla \cdot (v_Z Z) = D_Z \Delta Z + \beta S Z + \zeta R - \alpha S Z \tag{15.4b}$$

$$\dot{R} = \alpha S Z - \zeta R, \tag{15.4c}$$

where S, Z and R are functions of x and t and are now interpreted as population densities. The parameters D_S and D_Z are the diffusion constants measuring the

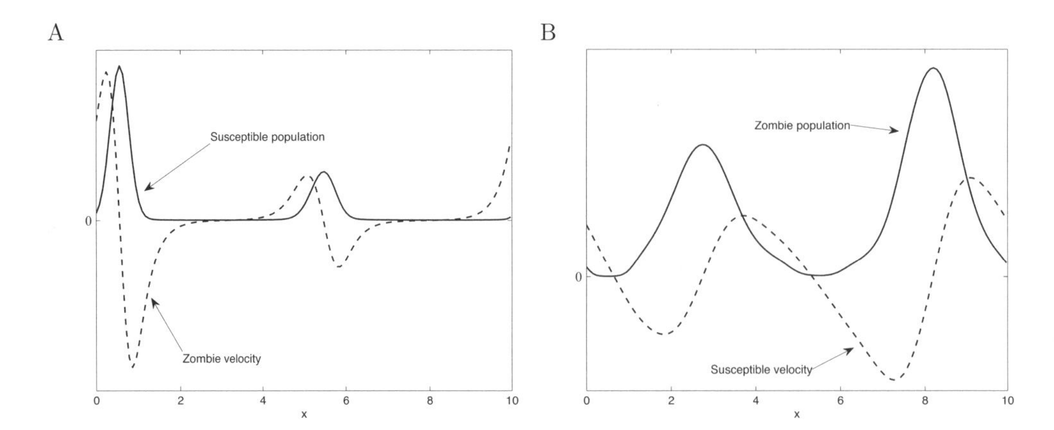

Figure 15.1: Social advection velocities. Schematic of typical populations plotted with the corresponding nonlocally determined advective velocities; see equations (15.5) and (15.6). A: Zombies are attracted to susceptibles and move towards maxima of S. B: Susceptibles are repelled from zombies and move outwards from maxima of Z. Though the population-density fields have different units from the velocity fields, we have plotted them on the same axes for a qualitative schematic comparison.

random motion of susceptibles and zombies. For simplicity, we restrict attention to one-dimensional spatial domains so that $\Delta \equiv \partial_{xx}$ and $\nabla \cdot \equiv \partial_x$.

Since susceptible individuals flee from zombies, we describe susceptibles' velocity due to social interaction as

$$v_S = K_S * Z, \quad K_S = \operatorname{sgn}(x) \cdot F_S e^{-|x|/\ell_S}. \tag{15.5}$$

Here, K_S is a repulsive kernel, where F_S and ℓ_S describe the characteristic strength and length scale of repulsion. We have chosen K to be a simple kernel capturing the essential property that a susceptible's reaction to a zombie should decay with the distance between the two, due to the limitations of human sensing. Similarly, we model the zombies' advective velocity as

$$v_Z = K_Z * S, \quad K_Z = -\operatorname{sgn}(x) \cdot F_Z e^{-|x|/\ell_Z}, \tag{15.6}$$

so that zombies are attracted to susceptibles. Figure 15.1 shows an example of the velocities v_S and v_Z for given population-density profiles. We assume that $F_Z \leq F_S$, so that alert humans are more violently repelled by oncoming zombies than zombies are attracted to humans. The opposite assumption would result in a faster velocity of zombies than susceptibles, and in turn the speedy annihilation of the entire susceptible population. We also assume that $\ell_Z \leq \ell_S$; that is, zombies generally sense with less acuity than susceptibles due to their lower level of brain function.

Plugging (15.5) and (15.6) into (15.4), we have our full equation

$$\dot{S} + (SK_S * Z)_x = D_S S_{xx} - \beta SZ \tag{15.7a}$$

$$\dot{Z} + (ZK_Z * S)_x = D_Z Z_{xx} + \beta SZ + \zeta R - \alpha SZ \tag{15.7b}$$

$$\dot{R} = \alpha SZ - \zeta R. \tag{15.7c}$$

Nondimensionalization. Nondimensionalization is a standard mathematical technique that involves making a change of variables to remove the physical dimensions (such as length, time and mass) from an equation in order to obtain a simpler model, possibly by reducing the number of model parameters. This reduction is convenient for analysis and for numerical simulation. One can always recast results in dimensional form simply by reversing the change of variables, without any loss of information. The nondimensionalization procedure for our model appears in Appendix A. The dimensionless model is

$$\dot{S} + (SK_S * Z)_x = D_S S_{xx} - \beta SZ \tag{15.8a}$$

$$\dot{Z} + (ZK_Z * S)_x = D_Z Z_{xx} + \beta SZ + R - \alpha SZ \tag{15.8b}$$

$$\dot{R} = \alpha SZ - R. \tag{15.8c}$$

K_S and K_Z have been redefined as

$$K_S = \text{sgn}(x) \cdot e^{-|x|} \tag{15.9}$$

$$K_Z = -\text{sgn}(x) \cdot Fe^{-|x|/\ell}. \tag{15.10}$$

The dimensionless parameters $F \leq 1$ and $\ell \leq 1$ measure, respectively, the relative strength and length scale of attraction to repulsion. The parameters D_S, D_Z, α and β are dimensionless versions of the corresponding parameters from the original dimensioned model.

Boundary conditions. Since the governing equations (15.8) describe a spatial model, one must specify a spatial domain and set appropriate boundary conditions, which dictate the behaviour of the model at the boundary of the domain. To facilitate our analysis and our computation, we assume that the domain is a finite interval with periodic boundary conditions. To get insight into the behaviour for very large domains, one can always choose L very large and focus on the middle of the domain where the finite-size and boundary effects are presumed to be small. For specificity, let the domain be the interval $[0, L]$. Then solutions for S have $S(0) = S(L)$, and $S_x(0) = S_x(L)$, with similar conditions for Z and F. In addition, the length scale of the interaction is assumed extremely small compared to L. Thus we modify our kernel functions K_S and K_Z to be periodic and this has little effect on their values. That is, outside a small region near their centres, K_S and K_Z are very close to zero.

15.3 Basic Model Properties

We now examine some basic properties of (15.8), seeking mass-balanced states and spatially homogeneous steady states; we also calculate the local stability of the latter and show that these spatially homogeneous steady states are the only possible steady states.

Conservation of total population. By construction, the total population size (alert humans, the dead and the undead all combined) should remain unchanged under the dynamics of (15.8). If we denote the population size at each point by $n(x,t)$ and the total population size as $N(t)$,

$$n(x,t) = S + Z + R \tag{15.11a}$$

$$N(t) = \int_0^L n(x,t)\, dx, \tag{15.11b}$$

then conservation of population means that N is constant, as in Munz *et al.* [1]. Appendix B derives the conservation of population—or as we will sometimes say, conservation of mass—for the model (15.8). For the remainder of this chapter, we treat the population size N as an additional parameter in the model.

Centre of mass and travelling solutions. One question of interest throughout the rest of this chapter pertains to the ability of the population to migrate under the dynamics of the model. One simple way to touch upon this question is to determine whether the centre of mass of the population can move or must be stationary. In qualitative terms, the centre of mass of the population is defined as the average of the positions of all of the individuals within the population. A proof that the centre of mass remains stationary would mean that solutions consisting of the population translating in one direction would be impossible. We show in Appendix C that, in fact, the centre of mass can migrate. This means that travelling solutions are not *a priori* ruled out, although in our numerical simulations of (15.8) we have only found transient travelling solutions; see Figures 15.2A and 15.3A–C. In Appendix C, we calculate the centre of mass to show that it can travel with a nonzero speed.

Mass-balanced solutions, steady states and stability. We now turn our attention to different types of equilibria of (15.8). Though a system such as (15.8) could, in theory, have long-term behaviour more complicated than an equilibrium, equilibria are some of the most fundamental long-term behaviours one might wish to study.

We are first interested in mass-balanced states. By mass-balanced states, we mean states where the number of individuals within each class is not changing in time. It is important to note that, even in such a state, the solution itself could still be changing (for instance, imagine a fixed population travelling in space). However, finding mass-balanced states would still provide useful information for epidemiological tallies. In Appendix D, we show that, at mass balance, $R = 0$, meaning there are no deceased individuals. All individuals are alert humans or zombies. Furthermore, we show that either $S = 0$ or $Z = 0$ at every point in space. This means that alert humans and zombies cannot coexist at the same spatial location.

Now consider steady states of (15.8). Steady states are states for which the time derivatives vanish, meaning that the solutions do not change in time. This is a more restricted class of solutions than mass-balanced states. That is, while every steady state is a mass-balanced state, not every mass-balanced state need be a steady state.

In Appendix E, we show that the only possible steady states are spatially homogeneous ones; that is, solutions where the mass is balanced, the solutions are no longer changing in time and there is no spatial variation in the profile. The steady-state solutions have population density within each S, Z, R class that is constant in space. The two steady-state solutions are the susceptible state

$$(S^*, Z^*, R^*) = (N/L, 0, 0) \tag{15.12}$$

and the doomsday state

$$(S^*, Z^*, R^*) = (0, N/L, 0), \tag{15.13}$$

in which all individuals have become zombies. To obtain the mass balance for each steady state (that is, the number of individuals within each class), one must integrate over the spatial domain. For (15.12), the mass balance is

$$\int_0^L S^*\, dx = N, \quad \int_0^L Z^*\, dx = 0, \quad \int_0^L R^*\, dx = 0,$$

and for (15.13) it is

$$\int_0^L S^*\, dx = 0, \quad \int_0^L Z^*\, dx = N, \quad \int_0^L R^*\, dx = 0.$$

These are the same mass balances as the steady states of the non-spatial model of Munz *et al.* [1]. We have not established whether (15.8) has other attractors such as inhomogeneous steady states or travelling waves. Our numerical investigations show transient travelling solutions, but have not revealed any long-term behaviour other than the spatially homogeneous states. See Section 15.4 for details.

We conclude this section with a study of the local stability of the two spatially homogeneous steady states (15.12) and (15.13). We review the concept of local stability of solutions to partial differential equations with the following (nonmathematical) thought experiment. Imagine taking an equilibrium solution for (S, Z, R) and applying a very small spatial perturbation to the solution profile. If all such small (in fact, infinitesimal) perturbations decay over time so that the system again reaches the equilibrium, then we say the equilibrium is locally stable. If any such perturbations grow in time, then the equilibrium is unstable. The stability properties of equilibria are crucial for us because they give hints about the possible long-time consequences of zombie epidemics. A stable equilibrium has the hope of being reached in the long term if the system begins at some non-equilibrium state (though it is not guaranteed). An unstable equilibrium can never in practice be reached (since tiny perturbations and fluctuations will always be present in real systems).

The calculation of the local stability of the equilibria (15.12) and (15.13) appears in Appendix F. We show that (15.12) is unstable with respect to generic noisy perturbations (at least, for L sufficiently large). This means that the state in which all individuals are healthy humans cannot be regained if any zombies are introduced into the population. On the other hand, the doomsday state (15.13) is locally stable. These stability properties are identical to the non-spatial model in Munz *et al.* [1].

15.4 Numerical Simulations and Dissipating Travelling Pulses

In this section, we present results of numerical simulations of (15.8). Our numerical scheme is spectral in space, taking advantage of the periodic domain. We treat the nonlocal (convolution) terms via multiplication in Fourier space and we compute the reaction terms pseudospectrally in real space. Time integration in Fourier space uses MATLAB's built-in timestepper ode45.[1]

Figure 15.2 compares the total number of individuals in each S, Z and R class as a function of time for three different models. Figure 15.2A corresponds to the full spatial model (15.8). The epidemiological parameters are $\alpha = 0.005$ and $\beta = 0.0095$ as in Munz *et al.* [1], and we take $F = 0.01$, $\ell = 0.02$, $D_S = 0.1$, $D_Z = 0.05$ and $L = 10$. Figure 15.2B is similar, but has the social interaction terms turned off; i.e., $K_S = K_Z = 0$. Both simulations begin with initial populations

$$\int_0^L S\,dx = 500, \quad \int_0^L Z\,dx = 3, \quad R = 0,$$

with S and Z distributed randomly in space. Despite the difference in the spatial effects for these two situations, the time evolution of the total number in each compartment is strikingly similar. Both models approach the doomsday equilibrium (15.13) and take equally long to do so. This can be understood by examining local stability results in Appendix F. From (15.34) and (15.36), diffusive terms affect the linearization around the susceptible and doomsday equilibria, but the social terms do not. Our simulations begin close to the susceptible equilibrium and conclude at the doomsday one. This suggests that the system spends most of its time in regimes where the linearizations are good descriptions, and hence that social interactions are only transiently important. Despite the similarity of Figures 15.2A and 15.2B, the

[1]For the zombie epidemiologist interested in conducting her own simulations, our MATLAB code is available at http://www.macalester.edu/~ctopaz/zombieswarm.m.

actual solution profiles for the two cases are markedly different; we will discuss these in a moment. Before doing so, we consider the non-spatial model of Figure 15.2C. The decay of the susceptible population is much faster. In fact, by the time the susceptible population has been extinguished in the non-spatial model, susceptibles within the spatial model have barely been impacted by the zombies.

Figure 15.3 shows S, Z and R profiles in space-time corresponding to the simulations of Figures 15.2A and 15.2B. For the full model, shown in Figure 15.3A–C, the solution shows transient travelling pulses, which are apparent as streakiness in the picture. These pulses are absent in the diffusion-only case, shown in Figure 15.3D–F.

We conclude from our numerical explorations that persistence of the healthy human population is impossible within the framework of (15.8). In all variations of the model considered here, the zombie population eventually overtakes the susceptible population, leading to the doomsday state. However, the manner of approach is important. Spatial effects can slow the decline of the healthy human population; furthermore, social attraction and repulsion provide for spatial localization of susceptibles and zombies. In the event of a zombie attack, the added time to extinction might allow the healthy, alert humans to seek a cure for zombie-ism. We consider the effect of such a cure in the next section.

15.5 Epidemics with Treatment

Now consider the case that there exists a treatment for zombie-ism. We repeat for this modified case the analytical studies presented in Section 15.3. We run numerical simulations in the next section.

Model construction, conservation of population and centre of mass. Assume that the treatment for zombie-ism moves zombies back to the susceptible class with rate c. Modifying (15.8) appropriately, we have

$$\dot{S} + (SK_S * Z)_x = D_S S_{xx} - \beta SZ + cZ \tag{15.14a}$$

$$\dot{Z} + (ZK_Z * S)_x = D_Z Z_{xx} + \beta SZ + R - \alpha SZ - cZ \tag{15.14b}$$

$$\dot{R} = \alpha SZ - R. \tag{15.14c}$$

This model is similar to that in Section 5 of Munz *et al.* [1], except that we do not include a class of latently infected zombies; the introduction of treatment even without latent infection introduces new solution types into the model as we show below. Calculations identical to those of Section 15.3 establish that (15.14) conserves the total population size but not the centre of mass.

Mass balance. We now seek mass-balanced solutions. There are two different cases of mass-balanced solutions for (15.14). The calculation of these is presented in Appendix D. The first mass-balanced state is the spatially homogeneous steady state

$$(S^*, Z^*, R^*) = (N/L, 0, 0). \tag{15.15}$$

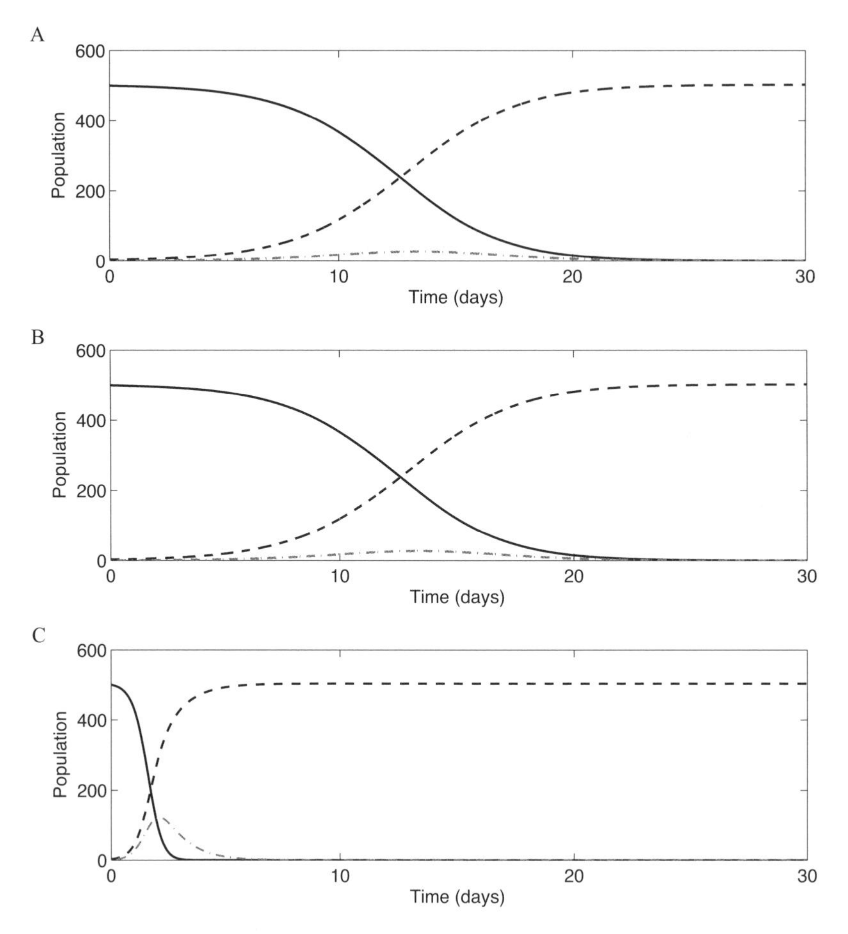

Figure 15.2: A comparison of spatial and non-spatial models. The total susceptible (solid), zombie (dashed) and removed (dot-dashed) populations plotted as a function of time. Decay to the doomsday state (no susceptibles) is much slower for the spatial models of A and B than the non-spatial model of C. Though the evolution of total mass in each compartment is similar for A and B, the actual solution profiles are different; see Figure 15.3. The epidemiological parameters are $\alpha = 0.005$ and $\beta = 0.0095$ as in Munz *et al.* [1]. We begin with a total susceptible mass of 500 and a zombie mass of 3 (distributed randomly in space). A: The full spatial model (15.8) with $F = 0.01$, $\ell = 0.02$, $D_S = 0.1$, $D_Z = 0.05$ and $L = 10$. B: Like A but with the social interaction terms turned off (i.e., $K_S = K_Z = 0$). C: The non-spatial version of the model.

The second mass-balanced state is actually a family of states parameterized by the quantity $\langle S, Z\rangle$, defined as

$$\langle S, Z\rangle \equiv \int_0^L SZ\,dx.$$

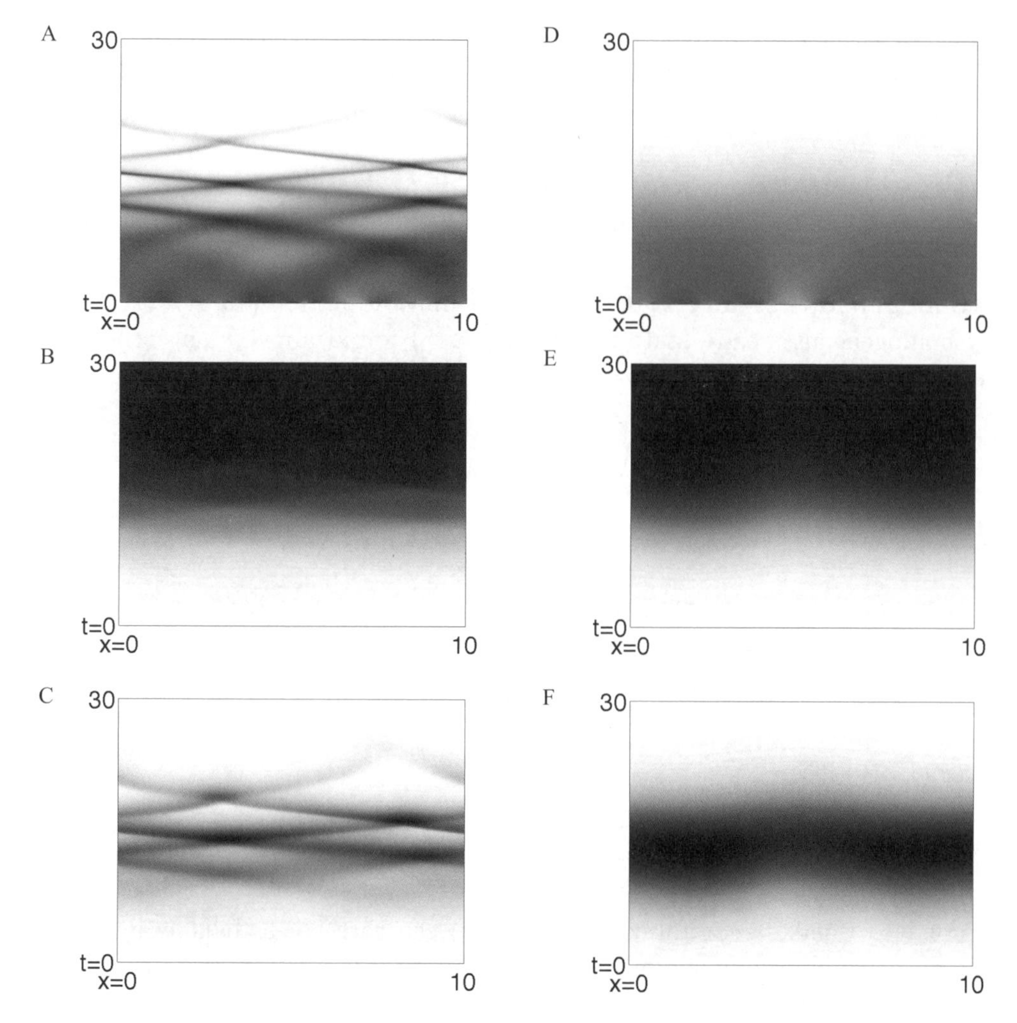

Figure 15.3: Space-time profiles of spatial models. A–C: S, Z and R (respectively) for the full spatial model simulation of (15.8) summarized in Figure 15.2A. Darker shading corresponds to higher population density. Note the presence of transient travelling pulses, which are apparent as streakiness in the picture. The system approaches the doomsday equilibrium (15.13). D–F: Like A–C, but for the diffusion-only (no social interaction) model summarized in Figure 15.2B. The system approaches the same equilibrium as in A, but transient pulses do not occur.

These states are

$$\int_0^L S\,dx = N - \left(\frac{\beta}{c} + \alpha\right)\langle S, Z\rangle$$

$$\int_0^L Z\,dx = \frac{\beta}{c}\langle S, Z\rangle$$

$$\int_0^L R\,dx = \alpha\langle S, Z\rangle.$$

Nothing in the conditions for mass balance selects a particular member of this family. For this family of solutions, the ratio of total members in the Z class to those in R is

$$\frac{\int_0^L Z\,dx}{\int_0^L R\,dx} = \frac{\beta}{\alpha c}.$$

Homogeneous steady states and stability. Equation (15.14) has two spatially homogeneous steady states. One is (15.15) found above. The other is the endemic state

$$(S^*, Z^*, R^*) = \left(\frac{c}{\beta}, \frac{N\beta - Lc}{L(\beta + \alpha c)}, \frac{\alpha c}{L\beta}\frac{N\beta - Lc}{\beta + \alpha c}\right), \tag{15.16}$$

which exists only for

$$c/\beta < N/L. \tag{15.17}$$

The mass balance for (15.15) is

$$\int_0^L S^*\,dx = N, \quad \int_0^L Z^*\,dx = 0, \quad \int_0^L R^*\,dx = 0, \tag{15.18}$$

and for (15.16) it is

$$\int_0^L S^*\,dx = \frac{cL}{\beta}, \quad \int_0^L Z^*\,dx = \frac{N\beta - Lc}{\beta + \alpha c}, \quad \int_0^L R^*\,dx = \frac{\alpha c}{\beta}\frac{N\beta - Lc}{\beta + \alpha c}. \tag{15.19}$$

It is crucial to make a careful comparison between these mass balances and those of the corresponding non-spatial model. We showed in Section 15.3 that, for the model without cure (15.8), the mass balance of steady states is the same for the spatial model as for the non-spatial model. This is not the case for the model with cure. The non-spatial version of the model,

$$\dot{S} = -\beta SZ + cZ \tag{15.20a}$$

$$\dot{Z} = \beta SZ + R - \alpha SZ - cZ \tag{15.20b}$$

$$\dot{R} = \alpha SZ - R, \tag{15.20c}$$

has susceptible steady state

$$(S^*, Z^*, R^*) = (N, 0, 0),$$

which does correspond to the mass balance (15.18) of the susceptible state in the spatial model. However, the non-spatial model has endemic steady state

$$(S^*, Z^*, R^*) = \left(\frac{c}{\beta}, \frac{N\beta - c}{\beta + \alpha c}, \frac{\alpha c}{\beta}\frac{N\beta - c}{\beta + \alpha c}\right), \tag{15.21}$$

which does not correspond to the mass balance (15.19) of the endemic state in the spatial model. Thus the introduction of spatial structure inherently shifts the mass balance of the endemic equilibrium in a manner dependent on the size of the domain.

Focusing now on steady states of the spatial model (15.14), we compute the local stability of the susceptible steady state (15.15). The full calculation is given in Appendix F, where we derive the stability condition

$$c/\beta > N/L \tag{15.22}$$

(which we have also verified numerically for sample parameters by measuring the evolution of small perturbations to the steady state). This condition is complementary to the condition (15.17) for existence of the endemic state.

Calculation of the local stability of (15.16) is more complex. The full analysis appears in Appendix F. A primary result is that the conditions

$$\beta > \alpha, \quad D_S > 2Z^* F \ell^2 \quad \text{and} \quad D_Z > 2S^* \tag{15.23}$$

are sufficient (though not necessary) for the local stability of the endemic state. Instability of the endemic equilibrium is possible as well. For instance, the parameters

$$\begin{gathered} \alpha = 1, \quad \beta = 0.1, \quad D_S = 0.1, \quad D_Z = 0.05 \\ c = 0.2, \quad F = 0.3, \quad \ell = 0.3, \quad N = 70 \quad \text{and} \quad L = 10 \end{gathered} \tag{15.24}$$

provide a numerical example of instability. For this set of parameters, note that $c/\beta = 2$ and $N/L = 7$. Hence, according to (15.22), the susceptible equilibrium (15.15) is unstable as well. Thus we expect that the system must have long-term behaviour other than a spatially homogeneous steady state. Our numerical simulations will demonstrate such behaviour (see Figures 15.4 and 15.5).

In summary, we have shown that the susceptible equilibrium (15.15) is stable when the cure rate c is sufficiently large (see (15.22)). The endemic state (15.16) can also be stable or unstable. We have demonstrated (Appendix F) the possibility of instability using the numerical example (15.24) and have derived the conditions (15.23) that guarantee stability for strong enough diffusion and transmission.

15.6 Numerical Simulations and Persistent Travelling Pulses

We now consider numerical results for three different models related to (15.14): the full model; the model with diffusion but no social interaction (i.e., $K_S = K_Z = 0$); and the version of the model with no spatial structure. The epidemiological parameters are $\alpha = 1$, $\beta = 0.1$ and $c = 0.2$, and we take $F = 0.03$, $\ell = 0.3$, $D_S = 0.1$, $D_Z = 0.05$ and $L = 10$. We take the initial condition to satisfy

$$\int_0^L S\,dx = 50, \quad \int_0^L Z\,dx = 50, \quad R = 0,$$

with the nonzero populations distributed randomly in space.

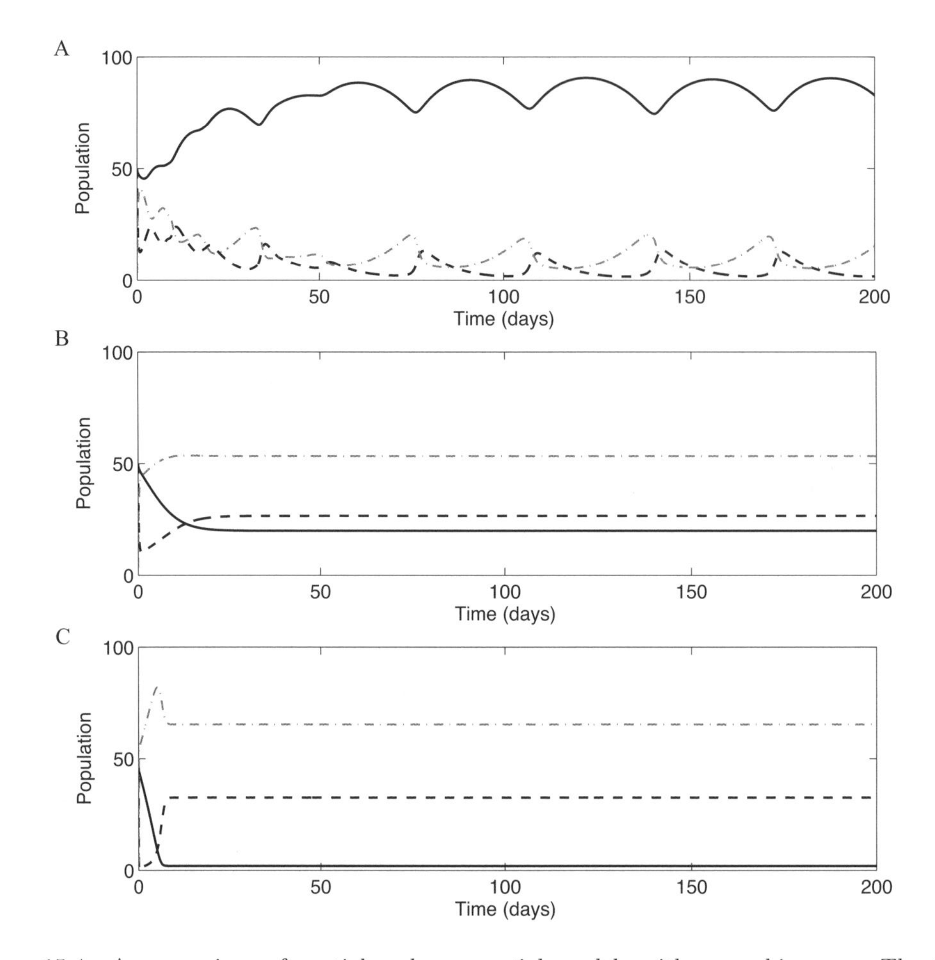

Figure 15.4: A comparison of spatial and non-spatial models with a zombie cure. The total susceptible (solid), zombie (dashed) and removed (dot-dashed) populations plotted as a function of time for $\alpha = 1$, $\beta = 0.1$, $c = 0.2$ and an initial condition of 50 susceptibles and 50 zombies. A: The full model (15.14) produces a persistent susceptible population. Here, $F = 0.03$, $\ell = 0.3$, $D_S = 0.1$, $D_Z = 0.05$ and $L = 10$. B: The reduced model (15.14) with no social forces ($K_S = K_Z = 0$) reaches the endemic equilibrium with mass balance (15.19). Asymptotically, more susceptibles survive for the case shown in A with social advection. C: The compartmental model reaches an endemic equilibrium with mass balance (15.21), in which even fewer susceptibles survive.

Figure 15.4A shows the total number of individuals in each S, Z and R class for the full spatial model (15.14). The system approaches neither a doomsday state nor an endemic steady state. Rather, the total mass in S, Z and R fluctuates around nonzero values. In contrast to the cure-less model (15.8), the population of healthy humans survives. The space-time plots of Figures 15.5A–C reveal that the susceptibles survive as persistent travelling pulses that are pursued by groups of zombies. Due to the finite size of the domain and periodic boundary conditions, there is a merging and splitting of pulses, and this effect is responsible for the slight oscillation in the long-term population count in Figure 15.4A.

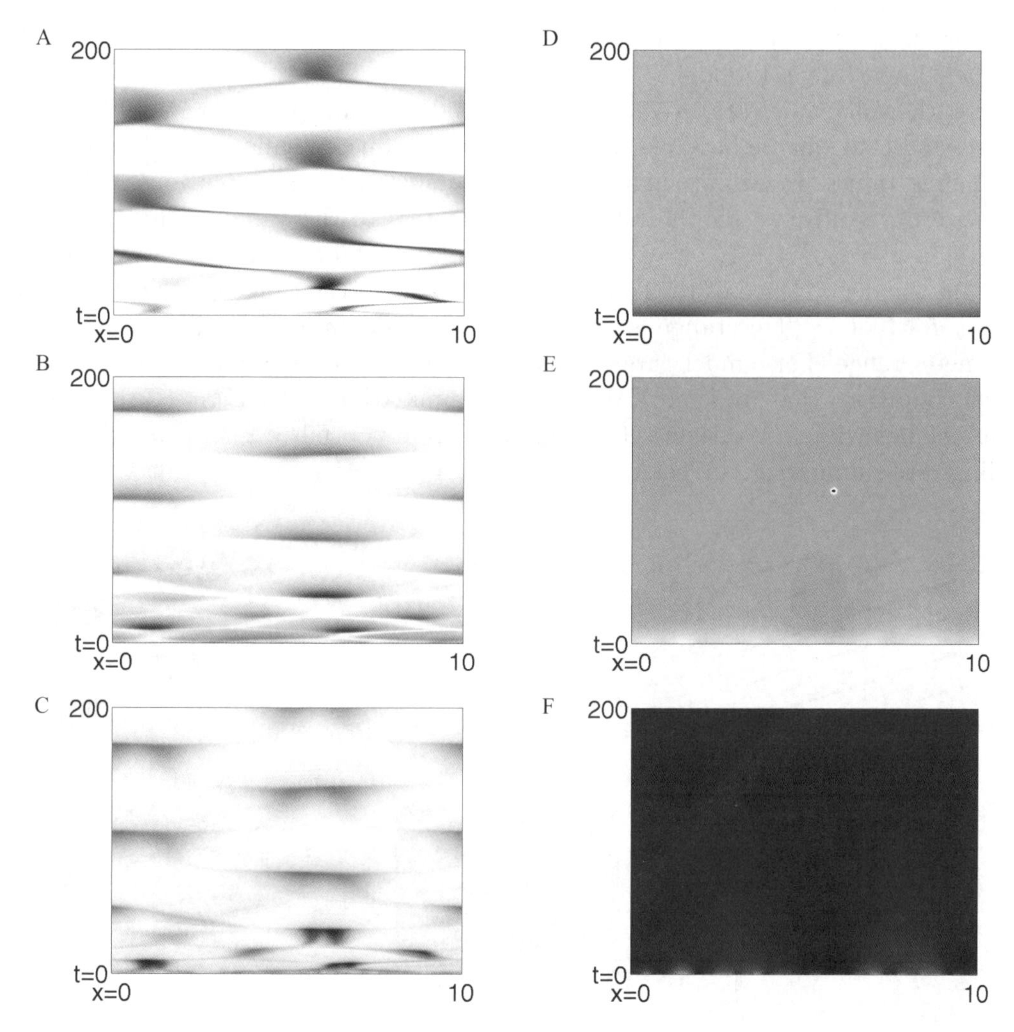

Figure 15.5: Space-time profiles of spatial models with zombie cure. A–C: S, Z and R (respectively) for the full spatial model simulation summarized in Figure 15.4A. Darker shading corresponds to higher population density. Note the presence of splitting and merging travelling pulses. D–F: Like A–C, but for the diffusion-only (no social interaction) model summarized in Figure 15.4B. Pulses do not occur, and the system reaches the spatially homogeneous steady state (15.16).

Figures 15.4B and 15.5D–F correspond to the reduced spatial model without social interaction. In this case, the system reaches the spatially homogeneous endemic equilibrium with mass balance (15.19). Although some healthy, alert humans survive (perhaps in shopping malls or village pubs), their total mass is less than that contained in the travelling pulse of susceptibles discussed above. Hence, for at least some parameters, social effects can enhance the survival of the healthy human population. Figure 15.4C corresponds to the compartmental model (15.20). The system again reaches an endemic equilibrium, albeit with different levels of S, Z and R. As discussed above, the mass balance (15.19) for the endemic equilibrium in the spatial model necessarily differs from the mass balance (15.21) for the compartmental model.

For the full model (15.14), the splitting and merging pulses are not the only long-term behaviour we have observed. We have conducted simulations at different parameter values, revealing slightly different non-steady-state solutions. To systematize our study, we fix the following parameters to the values in Figures 15.4 and 15.5: $\alpha = 1$, $\beta = 0.1$, $\ell = 0.3$, $D_S = 0.1$, $D_Z = 0.05$ and $L = 10$. We concentrate on the dependence of the solutions on c and F in the range $0 \leq c \leq 0.21$ and $0.01 \leq F \leq 0.05$. (This range is chosen based on preliminary studies on a larger parameter range. For c much larger, the susceptible steady state attracts all of the initial conditions that we tried.) We run each simulation to time $t = 60$ to remove transient behaviour. We then retain the data from $t = 60$ to 120 for study. Our findings are summarized in Table 15.1.

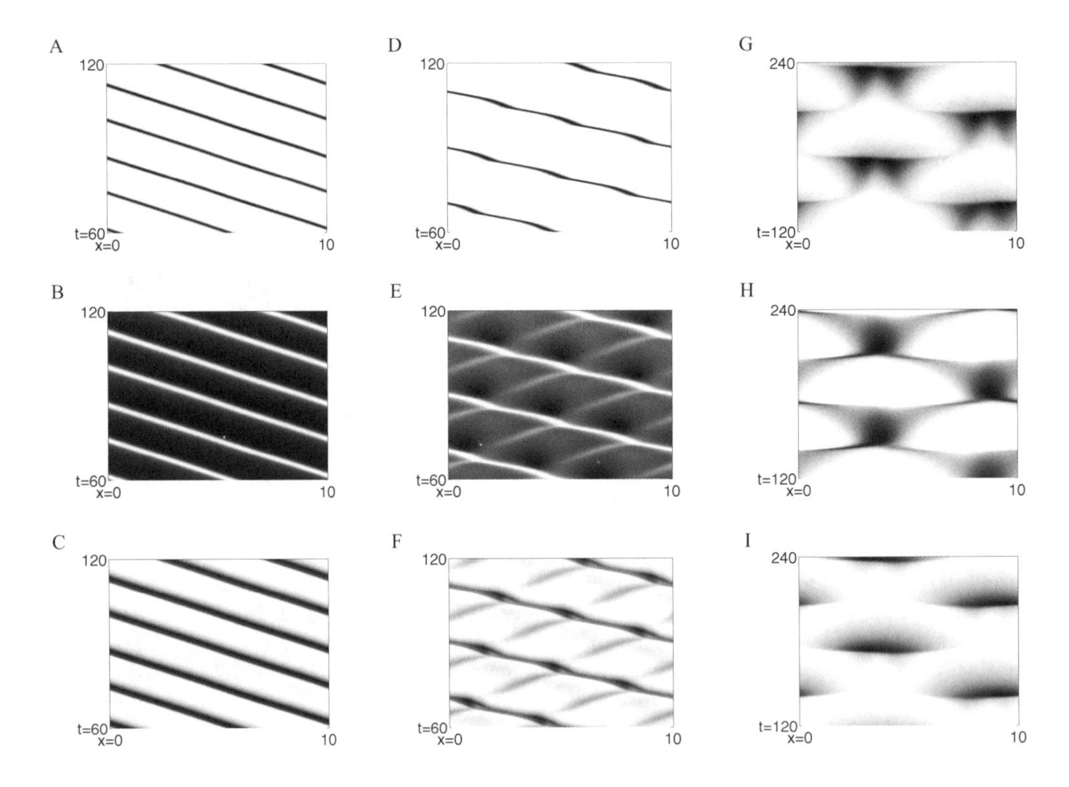

Figure 15.6: Different persistent travelling solutions in the case of a zombie cure. Space-time diagrams of susceptibles (top), zombies (middle) and removed (bottom) in simulations of (15.14) with $\alpha = 1$, $\beta = 0.1$, $\ell = 0.3$, $D_S = 0.1$, $D_Z = 0.05$ and $L = 10$, and initial conditions of 50 susceptibles and 50 zombies distributed randomly in space. For all figures, we show only the latter portion of the simulation to eliminate any transients. A–C: A travelling pulse with constant velocity, for $c = 0.054$ and $F = 0.032$. D–F: A travelling pulse with oscillating velocity, for $c = 0.054$ and $F = 0.037$. G–I: Merging and splitting pulses, for $c = 0.21$ and $F = 0.046$.

Table 15.1: Non-steady-state attracting solutions and their velocities for the spatial model with zombie cure. This table summarizes the type of solutions found in simulations of (15.14) for varying c and F values in the range $0 \leq c \leq 0.21$ and $0.01 \leq F \leq 0.05$ where we have fixed $\alpha = 1, \beta = 0.1, \ell = 0.3, D_S = 0.1, D_Z = 0.05$ and $L = 10$. The initial conditions have total mass of the S and Z populations both equal to 50, distributed randomly in space. We have performed one simulation per parameter value. If the solution is a travelling pulse, its velocity is stated. Oscillatory and highly oscillatory travelling pulses are denoted with a P and a PP respectively after the value of the average velocity. Finally, HC denotes solutions in which pulses periodically split and rejoin, giving a honeycomb-like appearance to the space-time diagram. A velocity no longer makes sense for such solutions.

c \ F	0.01	0.014	0.019	0.023	0.28
0.01	Transient Pulses				
0.03	2.43	2.41	2.26	2.30	2.15
0.05	2.72	−2.30	−2.75	−2.13	2.08
0.08	3.13	3.07	−3.09	2.28	−2.35
0.10	−3.84	−2.80	2.47	2.62	3.60
0.12	−3.23	−3.10	3.09	−2.76	−3.14
0.14	4.76 P	3.73	HC	−4.04 P	3.71
0.17	−4.36	HC	−4.75	4.01	−4.06
0.19	5.02 P	HC	HC	HC	−4.58 PP
0.21	5.17 PP	HC	HC	4.92 P	HC
c \ F	**0.032**	**0.037**	**0.041**	**0.046**	**0.05**
0.01	Transient Pulses				
0.03	2.13	−2.10	2.12	2.04	−1.97
0.05	−2.57	1.99	1.94	1.89	1.87
0.08	2.19	−2.41	−2.08	−2.14	−2.07
0.10	−2.41	3.19	−2.72	−2.37	2.41
0.12	2.85	−2.71	−2.83	−2.71	HC
0.14	3.76 P	3.76 P	−3.15	−3.21	3.38
0.17	−4.05	4.04	4.39 P	4.36 P	−4.25
0.19	−5.02 PP	4.48	4.82 P	−4.77 P	4.19 P
0.21	−5.80 PP	5.48 PP	−5.38 PP	HC	HC

We find three distinct long-term behaviours that are not spatially homogeneous steady states. For c slightly greater than zero, there are travelling pulse solutions with constant velocity; Figure 15.6A–C shows an example. These types of solutions are indicated in the table by entries consisting only of a numerical entry, which is the travelling velocity of the pulse. The speed depends on c and F, but the simulations indicate that the direction is dependent only on the initial conditions (which is consistent with the equations themselves having no inherent direction preference). As c grows larger, the velocity increases. For larger c values, the solutions still are travelling pulses, but with accompanying oscillation in the pulse speed; Figure 15.6D–F shows an example. These solutions are indicated in the table by entries marked with a P or PP (to denote low or high degrees of oscillation). The number given is the average velocity, which generally appears to be close to the speed of non-oscillating pulses nearby in parameter space. Finally, for certain large c values, pulses periodically join and split. We have already shown this type of solution in Figure 15.5A–C and we give an additional example in Figure 15.6G–I for comparison with the other solution types. The solutions look much like honeycombs in space-time, and we have marked them as HC in the table.

Theoretically, at certain parameter values, there could be coexisting stable solutions. In fact, based on preliminary numerical experiments, we anticipate that, for the larger c values for which we start seeing oscillatory pulses and merging/ splitting pulses, the individual pattern selection (i.e., long term behaviour of splitting/merging pulses versus oscillatory pulses) depends only on the initial condition, and this dependence is probably responsible for the nonmonotonic dependence of solution type on the parameters c and F in the table.

15.7 Conclusion

In this chapter, we have considered spatial models of zombie epidemics. Our model is based on the Munz model [1]. The spatial elements that we introduce capture social forces between zombies and alert, healthy humans. In particular, we have incorporated the tendency of healthy humans to move away from zombies, and the tendency of zombies to move towards healthy humans.

For our spatial model (15.8), these are our main analytical results:

- We show explicitly that the centre of mass of the population may travel.
- Mass-balanced solutions have no removed individuals. They have spatial separation between zombies and susceptibles.
- The only possible steady-state solutions are the spatially homogeneous susceptible and doomsday states.
- The susceptible state is unstable, and the doomsday state is locally stable.

Numerical simulations provided further information. For the non-spatial version of the Munz model [1], the only stable equilibrium is the doomsday state in which all individuals are zombies. In the spatial model, not only is this state locally stable, but it is also the only long-term behaviour observed in numerical simulations. However, in the spatial model, convergence to the equilibrium can occur more slowly than in the non-spatial model, and this convergence occurs when a self-organized cluster of susceptibles is pursued by a group of zombies. The additional time to the annihilation of alert humans might allow for the development of a cure.

In this vein, we considered a modification of our spatial model to include a cure for zombie-ism. For (15.14), we found that:

- Mass-balanced solutions do not necessarily have spatial separation between zombies and susceptibles.
- The model has two spatially homogeneous steady states. One is the susceptible state, which has the same mass balance as the susceptible state in the non-spatial analogue of the model. The other is the endemic state. The mass balance of the endemic state differs from that of the endemic state in the non-spatial model.
- The susceptible steady state exists for all parameter values, and is locally stable for a sufficiently large cure rate.
- The endemic steady state exists for a sufficiently small cure rate, and is stable if diffusion and transmission are sufficiently large.
- There exist parameter regimes in which both homogeneous steady states are unstable.

Using numerical simulations, we saw the existence of three qualitatively distinct types of solutions that are not homogeneous steady states. Each consisted of self-organized pulses of healthy, alert humans pursued by groups of zombies. In the first type, the pulse travels at constant velocity. In the second, the pulse travels at a nonzero average velocity, but is oscillatory. In the third type, there is a periodic splitting and rejoining of the susceptible population. In contrast to the first model (15.8) in which a cure was absent, these travelling pulses of susceptibles in (15.14) are all persistent, signalling hope of survival for the healthy population. Furthermore, in many of our example simulations, the total mass contained in the susceptible pulse was greater than that of the susceptibles within the spatially homogeneous endemic equilibrium.

We conclude with some open questions for the community of mathematicians undertaking zombie research. We show in Appendix E that homogeneous steady states are the only steady states of (15.8). However, this argument assumes continuity of the functions S and Z. A proof of regularity of the solutions is needed to justify this assumption. On a related topic, our numerics indicate that homogeneous steady states are not just the only steady states, but are the only mass-balanced states; however, a formal proof of this statement does not exist. In addition, one could imagine attractors that are not steady states, and are not mass balanced, but perhaps are time-periodic. Based on the variety of numerical simulations we have conducted, we conjecture that the doomsday state is the only attractor, but again no definitive formal proof exists. Such a proof would also serve, among other things, to rule out the possibility of persistent travelling pulses for this model.

For the model (15.14) with a cure, we have shown analytically that, depending on parameter choices, there is a susceptible steady state and an endemic steady state in which zombies and alert humans coexist. Our numerical simulations indicate that there exist attractors other than homogeneous steady states. In fact, we observed three different types of travelling pulses. We pose as an open problem the analytical proof for the existence of travelling pulses. Numerical simulations also indicate that bistability of the different types of solutions is possible. There is a need for more extensive numerical examination of these coexisting pulse types, followed by the development of early detection methods to distinguish between them. The early detection methods are of high priority, since they would allow more time for the implementation of the distinct corresponding strategies which would be needed for surviving a zombie epidemic.

Appendices

A Nondimensionalization

To reduce the number of parameters in the original dimensioned model (15.7), we nondimensionalize as follows. Let

$$\begin{aligned} x &\to \ell_S \hat{x} \\ t &\to \frac{1}{\zeta}\hat{t} \\ \{S, Z, R\} &\to \frac{\ell_S \zeta}{F_S}\{\hat{S}, \hat{Z}, \hat{R}\}. \end{aligned}$$

Substituting, cleaning up and dropping the hats, we have

$$\begin{aligned} \dot{S} + \left(S\left\{\mathrm{sgn}(x)\cdot e^{-|x|} * Z\right\}\right)_x &= \frac{D_S}{\ell_S^2\zeta}S_{xx} - \frac{\beta\ell_S}{F_S}SZ \\ \dot{Z} + \left(Z\left\{-\mathrm{sgn}(x)\cdot F_Z/F_S e^{-|x|\ell_S/\ell_Z} * S\right\}\right)_x &= \frac{D_Z}{\ell_S^2\zeta}Z_{xx} - \frac{\beta\ell_S}{F_S}SZ \\ &\quad + R - \frac{\alpha\ell_S}{F_S}SZ \\ \dot{R} &= -\frac{\alpha\ell_S}{F_S}SZ - R, \end{aligned}$$

where the overdot and the subscript x are now understood to represent derivatives with respect to dimensionless variables. Now define

$$F = F_Z/F_S$$
$$\ell = \ell_Z/\ell_S.$$

For convenience, redefine

$$K_S = \operatorname{sgn}(x) \cdot e^{-|x|} \tag{15.25a}$$
$$K_Z = -\operatorname{sgn}(x) \cdot F e^{-|x|/\ell} \tag{15.25b}$$

and

$$\frac{\alpha \ell_S}{F_S} \to \alpha$$
$$\frac{\beta \ell_S}{F_S} \to \beta$$
$$\frac{D_S}{\ell_S^2 \zeta} \to D_S$$
$$\frac{D_Z}{\ell_S^2 \zeta} \to D_Z.$$

Then the model reads

$$\dot{S} + (SK_S * Z)_x = D_S S_{xx} - \beta SZ$$
$$\dot{Z} + (ZK_Z * S)_x = D_Z Z_{xx} + \beta SZ + R - \alpha SZ$$
$$\dot{R} = \alpha SZ - R,$$

which is (15.8). Here, K_S and K_Z are given by (15.25). The dimensionless parameters $F \leq 1$ and $\ell \leq 1$ measure, respectively, the relative strength and length scale of attraction to repulsion. The parameters D_S, D_Z, α and β are dimensionless versions of the corresponding parameters in the original dimensioned model.

B Conservation of Population

Here we show that the total population size remains constant under the dynamics of (15.8). First, we add the three equations in (15.8) together, rearrange and expand spatial derivatives to obtain

$$\frac{\partial}{\partial t}(S + Z + R) = D_S S_{xx} + D_Z Z_{xx} - (SK_S * Z)_x - (ZK_Z * S)_x. \tag{15.26}$$

Now integrate over the domain $[0, L]$ and assume that this operation commutes with time differentiation. We recall the definitions (15.11) and we also define the inner product on two functions a and b as

$$\langle a, b \rangle \equiv \int_0^L a \cdot b \, dx.$$

Then (15.26) is

$$\dot{N} = D_S\langle S_{xx}, 1\rangle + D_Z\langle Z_{xx}, 1\rangle - \langle (SK_S * Z)_x, 1\rangle - \langle ZK_Z * S, 1\rangle.$$

Using integration by parts, and recalling that the convolution of two L-periodic functions is L-periodic, we have

$$\begin{aligned}\langle S_{xx}, 1\rangle &= S_x(L) - S_x(0) = 0\\ \langle (SK_S * Z)_x, 1\rangle &= S(L)(K_S * Z)(L) - S(0)(S_S * Z)(0) = 0.\end{aligned}$$

Similar statements hold for the Z terms. Hence

$$\dot{N} = 0$$

and the total population is conserved, as with the compartmental model from Munz *et al.* [1]. The calculation showing conservation of mass for (15.14) is essentially identical.

C Centre of Mass

In this appendix, we show that the population centre of mass is not conserved under (15.8), and we derive a time-dependent upper bound on the speed at which the centre of mass can travel.

Begin with the first moment $\langle n, x\rangle$ of $n(x,t)$ (proportional to the centre of mass) by taking an inner product of $\dot{n}$ with x. We obtain

$$\begin{aligned}\langle \dot{n}, x\rangle &= D_S\langle S_{xx}, x\rangle + D_Z\langle Z_{xx}, x\rangle && (15.27a)\\ &\quad - \langle (SK_S * Z)_x, x\rangle - \langle (ZK_Z * S)_x, x\rangle \\ &= H(t) + \langle SK_S * Z, 1\rangle + \langle ZK_Z * S, 1\rangle && (15.27b)\\ &= H(t) + \langle S, K_S * Z\rangle + \langle K_Z * S, Z\rangle && (15.27c)\\ &= H(t) + \langle S, K_S * Z\rangle + \langle S, K_Z * Z\rangle && (15.27d)\\ &= H(t) + \langle S, K * Z\rangle, && (15.27e)\end{aligned}$$

where $K = K_S + K_Z$ and $H(t) = L(D_s S_x(L,t) + D_z Z_x(L,x) - (Sv_s(L,t) + Zv_z(L,t)))$. The second equation follows from integration by parts (twice for the diffusive terms and once for the advective ones), the third follows from properties of multiplication and the fourth follows from the fact that convolution may be moved across the inner product. In the integration by parts leading to the second equation, note that the boundary terms that would arise all collapse to $H(t)$ due to the periodic boundary conditions. Since the quantity in (15.27e) is generically nonzero, it follows that the centre of mass of the solution is not conserved, which means travelling solutions are not *a priori* ruled out, although in our numerical simulations of (15.8) we have only found transient travelling solutions; see Figures 15.2A and 15.3A–C.

Using our previous calculation, we can bound the speed at which the centre of mass of n (equivalently, the first moment divided by the total mass) is permitted to move. Define $x_c(t) = \langle n, x\rangle/N$, the centre of mass of n. Let $G(t) = H(t)/N$. From our calculation above,

$$\begin{aligned}
|\dot{x}_c| &= |\langle S, K * Z\rangle|/N + G(t) \\
&\leq \langle S, |K| * Z\rangle/N + G(t) \\
&= \|S|K| * Z\|_1/N + G(t) \\
&\leq \|S\|_1 \cdot \||K| * Z\|_\infty/N + G(t) \\
&\leq \|S\|_1 \cdot \|K\|_\infty \cdot \|Z\|_1/N + G(t) \\
&\leq \frac{1}{4}\|K\|_\infty N^2/N + G(t) \\
&\leq \frac{1}{4}\|K\|_\infty N + G(t),
\end{aligned}$$

where the second line follows from the property of absolute value, the third from the definition of the L^1 norm, the fourth from Hölder's inequality, the fifth from Young's inequality for convolutions and the sixth from the fact that $\|S\|_1 + \|Z\|_1 \leq N$. Thus the maximum value of $\|S\|_1 \cdot \|Z\|_1$ is $N^2/4$. Therefore

$$\frac{1}{4}\|K\|_\infty N + G(t) \tag{15.28}$$

is an upper bound on the speed of the centre of mass of the population.

We can find $\|K\|_\infty$ using simple calculus. From the antisymmetry of K, it suffices to consider K on $x \geq 0$, where $K \geq 0$. The global maximum on this domain must occur either at $x = 0$ or at the critical point where $K' = 0$. The critical point is

$$x_{crit} = \frac{\ell \ln(\ell/F)}{\ell - 1}.$$

If $F < \ell$, then the critical point is outside of the domain. The global maximum, therefore, is $K(0) = 1 - F$. If $F = \ell$ then the critical point coincides with the domain boundary and $1 - F$ is still the global maximum. On the other hand, if $F \geq \ell$, the critical point is on the interior of the domain, and the global maximum is $K(x_{crit})$. In summary, we have

$$\|K\|_\infty = \begin{cases} 1 - F & F \leq \ell \\ \left(\frac{\ell}{F}\right)^{\frac{\ell}{1-\ell}} - F\left(\frac{\ell}{F}\right)^{\frac{1}{1-\ell}} & F > \ell, \end{cases}$$

where in the bottom expression we have evaluated $K(x_{crit})$.

Note that, as $F, \ell \to 1$, the upper bound (15.28) implies that the maximum speed of a travelling solution limits to $L/N(D_s S_x(L,t) + D_z Z_x(L,t))$.

For the cure model (15.14), the demonstration that the centre of mass is not conserved (as well as the provision of a bound for the speed of the centre of mass) is essentially identical to what we have presented in this appendix.

D Mass-Balanced States

We seek mass-balanced states of (15.8); that is, states where the number of individuals in each class is not changing in time, even if the solution profiles themselves are. We integrate each of (15.8) on the real line to obtain

$$\langle \dot{S}, 1\rangle + \langle (SK_S * Z)_x, 1\rangle = \langle D_S S_{xx}, 1\rangle - \langle \beta SZ, 1\rangle \tag{15.29a}$$

$$\langle \dot{Z}, 1\rangle + \langle (ZK_Z * S)_x, 1\rangle = \langle D_Z Z_{xx}, 1\rangle + \langle \beta SZ, 1\rangle + \langle R, 1\rangle - \langle \alpha SZ, 1\rangle \tag{15.29b}$$

$$\langle \dot{R}, 1\rangle = \langle \alpha SZ, 1\rangle - \langle R, 1\rangle. \tag{15.29c}$$

At mass balance, the first term of each equation will be zero by definition. Using integration by parts, we find that the terms involving spatial derivatives are zero. Then (15.29) is simplified to

$$\langle S, Z\rangle = 0 \tag{15.30a}$$

$$(\beta - \alpha)\langle S, Z\rangle + \langle R, 1\rangle = 0 \tag{15.30b}$$

$$\alpha\langle S, Z\rangle - \langle R, 1\rangle = 0. \tag{15.30c}$$

Substituting (15.30a) into (15.30c) shows that R must be identically zero. Equation (15.30a) itself implies that, in a point-wise sense, either S or Z is 0. Thus, at mass balance (and therefore at any steady states), there is physical separation between susceptibles and zombies. They cannot coexist at the same spatial location.

We now consider mass-balanced states of (15.14). The conditions (15.30) are now modified to

$$-\beta\langle S, Z\rangle + c\langle Z, 1\rangle = 0$$

$$(\beta - \alpha)\langle S, Z\rangle + \langle R, 1\rangle - c\langle Z, 1\rangle = 0$$

$$\alpha\langle S, Z\rangle - \langle R, 1\rangle = 0.$$

The solution may be written in terms of $\langle S, Z\rangle$ as

$$\int_0^L S\,dx = N - \left(\frac{\beta}{c} + \alpha\right)\langle S, Z\rangle$$

$$\int_0^L Z\,dx = \frac{\beta}{c}\langle S, Z\rangle$$

$$\int_0^L R\,dx = \alpha\langle S, Z\rangle.$$

We consider separately two cases for $\langle S, Z\rangle$. If $\langle S, Z\rangle = 0$, then the only mass-balanced state is the spatially homogeneous steady state (15.15). If $\langle S, Z\rangle \neq 0$, then there is a family of mass-balanced states parameterized by $\langle S, Z\rangle$, as discussed in Section 15.5.

E Steady States of Equation (15.8)

We show that the only possible steady states of (15.8) are spatially homogeneous. The steady-state problem corresponding to (15.8) is

$$(SK_S * Z)_x = D_S S_{xx} - \beta SZ \tag{15.31a}$$

$$(ZK_Z * S)_x = D_Z Z_{xx} + \beta SZ + R - \alpha SZ \tag{15.31b}$$

$$0 = \alpha SZ - R \tag{15.31c}$$

Since steady states are automatically mass-balanced states, we know from Appendix D that R is identically zero, and the product SZ is identically zero.

Assume that $Z(a) \neq 0$ for some $a \in [0, L]$ and that Z is continuous. We now show that the solution is a spatially homogeneous steady state. By the continuity of Z, $Z \neq 0$ in a neighbourhood of a. By the fact that $SZ = 0$, $S(x) = S_x(x) = 0$ in this neighbourhood. By (15.31),

$$(SK_S * Z)_x = D_S S_{xx}.$$

Integrating from a to x, we get

$$S_x = \left(\frac{K_S * Z}{D_S}\right) S.$$

The solution to this equation is

$$S(x) = S(a) \exp\left(\frac{\int_a^x K_S * Z ds}{D_S}\right) \equiv 0.$$

Therefore we have shown that the solution is a spatially homogeneous steady-state solution. A similar calculation shows that if $S(a) \neq 0$, then $Z \equiv 0$. Thus the only steady states are spatially homogeneous ones: namely (15.12) and (15.13).

F Linear Stability

We compute the local stability of the steady states (15.12) and (15.13) of (15.8) by allowing small perturbations, which for convenience we write in a normal mode expansion; that is,

$$S = S^* + \sum_k S_k(t)e^{ikx}$$

$$Z = Z^* + \sum_k Z_k(t)e^{ikx}$$

$$R = R^* + \sum_k R_k(t)e^{ikx},$$

where the wave number k is

$$k = \frac{2j\pi}{L}, \quad j \in \mathbb{Z},$$

to be commensurate with the periodic domain.

In the analysis below, since the ranges of K_S and K_Z are small compared to the size of the domain (i.e., $L \gg 1$), we approximate the Fourier transforms of the repulsive and attractive kernels by their values on the infinite domain:

$$\widehat{K}_S(k) \approx -\frac{2ik}{1+k^2} \tag{15.32a}$$

$$\widehat{K}_Z(k) \approx \frac{2iF\ell^2 k}{1+\ell^2 k^2}. \tag{15.32b}$$

Let $U_k = (S_k, Z_k, R_k)^T$. We substitute (15.12) and (15.13) into a linearized version of (15.8), giving us the system

$$\frac{d}{dt}U_k = \mathbf{L}U_k, \tag{15.33}$$

where $\mathbf{L}$ consists of terms arising from advection, diffusion and the original compartmental model of Munz *et al.* [1]:

$$\mathbf{L} = \mathbf{L}_{\text{adv}} + \mathbf{L}_{\text{diff}} + \mathbf{L}_{\text{comp}}.$$

We have

$$\mathbf{L}_{\text{comp}} = \begin{pmatrix} -\beta Z^* & -\beta S^* & 0 \\ (\beta-\alpha)Z^* & (\beta-\alpha)S^* & 1 \\ \alpha Z^* & \alpha S^* & -1 \end{pmatrix}$$

and

$$\mathbf{L}_{\text{diff}} = \begin{pmatrix} -D_S k^2 & 0 & 0 \\ 0 & -D_Z k^2 & 0 \\ 0 & 0 & 0 \end{pmatrix}.$$

For the linearized advection operator, using the fact that x derivatives transform as ik and convolution with a kernel K transforms as multiplication by $\widehat{K}$, we have

$$\mathbf{L}_{\text{adv}} = \begin{pmatrix} 0 & -ik\widehat{K}_S(k)S^* & 0 \\ -ik\widehat{K}_Z(k)Z^* & 0 & 0 \\ 0 & 0 & 0 \end{pmatrix},$$

but use (15.32) to write

$$\mathbf{L}_{\text{adv}} \approx \begin{pmatrix} 0 & -\dfrac{2k^2}{1+k^2}S^* & 0 \\ \dfrac{2F\ell^2k^2}{1+\ell^2k^2}Z^* & 0 & 0 \\ 0 & 0 & 0 \end{pmatrix}.$$

Using this general form of the linearization, we can analyze stability for each of the steady-state solutions.

For (15.12), the linearization is

$$\mathbf{L} = \begin{pmatrix} -D_S k^2 & -\dfrac{2(N/L)k^2}{1+k^2} - \beta N/L & 0 \\ 0 & -D_Z k^2 + (\beta - \alpha)N/L & 1 \\ 0 & \alpha N/L & -1 \end{pmatrix}.$$

Linear (in)stability will depend on the eigenvalues of the matrix **L**. Though the eigenvalues may be computed directly, the expressions are inconvenient to analyze. Instead, recall that the characteristic polynomial may be written down in terms of the eigenvalues $\lambda_{1,2,3}$ as

$$\lambda^3 - \tau\lambda^2 + \gamma\lambda - \Delta = 0,$$

where

$$\tau = \text{trace} \quad = \lambda_1 + \lambda_2 + \lambda_3 \tag{15.34a}$$
$$\gamma = \text{cross terms} \quad = \lambda_1\lambda_2 + \lambda_1\lambda_3 + \lambda_2\lambda_3 \tag{15.34b}$$
$$\Delta = \text{determinant} = \lambda_1\lambda_2\lambda_3. \tag{15.34c}$$

From (15.33), it follows that

$$\tau(k) = -(D_S + D_Z)k^2 + (\beta - \alpha)N/L - 1 \tag{15.35a}$$
$$\gamma(k) = D_S D_Z k^4 + [D_S(\alpha - \beta)N/L + D_S + D_Z]\, k^2 - \beta N/L \tag{15.35b}$$
$$\Delta(k) = -D_S D_Z k^4 + (D_S \beta N/L)k^2. \tag{15.35c}$$

For the equilibrium to be locally stable, the eigenvalues $\lambda_{1,2,3}(k)$ must have negative real parts for all admissible k. It follows from (15.34) that necessary conditions are $\tau < 0$, $\gamma > 0$, $\Delta < 0$ for all k. An examination of (15.35) shows that it is impossible to satisfy these necessary conditions. For instance, the opposing signs on the k^2 and k^4 terms in $\Delta(k)$ mean that $\Delta > 0$ for some k, and hence the equilibrium is locally unstable to generic noisy perturbations as long as L is sufficiently large. This result of linear instability is the same as for the non-spatial model in Munz *et al.* [1].

For the susceptible equilibrium (15.12), the linear stability matrix is

$$\mathbf{L} = \begin{pmatrix} -D_S k^2 - \beta N/L & 0 & 0 \\ \dfrac{2NF\ell^2 k^2}{1+\ell^2 k^2} + (\beta - \alpha)N/L & -D_Z k^2 & 1 \\ \alpha N/L & 0 & -1 \end{pmatrix},$$

which has eigenvalues with simple expressions that are easy to analyze. We have

$$\lambda_1 = -D_S k^2 - \beta N/L, \quad \lambda_2 = -D_Z k^2 \quad \text{and} \quad \lambda_3 = -1. \tag{15.36}$$

Since $\lambda_{1,2,3} < 0$, we see that the population is locally stable, as in the non-spatial model. The only exception is for a perturbation with $k = 0$, in which case there is one eigenvalue equal to zero, with corresponding eigenvector $(S_0, Z_0, R_0) = (0, 1, 0)$, which tells us that a perturbation that changes an all-zombie state to another all-zombie state (by changing the total population size) is neutrally stable.

We now turn our attention to the local stability of homogeneous steady states of the cure model (15.14). The calculation is similar to those presented above. Stability of the susceptible equilibrium (15.15) is determined by the matrix

$$\mathbf{L} = \begin{pmatrix} -D_S k^2 & -\dfrac{2(N/L)k^2}{1+k^2} - \beta N/L + c & 0 \\ 0 & -D_Z k^2 + (\beta - \alpha)N/L - c & 1 \\ 0 & \alpha N/L & -1 \end{pmatrix}.$$

One eigenvalue has a simple expression, $\lambda_1 = -D_S k^2 \leq 0$. It follows from above that the remaining eigenvalues satisfy

$$\begin{aligned} \lambda_2 + \lambda_3 &= -D_Z k^2 + (\beta - \alpha)N/L - 1 - c \\ \lambda_2 \lambda_3 &= D_Z k^2 - \beta N/L + c. \end{aligned}$$

Necessary and sufficient conditions for the remaining two eigenvalues to have negative real parts are $\lambda_2 + \lambda_3 \leq 0$ and $\lambda_2\lambda_3 \geq 0$, which lead to stability condition (15.22).

The local stability of the endemic state (15.16) is determined by the matrix

$$\mathbf{L} = \begin{pmatrix} -D_S k^2 - \beta Z^* & -\dfrac{2S^* k^2}{1+k^2} & 0 \\ \dfrac{2Z^* F \ell^2 k^2}{1+\ell^2 k^2} + (\beta - \alpha)Z^* & -D_Z k^2 - \alpha S^* & 1 \\ \alpha Z^* & \alpha S^* & -1 \end{pmatrix}, \tag{15.37}$$

where we have used the equilibrium value for S^* to simplify parts of some entries in the matrix in order to have convenient expressions. It is straightforward to compute eigenvalues of (15.37) numerically. We also make some analytical statements. Recall from (15.34) that necessary conditions for stability are $\tau < 0$, $\gamma > 0$ and $\Delta < 0$ for all k. For (15.37),

$$\begin{aligned} \tau(k) &= -(D_S + D_Z)k^2 - 1 - Z^*\beta - \frac{\alpha c}{\beta} && < 0 \\ \Delta(k) &= -D_Z k^2 \left(\beta Z^* + D_S k^2\right) - \frac{2cZ^* k^2}{1+k^2} - \frac{4F\ell^2 c Z^* k^4}{\beta\left(1+\ell^2 k^2\right)\left(1+k^2\right)} && \leq 0, \end{aligned}$$

with $\Delta(k) = 0$ only for $k = 0$. However, the lengthy expression for γ (omitted here) is of indeterminate sign. Negativity of γ would guarantee instability, but positivity of γ does not guarantee stability as these are necessary conditions. The two conditions $\tau < 0$ and $\Delta < 0$ from (15.34) combined with the condition $\Delta - \tau\gamma >$

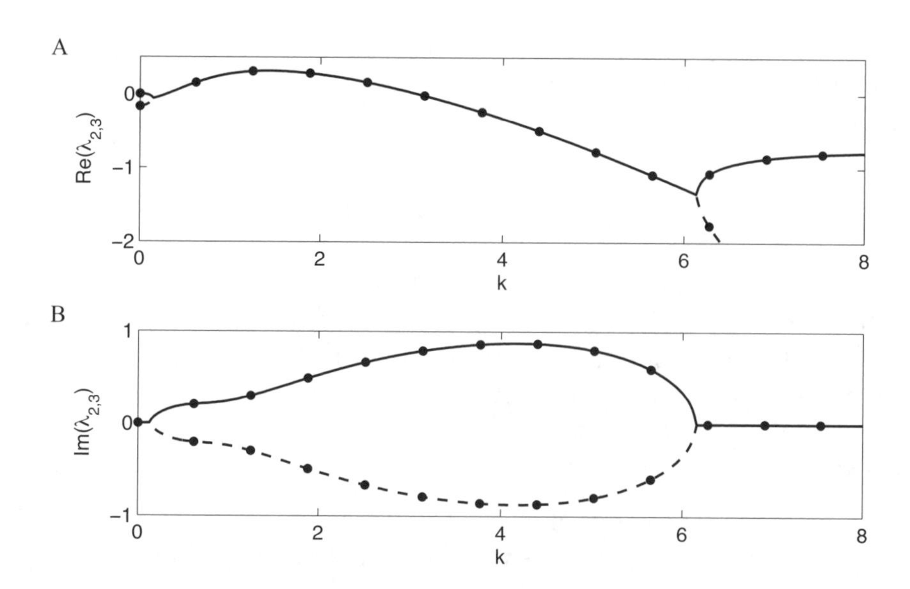

Figure 15.7: Example instability of the endemic state. A: Real part of two eigenvalues $\lambda_{2,3}$ for the linearization (15.37) of the endemic steady state (15.16) of the model (15.14) with parameters (15.39). As the spectrum is discrete, data are given by the filled circles; curves are plotted to guide the eyes. The first eigenvalue λ_1 (not shown) is always negative. Since there exist values of the wave number k for which $\mathrm{Re}(\lambda_{2,3}) > 0$, the equilibrium is unstable. B: Like A, but depicting $\mathrm{Im}(\lambda_{2,3})$. The eigenvalues are complex for a range of k, including the subset for which $\mathrm{Re}(\lambda_{2,3}) > 0$, indicating an oscillatory instability.

0, which results from the Routh–Hurwitz criteria (see Appendix B of Murray [21]), constitute a set of necessary and sufficient conditions for stability. Unfortunately, the Routh–Hurwitz conditions for (15.37) are difficult to analyze due to the large number of parameters and the rational dependence on k.

Elegant conditions sufficient for stability can be obtained by using the Gerschgorin Circle Theorem [22] applied to the columns of $\mathbf{L}$. Application of the theorem guarantees that the eigenvalues of $\mathbf{L}$ lie in the union of three discs in the plane of complex numbers w:

$$\left|w + \left(D_S k^2 + \beta Z^*\right)\right| \leq \left|\frac{2Z^* F \ell^2 k^2}{1 + \ell^2 k^2} + (\beta - \alpha)Z^*\right| + \alpha Z^* \tag{15.38a}$$

$$\left|w + \left(D_Z k^2 + \alpha S^*\right)\right| \leq \frac{2S^* k^2}{1 + k^2} + \alpha S^* \tag{15.38b}$$

$$|w + 1| \leq 1. \tag{15.38c}$$

By forcing these discs to lie entirely in the left half-plane, we can guarantee stability. By simple algebra, (15.38a) lies in the left half-plane for all k if $\beta > \alpha$ and $D_S > 2Z^* F \ell^2$. The second disc (15.38b) lies in the left half-plane if $D_Z > 2S^*$. The third disc (15.38c) lies in the left half-plane no matter what. Hence the three conditions

$$\beta > \alpha, \quad D_S > 2Z^* F \ell^2 \quad \text{and} \quad D_Z > 2S^*$$

are sufficient for stability of the endemic state.

Instability of the endemic equilibrium is possible as well. As mentioned in Section 15.5, the parameters

$$\begin{aligned} &\alpha = 1, \quad \beta = 0.1, \quad D_S = 0.1, \quad D_Z = 0.05, \\ &c = 0.2, \quad F = 0.3, \quad \ell = 0.3, \quad N = 70 \quad \text{and} \quad L = 10 \end{aligned} \tag{15.39}$$

provide a numerical example of instability. One eigenvalue λ_1 is negative for all k. For the remaining eigenvalues, $\mathrm{Re}(\lambda_{2,3})$ are shown in Figure 15.7B and $\mathrm{Im}(\lambda_{2,3})$ in Figure 15.7B. For an intermediate range of k, $\mathrm{Re}(\lambda_{2,3}) > 0$ and $\mathrm{Im}(\lambda_{2,3}) \neq 0$, rendering the endemic equilibrium unstable to an oscillatory solution.

G Glossary

Advective velocity. The portion of a population's velocity that results from bulk movement; in contrast to diffusive velocity, describing random motion.

Centre of mass. A point representing the average position of all organisms in a given group.

Chemotaxis. The movement (usually of biological organisms) toward or away from chemicals in the environment.

Compartment model. A type of mathematical model convenient for studying the transfer of some quantity between different compartments or states of a system; a classic example is the SIR (susceptible-infected-removed) model from epidemiology, in which sick individuals pass through these three compartments over the course of their illness.

Conservation equation. An equation describing a system in which some quantity does not change as the system evolves; in the context of the zombie model, the total number of individuals is conserved.

Diffusion. A transport phenomenon in nature that describes macroscopic spreading of a group of objects due, usually, to underlying random motion.

Doomsday state. A state in which all of humanity has become zombified.

Endemic state. A state in which a disease persists within a population without input; for instance, chickenpox or the undead.

Equilibrium. A state in which the spatiotemporal distributions of humans and zombies remain fixed for all time.

Mass-balanced state. A state of the system for which the number of individuals in each class is not changing in time; every equilibrium is a mass-balanced state, but not every mass-balanced state is an equilibrium.

Mean field hypothesis. The hypothesis that the behaviour of individuals can be understood by considering their average behaviour; this is most frequently used for fluids, but it is useful in many other situations, including human population models.

Nondimensionalization. A standard mathematical technique that simplifies a model by using changes of variables to replace physical dimensions with problem-specific dimensions.

Social interaction kernel. A function describing the effect that one organism has on a second as a result of the second organism being able to sense the first one through sight, sound, smell, touch or other method.

Stability. An equilibrium state is locally stable if all states starting nearby stay near the equilibrium state forever.

Swarming model. A model for biological organisms, describing the ability for individuals to sense each other over finite distances and react accordingly. Such models have been successfully used to understand fish schools, bird flocks and insect interaction, as well as collective human behaviour.

Travelling pulse. A wave with a single crest that moves through space without dissipating.

References

[1] P. Munz, I. Hudea, J. Imad, R.J. Smith?. When zombies attack!: Mathematical modelling of an outbreak of zombie infection. In *Infectious Disease Modelling Research Progress*, J.M. Tchuenche and C. Chiyaka, eds., Nova Science, Happuage, NY 2009, pp. 133–156.

[2] T. Prentice, L.T. Reinders. *Technical Report.* World Health Organization, 2007.

[3] E.F. Keller, L.A. Segel. Model for Chemotaxis. *J. Theor. Biol.*, 30:225–234, 1971.

[4] A. Okubo. *Diffusion and Ecological Problems.* Springer, New York, NY, 1980.

[5] D. Bernoulli, Essai d'une nouvelle analyse de la mortalité causée par la petite vérole. *Mem. Math. Phy. Acad. Roy. Sci. Paris*, 1–45, 1760.

[6] D. Bernoulli, S. Blower. An attempt at a new analysis of the mortality caused by smallpox and of the advantages of inoculation to prevent it. *Rev. Med. Virol.*, 14:275–288, 2004.

[7] H. Hethcote. The mathematics of infectious diseases. *SIAM Rev.*, 42:299–653, 2000

[8] J.D. Murray. *Mathematical Biology II: Spatial Models and Biomedical Applications.* Springer, New York, NY, 2002.

[9] S. Ruan. Spatial-Temporal Dynamics in Nonlocal Epidemiological Models. In *Mathematics for Life Sciences and Medicine*, Y. Takeuchi, K. Sato and Y. Isawa, eds., Springer-Verlag, New York, NY, 2006, 97–122.

[10] J.V. Noble. Geographic and temporal development of plagues. *Nature* 250:726–729, 1974.

[11] J.D. Murray, E.A. Stanley, D.L. Brown. On the spatial spread of rabies among foxes. *Proc. R. Soc. Lond. B* 299:111–450, 1986.

[12] S. Fedotov. Front propagation into an unstable state of reaction-transport systems. *Phys. Rev. Lett.*, 86:926–929, 2001.

[13] J. Medlock, M. Kot. Spreading disease: Integro-differential equations old and new. *Math. Biosci.*, 184:201–222, 2003.

[14] D.G. Kendall. Mathematical Models of the Spread of Infection. In *Mathematics and Computer Science in Biology and Medicine*, 1965, pp. 213–224.

[15] D. Mollison. Possible velocities for a simple epidemic. *Adv. Appl. Prob.*, 4:233–257, 1972.

[16] T. Hillen, K.J. Painter. A user's guide to PDE models for chemotaxis. *J. Math. Biol.*, 58:183–217, 2009.

[17] A. Mogilner, L. Edelstein-Keshet. A non-local model for a swarm. *J. Math. Biol.*, 38:534–570, 1999.

[18] A.J. Leverentz, C.M. Topaz, A.J. Bernoff. Asymptotic dynamics of attractive-repulsive swarms. *SIAM J. Appl. Dyn. Sys.*, 8:880–908, 2009.

[19] R. Eftimie, G. de Vries, M.A. Lewis. Complex spatial group patterns result from different animal communication mechanisms. *Proc. Natl. Acad. Sci.*, 104:6974–6979, 2007.

[20] S.R. Partan, P. Marler. Issues in the classification of multimodal communication signals. *Am. Nat.*, 166:231–245, 2005.

[21] J.D. Murrary. *Mathematical Biology I: An Introduction*, Springer, New York, NY, 2002.

[22] R.A. Brualdi, S. Mellendorf. Regions in the Complex Plane Containing the Eigenvalues of a Matrix. *Amer. Math. Month.*, 101:975–985, 1994

CONCLUSION

Robert Smith?

So what did we learn?

To answer this question, I'd like to return to the very roots of the current epidemic: that's right, the original 2009 Munz zombie paper. Or, more accurately, the reason it was published in the first place.

Although publishing a mathematical modelling paper on zombies was a fun thing to do (as the contributors to this volume will surely agree), there was reason it was published in the book it was. Specifically, the editors of that book requested chapters that showcased the way in which models were constructed and requested that each article include more than one mathematical model.

That's because mathematical modelling is a conversation. No single model can be the final word on reality, any more than a map can. The entire point of modelling is to strip out what's distracting, in order to focus on what's important. Your version of 'important' or 'distracting' is going to differ from mine.

One of the most obvious omissions from the original paper was that (in the first three models, anyway) we neglected to include a permanent state of death for the zombies. This was a deliberate choice: in *Shaun of the Dead*, Shaun kills one with his car and it gets up again. However, it's true that if you shoot a zombie in the head, it dies for good, and we didn't include that until the final model.

This omission made a lot of people very angry. Well, hey, it was the most popular thing on the internet at the time, and just about anything on the internet is going to make people very angry! If I had to do it again, I'd have included a 'permanent death' compartment in all four of those models. But you know what? I'm glad we didn't.

Why? Because if the original model had been perfect, then there'd be nothing to say. And you probably wouldn't be reading this book. (There's a reason why you aren't reading an edited collection of articles on Bieber Fever. Okay, several reasons.) Because that omission opened the door. It inspired people.

I've never had a problem with people criticizing my work. Indeed, I welcome it. Part of what academia is about is building on each other's work so that, as Isaac Newton said, we stand on the shoulders of giants. If you're precious about criticism of your research, then the only surefire conclusion is that you aren't in the running to be a giant.

So I see mathematical modelling as a conversation. My students and I had the opening word, but the book you've just finished is the result of a dynamic and interactive discussion about just how to model a new epidemic. And that's something we need more of.

In fact, this is how modelling finds its way into policy. Or good policy, at any rate. Rather than just accepting the conclusions of one research group, policy documents tend to look at what everyone is saying. If the entire modelling community points out that an HIV vaccine with low efficacy may cause more harm than good, then that's something worth paying attention to. So although we might have a dynamic conversation along the way, it's when those voices come into synch that we speak the strongest. That's when we sing.

So what are the melodies of this book, then? For one thing, we see the importance of spatial components to modelling. Zombies are much more homogenous than humans, yet we see many voices saying how important spatial effects and distributions are. We also learn that decision-making is critical and that we might just be stuck with having to make decisions with partial or imperfect information. That's unfortunate, but knowing that it's a necessary evil means we can do something about it. Oh, and we're probably going to get oscillations no matter what we do. History repeats itself, even with zombie invasions.

But this book isn't the last word on the subject, nor is it intended to be. One of the unexpected side effects I've found from the original zombie paper is that mathematics teachers have started teaching zombies. I've seen this done in universities, in high schools and even to nine-year-olds. There's a very good reason: it engages students immediately, sometimes when nothing else will. This is a very important piece of the process: if we are to be giants, then we need the next giants to stand on our shoulders.

And that means the next step is up to you. Take the lessons learned here and create your own zombie models. Tweak one of the models and see what happens. Use a technique illustrated here to see how it might apply to fast zombies, vampires or werewolves. Create your own model entirely from scratch, just to see how your instincts compare to ours.

The conversation hasn't ended. It's only just begun.

CONTRIBUTORS

The Undead

Daniel Ashlock (dashlock@uoguelph.ca) is a professor of mathematics holding the chair in bioinformatics of the department of mathematics and statistics at the University of Guelph in Guelph, Ontario, Canada. He is a senior member of the IEEE and a member of the IEEE Computational Intelligence Society. His research interests include bioinformatics, computational intelligence, combinatorics and, of course, applications to the modelling and prevention of zombie outbreaks.

Jennifer Badham (research@criticalconnections.com.au) is an applied mathematician interested in using models and simulations to assist in making policy decisions. She has contributed to health policy for government and industry associations for over twenty years. She is currently a research fellow at the Centre for Research in Social Simulation at the University of Surrey. Her other research focus is dynamic processes over social networks, including algorithms to generate networks with specific properties and the way in which network structure affects transmission of beliefs, ideas or diseases across the network.

Ruth Baker (baker@maths.ox.ac.uk) studied mathematics at the University of Oxford, before joining the Centre for Mathematical Biology in Oxford for her doctoral studies. She was subsequently appointed to an RCUK Fellowship in mathematical biology and more recently to a University Lectureship, both also in the Centre for Mathematical Biology. Her research interests include mathematical modelling of aspects of developmental biology, including self-organization, cell motility and domain growth. Her knowledge of zombie behaviour has increased exponentially during this project.

Nick Beeton (nick.beeton@utas.edu.au) has always enjoyed playing with numbers, so was inevitably trapped by the allure of applied mathematics during his science degree. After studying his Ph.D. on modelling the spread of festering infectious tumours on Tasmanian devils, modelling zombies seemed a natural next step. Currently, Nick is a research fellow at the University of Tasmania funded by the

National Environment Research Program, investigating ways to protect Australia's animal biodiversity from threats other than marauding zombies.

Joseph Alexander Brown (jb03hf@gmail.com) was born in Niagara-on-the-Lake, Ontario, Canada. He received his B.Sc. (Hons.) in computer science from Brock University, St. Catharines in 2007 and an M.Sc. in computer science in 2009. He is currently a doctoral student at the University of Guelph. Other than worrying about how to deal with the pressing matters of zombies, his academic studies deal with bioinformatics and evolutionary computation. He's currently examining how extinction can be used as an evolutionary aid for regression techniques. He's also a student member of the IEEE and the Association for Computing Machinery (ACM). He is the local Treasurer of the Student ACM Group and coach of the ACM International Collegiate Programming Competition Teams at the University of Guelph.

Ben Calderhead (b.calderhead@imperial.ac.uk) is a lecturer in the Department of Mathematics at Imperial College London. He obtained his Ph.D. from the University of Glasgow, funded by a Microsoft Research Scholarship. His thesis introduced the use of differential geometry within Markov chain Monte Carlo methods, leading to a paper that was discussed before the Royal Statistical Society. Subsequently, he was a 2020 Science Research Fellow at University College London before moving to Imperial. His current work involves the development of novel computational statistical methodology, in particular Monte Carlo methods, with applications in many fields, from numerical analysis to computational and systems biology.

Andrew Cartmel (andrew.cartmel.3@facebook.com) is a science-fiction writer, journalist and former script editor of *Doctor Who*. He was born in Britain and grew up in Canada. Seduced by a book on writing for television, he ended up as the show runner on three seasons of *Doctor Who*, from 1987 to 1989. Since then, he has freelanced, most recently with script commissions for *Torchwood* and *Midsomer Murders*. He has also written two plays for London's Fringe Theatre, had ten novels and a novella published, scripted four audio plays and performed stand-up comedy. He has two cats, two agents, a manager, a large collection of scrounging friends and an even larger collection of jazz recordings on vinyl to support. Consequently, he's a busy fellow.

Micael Santos Couceiro (micaelcouceiro@gmail.com) obtained his Ph.D. in Electrical and Computer Engineering (Automation and Robotics) at the University of Coimbra. Over the past six years, he has been conducting research in computer vision, sports engineering, economics, sociology and digital media at the Institute of Systems and Robotics. Currently, he is the CEO of the research and technological development company Ingeniarius.

Mike Delorme (mike.j.delorme@gmail.com) received his M.Sc. from McMaster University specializing in mathematical models of influenza. He received his B.Sc.

(Honours) from Queen's University specializing in mathematical biology. During the 2009/2010 influenza season, Mike was a Pandemic Influenza Research Intern at McGill University and the Direction de Santé Publique de Montréal. He then worked for three years in the Infectious Disease, Dental and Sexual Health Division at Region of Waterloo Public Health in Waterloo, Ontario, Canada. He now works for Manulife Financial Canada as the statistical lead for the applied analytics section.

Louis J. Dubé (Louis.Dube@phy.ulaval.ca) received his B.Sc. in physics from Université Laval (1973) and completed his graduate studies at Yale University (M.Sc. 1974, Ph.D. 1977). This was followed by years of directed random walks, first to Oxford, UK (1978–79) and then to Freiburg, Germany (1979–86) where he emerged with the German doctorate and the Habilitation (Dr.rer.nat.habil. 1986). At this point, he joined the professorial ranks at his *alma mater* Université Laval; this was the end of the first cycle. He has so far resisted other basins of attraction (Bordeaux, 1992–93; Paris, 2002–2008) only to return to the original attractor. He had the chance to reinvent himself several times from collision theory to complex dynamical systems. He embodies an *RSIR* model: he was born *ready* for, quite *susceptible* to, rapidly *infected* by and still not *recovered* from scientific research and education. In the course of this work, he learned that zombies were, by strict definition, not vampires.

Bard Ermentrout (ermentrout@gmail.com) is a professor of mathematics and amateur apethantologist at the University of Pittsburgh. Apethantology (coined by Stefanos Folias) is derived from the Greek root pethanos [it has died] along with 'a-' to negate it. Thus, the study of undead. One of the first dates with his future wife was to see a double feature featuring *Night of the Living Dead.* Upon arriving in Pittsburgh he made a pilgrimage the Monroeville Mall; he has 'The Time of the Season' on his iPod; and he always reminds his kids: "Rule number three: seat belts."

Kyle Ermentrout (kyle.ermentrout@gmail.com) received his B.A. in philosophy at the University of Pittsburgh and is currently employed with Clean Water Action. His interest in horror and monster movies developed at an early age, primarily due to the influences of his father. He particularly enjoys the zombie mythos both for its potential in comedy and for its analogues to the real world.

Carlos Manuel Clímaco Figueiredo (cfigueiredo@isec.pt) graduated from the Engineering Institute of Coimbra with an M.Sc. in Automation and Communications (2008). He worked at Institute Pedro Nunes at Coimbra developing new technologies for alternative transportation systems based on autonomous electric vehicles. He is currently responsible for electronics at SOLIEN (Integrated Engineering Solutions).

Eamonn Gaffney (gaffney@maths.ox.ac.uk) studied mathematics at the University of Cambridge, prior to a Ph.D. in Theoretical High Energy Physics and a Wellcome Trust post-doctoral fellowship at the Centre for Mathematical Biology, Oxford. After a faculty position at the School of Mathematics, University of Birmingham, he returned to a University Lectureship at Oxford and the Centre for Mathematical Biology in 2006. His research interests include microbiological continuum mechanics, biological self-organization and cell motility, signalling and interactions.

Mark Girolami (m.girolami@ucl.ac.uk) is a professor of computing and inferential science at the University of Glasgow. He was awarded an EPSRC Advanced Research Fellowship in 2007. He obtained his Ph.D. in 1998 with a thesis on Independent Component Analysis, and there are currently more than 1500 citations to the papers published during his Ph.D. studies. In 2009 he was awarded the SPIE Pioneer Award (International Society of Photo-Optical Engineers) for the impact his contributions to ICA have had on advancing neuro-imaging analysis technology. At Glasgow, he leads the cross-disciplinary Inference Research Group, where the main focus of research is methodological development in computational statistical inference with major applications in computational and systems biology. Other recent successful applications include statistical modelling of telecom fraud and counterfeit currency detection.

Jane M. Heffernan (jmheffer@mathstat.yorku.ca) is from an area in rural Ontario known for growing potatoes, which are great weapons in defending against zombie attacks. At an early age, Jane realized that she had a keen interest in math and she decided to become a high-school math teacher. Jane pursued her undergraduate degree in mathematics and computer science at Trent University, and then her bachelor of education at Queen's University. However, after teaching zombielike high-school students, she decided to pursue graduate studies in mathematical biology. After completing a master's and Ph.D. at the University of Western Ontario, she took up a post-doctoral fellowship at the University of Warwick in England and then became a professor at York University in 2007. In her work, she develops mathematical models that describe various aspects of HIV, measles, influenza, Hepatitis B, Hepatitis C, the immune system and now zombies and vampires. Jane and Derek Wilson met in their undergraduate days, married and are now parents. This chapter is Derek and Jane's first collaborative research project.

Des Higham (d.j.higham@strath.ac.uk) works in computational and applied mathematics at the University of Strathclyde, and has interests in biology and finance. He has collaborated with a wide range of colleagues across the life sciences. In 2005, he was awarded the Germund Dahlquist Prize from the Society for Industrial and Applied Mathematics (SIAM), an international award made every two years for research contributions in numerical methods for scientific computing. In 2006, his article "A matrix perturbation view of the small world phenomenon" was chosen to feature as a SIGEST article, representing an "exceptional" paper from SIAM's specialized research journals. Higham was invited to give the 2007 Magnus lectures at Colorado State University, including a public lecture on "Network Science: Joining the Dots." In 2009, he was one of only six UK mathematicians to be elected to an inaugural SIAM Fellowship, with a citation for "contributions to numerical analysis and stochastic computation." He has also published expository articles in SIAM Review's Education section, and has written undergraduate-level texts in mathematical finance and numerical analysis.

Alex Hoare (ahoare@gmail.com) holds a Ph.D. in infectious disease modelling from the University of New South Wales. He is an avid gamer and relishes any opportunity to bring mathematics, computer games and zombies together.

Clinton Innes (innesc@uoguelph.ca) is an undergraduate student in mathematics at the University of Guelph. In addition to understanding long walks on the beach, his research interests include the modelling and prevention of zombie outbreaks and the modelling of DNA replication. In the future, he plans to attend graduate school in mathematics or teachers college.

João Miguel Almeida Luz (miguel.luz@isec.pt) graduated from the Engineering Institute of Coimbra, Portugal, with a B.Sc. and a Teaching Licensure in Electrical Engineering and Communications. Since then, he has worked at WSBP Electronics, in Coimbra, in data and sensor fusion and for the research group RoboCorp. He is a Ph.D. student in Electrical and Computer Engineering at the University of Coimbra.

Philip Maini (maini@maths.ox.ac.uk) gained his D.Phil. in mathematical biology under the supervision of Jim Murray. He is professor of mathematical biology and director of the Centre for Mathematical Biology, Oxford, and a principal investigator at the Oxford Centre for Integrative Systems Biology in the Department of Biochemistry in Oxford. His research interests include the application of mathematical modelling in areas in developmental biology, wound healing and cancerous tumour growth.

Philip Munz (pmunz@wlu.ca) has been an avid fan of zombie media since *Night of the Living Dead* scared the popcorn out of him as a teenager. He recently completed a master's degree in applied mathematics at Carleton University, where he co-authored the original paper on mathematical modelling of zombies. He is currently employed at Wilfrid Laurier University teaching math and zombie survival to first and second year undergraduate students. He spends much of his spare time coming up with quirky applications for mathematical concepts and researching the undead—without the help of any government funding—and has dreams of becoming zombie fodder in some upcoming blockbuster zombie flick.

Pierre-André Noël (pierre-andre.noel.1@ulaval.ca), **Antoine Allard** (antoine.allard.1@ulaval.ca), **Laurent Hébert-Dufresne** (laurent.hebert-dufresne.1@ulaval.ca) and **Vincent Marceau** (vincent.marceau.2@ulaval.ca) were born and subsequently raised in the province of Québec. Unbeknownst to one another, they shared first-hand zombie encounters early in life (PAN: *The Evil Dead*; AA: Saw a bunch of them kicked out of a bar once; LHD: *Zombies Ate My Neighbors*, VM: Michael Jackson's *Thriller*). Unprepared at the time, they ignored their calling until their paths intertwined at Université Laval where they each received a B.Sc. in physics (PAN: 2005; AA: 2006; LHD and VM: 2009). They eventually chose to face the trials and tribulations of graduate studies (purification by fire) within the same holy halls, where they gathered under the Dynamics flag of Prof. Louis J. Dubé's research group on the dynamics of networks (actually PAN started with mosquitos, but, as they say, from sucking blood to eating flesh is a one-way trip). Undeterred and hardened by the growing challenges, they have prevailed against papers, degrees and diplomas (PAN: M.Sc. 2007, Ph.D. 2012; AA: M.Sc. 2008, Ph.D. 2014; LHD: M.Sc. 2011, Ph.D. 2014; and VM: M.Sc. 2011). They have now returned to face the zombies of their childhood nightmares for the single greatest modelling challenge of all times: the undead hordes . . . to save us all, or go mad trying.

Judy-anne Osborn (Judy-anne.Osborn@newcastle.edu.au) is a mathematician working in combinatorics and mathematics education. She is based at the University of Newcastle, Australia, where she took up a continuing position in 2013 following a post-doctoral fellowship in the Centre for Computer Assisted Research Mathematics and its Applications and a previous post-doctoral fellowship in the Mathematical Sciences Institute at the Australian National University where she retains a visiting fellowship. These post-doctoral roles followed a 2007 Ph.D. in Mathematics at the University of Melbourne, where she was a member of the ARC Centre of Excellence for Mathematics and Statistics of Complex Systems. She has a number of projects with associated grants in mathematics research and education, including a large Australian Office of Learning and Teaching grant "Inspiring Mathematics and Science in Teacher Education." She has presented her work on mathematics and mathematics education in Australia, the UK, France and Indonesia. She recently spoke of her zombie-related work in an invited talk at the Australian Academy of Science in their "Science Stars of Tomorrow" series, with a lecture entitled "Tipping the balance towards scientific thinking, via zombies and maths."

Gergely Röst (rost@math.u-szeged.hu) was born in western Hungary. He studied mathematics at the University of Szeged (southern Hungary) and also in Italy (Pisa, Potenza) and Germany (Giessen). He received his Ph.D. in 2006 with the special 'Golden Ring of Hungary' award, given personally by the president of the country. In his thesis, he developed a spectral projection technique to study bifurcations in time-periodic dynamical systems in infinite dimensions. After that he spent two years in Toronto as a post-doctoral research fellow at York University, where he started to apply his obscure mathematical tools to real-life problems arising in epidemiology, such as the emergence of antiviral resistance during pandemic influenza. Two years after his return to Hungary, he won the highly prestigious European Research Council's Starting Investigator Grant, and now he is organizing a new research group called EPIDELAY at the University of Szeged, but only upon his return from his Fulbright scholarship at Arizona State University. He is one of the toughest soccer players in Hungary Division 5.

Evelyn Sander (esandergmu@gmail.com) is a professor of Mathematics at George Mason University in Fairfax, Virginia, where she teaches and mentors both graduate and undergraduate students. Sander mainly does research on dynamical systems and differential equations. In addition to the distinction of having been the first paid electronic-only journalist, she also held first place at the national level in unicycle racing. She is attempting to print one of everything on a three-dimensional printer, be it living, dead or undead. For the preparation of this paper, Sander spent many diligent hours researching zombie video games.

Without really intending to, **Robert Smith?** (rsmith43@uottawa.ca) appears to have accidentally created the subdiscipline of mathematical modelling of zombies. Huh. By day, he's a professor of biomathematics at the University of Ottawa, studying infectious diseases such as HIV, human papillomavirus and various tropical diseases. By night, he's a writer, having written or edited *Who's 50: The 50 Doctor Who Stories to Watch Before You Die* (ECW Press, 2013), *Outside In: 160 New Perspectives on 160 Classic Doctor Who Stories by 160 Writers* (ATB Publishing, 2012), *Who is the Doctor: The Unofficial Guide to the New Series* (ECW Press, 2012), *Braaaiiinnnsss: From Academics to Zombies* (University of Ottawa Press, 2011), two volumes of *Time Unincorporated: The Doctor Who Fanzine Archive* (Mad Norwegian Press, 2010 and 2011) and *Modelling Disease Ecology with Mathematics* (American Institute of Mathematics Sciences, 2008). He was the senior author on the zombie paper that sparked all this madness and, for an encore, became the world's leading expert on the spread of Bieber Fever—although he's not sure whether to hope that this scientific discipline also takes off or not.

Ben Tippett (ben.tippett@ubc.ca) earned his doctorate studying black holes (a type of ravenous, undead star) at the University of New Brunswick. He received his master's degree in physics from Queen's University, and a B.Sc. in physics from the University of British Columbia. He developed an acute fear of the undead in 2005, and has been actively researching them since 2008. He currently resides in Kelowna, British Columbia, with his intrepid wife Bethany (who promises to save him in case of zombie attack). He spends his free time writing physics papers about Cthulhu and TARDISes. He hosts and produces The Titanium Physicists Podcast.

Chad Topaz (chad.topaz@gmail.com) is an associate professor of Mathematics at Macalester College in St. Paul, Minnesota, and he wants to eat your brains.

When not eating brains, he teaches undergraduate applied mathematics courses and mentors student research in dynamical systems and mathematical biology within his interdisciplinary XMAC (eXperiment, Modeling, Analysis and Computation) Laboratory. He has been a Kavli Frontiers Fellow of the National Academy of Sciences and a New Directions Research Professor at the Institute for Mathematics and its Applications. For his teaching efforts, and for refraining from eating his students' brains, he was awarded Macalester's Jack and Marty Rossmann Excellence in Teaching Award.

Brody Walker (brody.walker@education.tas.gov.au) has had a passion for all things zombie-related ever since he can remember. With this level of enthusiasm, it comes as no surprise that he has spent many hundreds of hours watching, playing and reading anything and everything involving zombies. When he was asked to provide a research base to a mathematical model of zombies, he jumped at the chance to put his extensive knowledge to the test. In his non-zombie life, Brody has a Bachelor of Arts (History Honours) and a Master of Teaching, both from the University of Tasmania. He now spends his days teaching zombie survival skills to the next generation of hunters, hoping to ensure the continued existence of humanity.

Derek J. Wilson (dkwilson@yorku.ca) was born in Toronto, Ontario, a city that has not been attacked by zombies (yet). His deep and abiding interest in the undead began at the age of four when he saw a light flashing on and off all by itself in a neighbour's garage and logically concluded that it must be a ghost. He would later discover that his neighbour had made the mistake of installing a light sensor in an enclosed space, which was a bit of a disappointment. Derek pursued his undergraduate degree at Trent University, majoring in Biochemistry. In third year, he met his future wife, Jane Heffernan, whom he wooed by lying to her and saying it was his intention to go to medical school. Instead, he went to graduate school at the University of Western Ontario where he completed a Ph.D. in Biochemistry and subsquently moved with his (now) wife to England to take up an NSERC post-doctoral fellowship at Cambridge. Shortly into that post-doctoral stint, he was hired as a professor at York University, starting in 2007. Since that time, he has published a number of primary research papers, review articles and book chapters, received tenure and, more importantly, has become a dad. As such, he is putting all zombies and vampires on notice that he is at least book-smart in counter-undead tactics and will defend his family to the (un)death.

Thomas Woolley (woolley@maths.ox.ac.uk) started at the University of Oxford as an undergraduate in 2004 and they are not rid of him yet. He achieved his Masters of Mathematics in 2008 and his doctorate in 2012. His research interests are stochastic pattern formation, nonlinear dynamics and partial differential equations. He hopes that one day humans and zombies will live peacefully together.

Daniel Zelterman (Daniel.Zelterman@yale.edu) spends many of his waking hours playing bassoon, a large, sinister-looking woodwind instrument that lurks in the back row of most symphony orchestras. At night, he dreams of completing his lifelong goal of hiking the Appalachian Trail. His day job is professor of biostatistics at Yale, a position he has managed to hold onto for over fifteen years. In this job, he is working toward cures for cancer, AIDS and statistical illiteracy. Among his publications he has shown that it is impossible for humans to live beyond 120 years. Not that you would want to. Just saying.

AFTERWORD

Robert Smith?

I didn't really mean to invent an entire mathematical subdiscipline. But I suppose no one ever does. I'd always kept my love of science fiction totally separate from my love of mathematics. But when I touched the two together briefly in 2009, the whole world went mad. And one of the nice consequences was discovering I wasn't the only one. As you can now tell, there are a surprisingly large number of people into mathematics and zombies. And if you made it this far, presumably you're one of them. Congratulations.

A book like this doesn't just throw itself together. All sorts of people helped make it better than it might have been, not least of whom are the authors themselves. Fortunately, you already know who they are, but there are others who also deserve a mention.

Every chapter (including my own) was peer reviewed. Peer reviewing isn't a task that gets a lot of of rewards, as it's entirely voluntary and, by definition, no one usually knows who you are. So, in alphabetical order, I'd like to thank Mo'tassem Al-arydah, Mayer Alvo, Chris Bauch, Shwetal Bansal, Catherine Beauchemin, Erin Bodine, Fred Brauer, Joseph Brown, Lucy Campbell, Chris DeHaan, Kristina Donato, Jozsef Farkas, Jonathan Forde, Andy Foster, Joshua Grant, Richard Gray, Des Higham, Ioan Hudea, Frithjof Lutscher, Arturo Madigin, Rachelle Miron, Jeff Musgrave, Tim Reluga, Shannon Patrick Sullivan, Arthur Vickers, David Vickers, Lindi Wahl, Derek Wilson and Thomas Woolley for their careful and thoughtful reviewing. I'd also like to thank Jane Heffernan for being the handling editor on my chapter and the two anonymous reviewers who reviewed my chapter. It amused me no end that my own chapter had the toughest reviews of all!

I'd also like to thank Bon Clarke, Justine Hall, Anne-Marie Hoskinson, Hans Petter Langtangen, Mano Manoranjan, Kent-Andre Mardal, Ryan Neighbour, Jose Chan, Byran Demianyk, Marek Laskowski, Shamir Mukhi, Marcia Friesen, Bob McLeod, Ola Skavhaug, Leon Tribe, Rebecca Tyson and Luis Zaman for discussions about potential submissions. Even though they didn't work out, I'm very grateful for the intense interest shown in zombie modelling that continues to grow.

I'd like to thank the University of Ottawa Press team for guiding this project through various bumps: Eric Nelson, Michael O'Hearn, Jessica Clarke, Rebecca Ross, Marie Clausen, Jessica Pearce, Lisa-Marie Smith, Lara Mainville, Bryn Harris, Trish O'Reilly-Brennan and especially Didier Pilon for some sterling work in the trenches and Dominike Thomas for going above and beyond. Thanks guys. And if you like what you see, check out *Braaaiiinnnsss: From Academics to Zombies* (University of Ottawa Press, 2011) for something a little bit similar. This book was all about mathematics and zombies; that was everything else.

On a personal note, I'd also like to thank Laura Collishaw, Sarah Hogenbirk and Kate Fleming. You know why. And none of this would have been possible without Phil Munz, who got the ball rolling back in 2009 (and who has now appeared in both zombie books, as well as the original paper).

In the interests of full disclosure, I should note that a version of the introduction originally appeared in *The Partner*, issue 47 (2012).

Finally, I'd like to thank you, the reader. Whether you're a mathematician who's interested to see how zombies can be used to illustrate modelling techniques or a zombie fan who took the plunge into mathematics, this book isn't your usual kind of thing. So thanks for giving it a shot.

Now go solve some equations. When the zombie apocalypse arrives, you'll thank me.